Basic Vehicle Dynamics and Suspension Design

Basic Vehicle Dynamics and Suspension Design

R Dale White

© Individual authors and College Publications 2023. All rights reserved.

ISBN 978-1-84890-430-9

College Publications
Scientific Director: Dov Gabbay
Managing Director: Jane Spurr

http://www.collegepublications.co.uk

Cover design by Laraine Welch

CONTENTS

LIST OF FIGURES

Chapter 5

LIST OF TABLES

Chapter 1:
INTRODUCTION

One of the more interesting fields of automotive engineering is the design of ground vehicles. This can be thought of as four general topics of study: (1) the mechanical configuration of the vehicle, (2) the dynamic behavior of the vehicle, (3) the chassis layout and structural design, and (4) the ergonomic design best suited for the driver. The objective of this book is to explore the first two topics. In particular, our first goal is to gain an understanding of the function of the vehicle's mechanical configuration, as determined by the engine and drive train systems, braking system, and suspension system, and how it influences vehicle performance. The second goal is to be able to employ relatively simple mathematical techniques to predict vehicle dynamic behavior under conditions of acceleration and cornering, along with the ability to estimate the quality of ride as perceived by the driver.

The dynamic behavior of the vehicle is governed entirely by its mechanical design, and not surprisingly, the mechanical design is influenced by the desired dynamic performance goals. Consequently, the two subject areas of vehicle design are not completely independent of each other. In this regard it is beneficial to start with a brief overview of the design process followed by a survey of suspension systems before a more thorough, in depth, analysis of vehicle dynamics is undertaken.

(1.1) STARTING THE DESIGN PROCESS
One of the more challenging aspects of predicting the performance of a new vehicle and suspension system design is acquiring numerical data such as the weight distribution, center of gravity location and dynamic index. These parameters are required to mathematically model the vehicle behavior, but specific data is unavailable because the car has yet to be built. The best we can do is to use our experience and engineering judgement to make reasonable estimates. More than likely as the design process evolves, and better information is obtained, a review of estimates will be warranted. In this regard it must be realized that design parameters are continuously updated and revised until the final vehicle design is achieved.

The first step in establishing a starting point is to specify written "foundation" criteria based on the intended use of the vehicle. These criteria include the wheelbase, the front and rear track widths, the design classification of the front and rear suspensions, and the front and rear spindle (axle) height. (The latter specification is needed later on the design process when tires are selected.) In some applications the

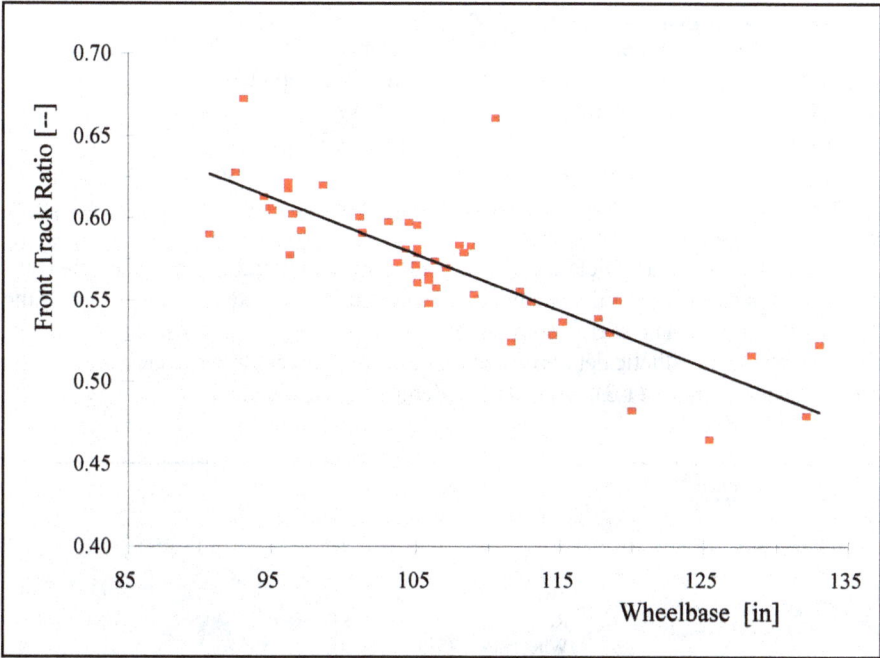

Figure (1.1): FRONT TRACK RATIO vs WHEELBASE

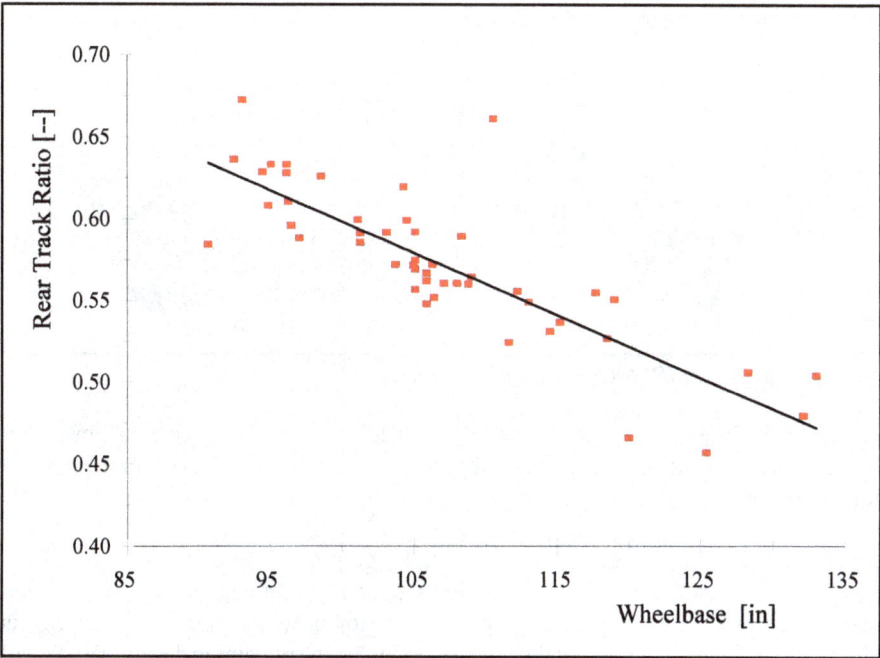

Figure (1.2): REAR TRACK RATIO vs WHEELBASE

number of passengers and/or cargo space may also be important.

Assistance, if necessary, in specifying front and rear track widths can be found in graphs shown in Figures (1.1) and (1.2). The data depicted in these graphs are a compilation of a variety of different vehicles that have been successfully produced over the years. As indicated, there is a general correlation between the track ratio (defined as the track width divided by the wheelbase) and the wheelbase. The front and rear track widths do not necessarily to have to be the same dimension, and, in deed, a slight difference is noted in the two graphs.

Envisioning a vehicle which might satisfy the foundation design criteria leads us towards accomplishing the second step; and that is to create a rough sketch of the "concept" vehicle. The sketch does not have to be of great artistic value and many of the geometric and esthetic details can be omitted. But, a well drawn sketch, such as the one shown in Figure (1.3), will clearly depict the mechanical layout and general

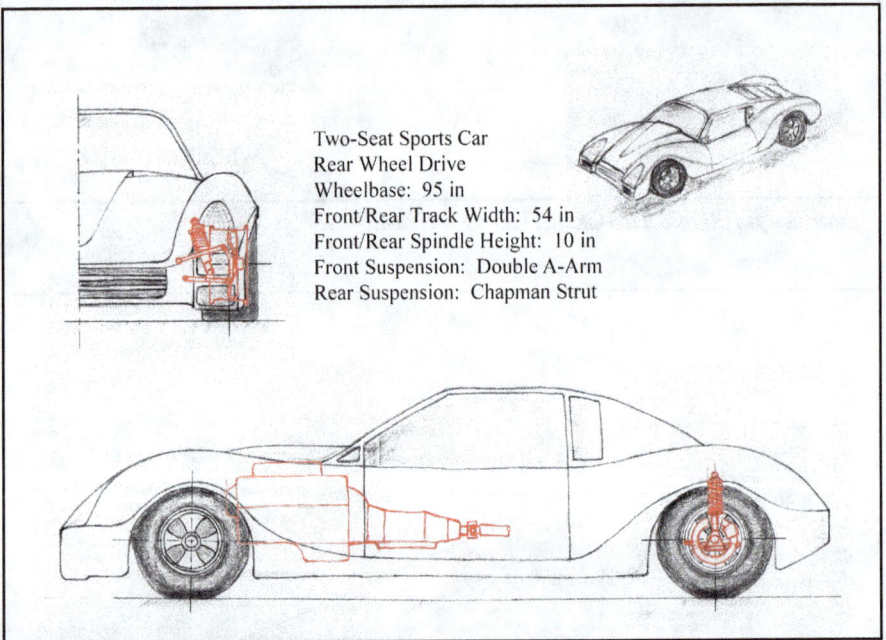

Two-Seat Sports Car
Rear Wheel Drive
Wheelbase: 95 in
Front/Rear Track Width: 54 in
Front/Rear Spindle Height: 10 in
Front Suspension: Double A-Arm
Rear Suspension: Chapman Strut

Figure (1.3): EXAMPLE CONCEPT CAR SKETCH

appearance of the vehicle, and be proportionally accurate, e.g., 1/8 scale, in dimension. The foundation criteria and sketch will give us a pretty good idea of (1) the general location of the engine/transmission, i.e., front, mid, or rear, and (2) whether the vehicle should be front wheel drive, rear wheel drive, or perhaps all wheel drive. We should point out that as the vehicle and suspension design progresses it will be necessary to specify secondary design criteria including size and type of engine/transmission, and suspension parameters, e.g., roll centers. But, for now, the foundation criteria is sufficient to begin an analysis of vehicle performance; the remaining design criteria will be introduced as the need arises.

As previously mentioned, specifying the classification of the front and rear suspension is necessary in the design process. In this regard it is helpful to have an understanding of various suspension systems.

(1.2) SURVEY OF BASIC SUSPENSION SYSTEMS

The function of a suspension system on a ground vehicle is twofold: (1) it maintains optimum tire contact with the road surface during cornering, acceleration and braking maneuvers, and (2) it provides an acceptable level of comfort, as perceived by the driver, through isolation of the vehicle chassis from irregularities in the road surface. It is interesting that while many different suspension systems have been devised over the years, only a few designs have been engineered well enough to achieve a certain degree of success. As might be expected, there is not any one particular suspension system configuration that has been universally accepted for all vehicles. All of the various suspension system designs have good and bad points with respect to vehicle cornering behavior and vehicle ride.

Suspension systems always fall into one of two categories: either dependent or independent. To understand this concept, consider the left tire of a rear suspension traveling over a bump. If the adjoining right tire experiences a change in camber or caster resulting from the movement of the left tire, then the suspension system is dependent. As you might guess, independent suspensions allow independent movement of adjoining tires. In general an independent suspension system offers substantial improvement over dependent suspension systems in road holding capability, but usually at the expense of increased tire wear. Both types of suspension systems have a place in car and truck suspension design.

(1.2.1) SOLID AXLE SUSPENSION SYSTEMS

The solid axle suspension system is a dependent suspension and is perhaps the simplest of all suspension systems. It is classified as a dependent suspension system. Rear solid axle systems are sometimes referred to as a "Hotchkiss drive", named after Benjamin B. Hotchkiss a manufacturer of munitions and automobiles. The drive mechanism, however was designed by Hotchkiss's chief engineer Georges Terrase in 1905 [1][1]. The concept of this early suspension system is still being used today.

The solid axle suspension shown in Figure (1.4) consists of a pair of semi-elliptic springs to locate an axle beneath the vehicle chassis. The axle component can be either (1) a single structural element for mounting wheels which are not driven by the engine/transmission, such as an I-beam configuration, or (2) a housing containing axle shafts with a differential unit for wheels under power. In addition, a steering knuckle component which provides for wheel rotation necessary to steer the vehicle can be incorporated into the solid axle housing, e.g., the front axle used on a heavy truck or an all wheel drive vehicle. Although the solid axle suspension is not generally found on performance road cars, it has been used on literally hundreds of different models of passenger cars and trucks. It is interesting to note that the illustration of Figure (1.4)

---

[1] *Numbers in brackets refer to References.*

shows the leaf springs mounted below the axle housing. This is generally the case for road cars where maintaining a low center of gravity for optimum cornering performance is desired. Truck suspensions normally anchor the springs on top of the axle housing to achieve a greater chassis ground clearance for better off-road mobility; however, the resulting high center of gravity greatly diminishes the cornering ability of the vehicle.

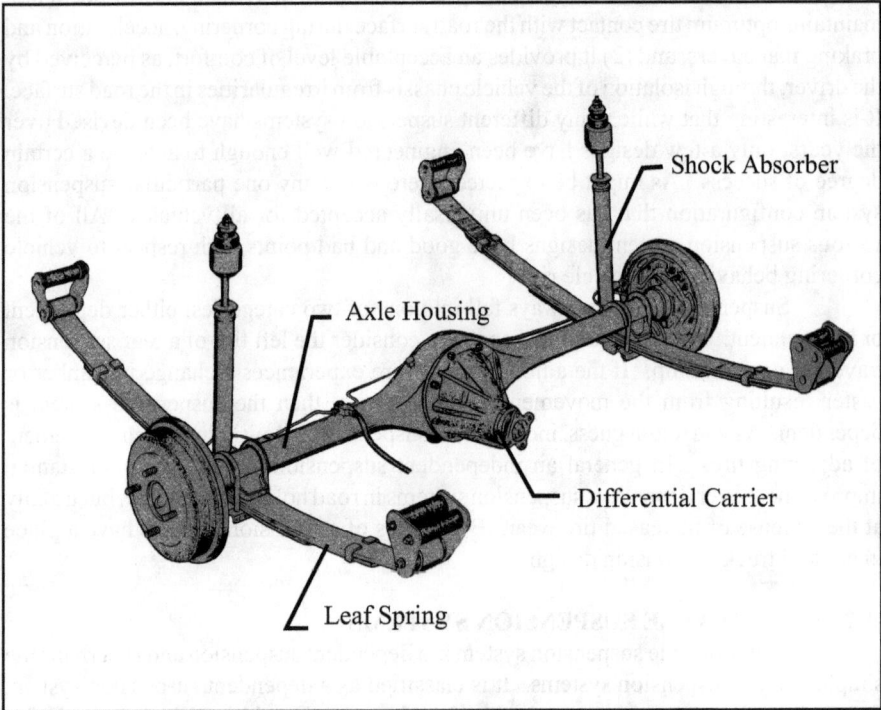

Figure (1.4): SOLID REAR AXLE SUSPENSION {Datsun B210}

The design and spacing of the leaf springs greatly influences other handling characteristics as well. In order to isolate the vehicle from minor irregularities in the road surface, the length of the main leaf in the leaf spring assembly must be relatively long. The long spring deflects in a lateral direction during cornering due to centrifugal forces acting on the vehicle. The lateral deflection in turn hinders the cornering performance. To reduce the lateral deflection, the basic leaf spring design is modified by skewing the individual leafs in a forward direction. This creates an asymmetric leaf spring. The axle locating pin is also moved forward from the spring center. The combination of securing the front spring eye to the chassis and the additional material in front of the axle locating pin resists lateral deflection. The back portion of the spring provides most of the flexing, or spring action, encountered during chassis bounce. The rear spring eye is shackled to accommodate the change in chord length resulting from spring deflection.

The roll stiffness of the vehicle is dependent on the magnitude of the spring base, i.e., the lateral distance between the left and right spring. A large spring base minimizes the deflection of the outside spring during cornering, thus producing only a small amount of body roll. This means that a solid axle suspension with a large spring base has high roll stiffness characteristics. Reducing the spring base decreases the roll stiffness, allowing more body roll. An anti-roll bar (torsion spring) is sometimes used in conjunction with the leaf springs to obtain desired roll stiffness characteristics.

Roll steer, that is, a change in vehicle direction resulting from body roll[2], occurs in solid axle systems when the height of the front spring eye is different from that of the rear eye. When common production rear suspensions are examined closely, it will be noticed that the rear spring eye is usually higher than the front. The affect of higher distance in the rear causes the vehicle to understeer. The effect of differences in spring eye heights can be understood by examining the geometrical behavior during cornering. When the outer spring is compressed to balance the influence of the centrifugal acceleration, the wheel moves up relative to the chassis, in a direction perpendicular to the imaginary line formed by connecting the center of the front spring eye with the center of the rear. If the rear eye is higher than the front, then the wheel moves forward of center during compression. Since the wheel is mechanically connected to the axle housing, the forward motion of one wheel causes the axle housing to rotate into the corner. A rear solid axle suspension would, therefore, steer the vehicle out of the turn, an understeering condition. Similarly, a condition of oversteer is created when spring eye distances are greater in the front.

The solid axle suspension has two desirable characteristics for good road handling: constant camber angle, and constant track width. An added benefit of these two characteristics is minimized tire wear. Because the solid axle is separate from the chassis, connected only by springs and shock absorbers, the wheels remain perpendicular (zero camber) to the road surface regardless of chassis position or movement. It should be recognized, however, that single wheel movement from variations in the road cause both wheels to change camber, which is undesirable.

A major disadvantage of solid axle systems applied to lightweight vehicles is the low sprung to unsprung weight ratio. The low ratio results in poor vehicle performance at higher speeds due to the suspension system's inability to follow the road contour. This phenomenon is referred to as "wheel hop" and is a problem which cannot easily be overcome in this type of suspension.

Solid axle suspension systems incorporating driving wheels have another severe disadvantage. The axle housing reacts, dynamically, to an externally applied torque. There are two sources of external torque: (1) the torque produced by the engine and subsequently transmitted by the transmission and drive shaft, and (2) the torque resulting from the tractive force of the ground on the wheel multiplied by the rolling radius of the tire. The axle housing responds to the torque applied by the engine/transmission by compressing the right spring. The amount of change in spring deflection from static ride conditions results in an unbalance of spring forces to counteract the applied torque. In some instances the torque produced by the engine is sufficient to induce extreme spring deflections, causing the right wheel to hop or lift off the ground. The effect of tractive torque, most noticeable during acceleration and

---

[2] *Standard vehicle dynamics terminology is discussed in Appendix A.*

braking, causes the springs to wind-up or deform into an "S" shape when viewed from the side. In extreme situations when accelerating through a corner, the torque reaction can vary the handling characteristics from over steering on right-hand turns to understeering on left-hand turns.

(1.2.2) FOUR LINK SOLID AXLE SYSTEMS

The four link solid axle suspension evolved from the solid axle system previously described. The goal of the suspension engineers in redesigning the solid axle system was to reduce the unsprung weight, and also reduce the effects of torque reaction. Both were accomplished by replacing the leaf spring components of the solid axle system with coil springs and control links.

The redesigned solid axle suspension system is called a "four link solid axle suspension" because it parallels the action of a four link mechanical mechanism. The lower control link has a shape and cross-section similar to that of a small I-beam. Both ends of the link have pinned connections such that it is only free to move in a vertical plane. A pair of links will then allow the axle to move up and down with respect to the

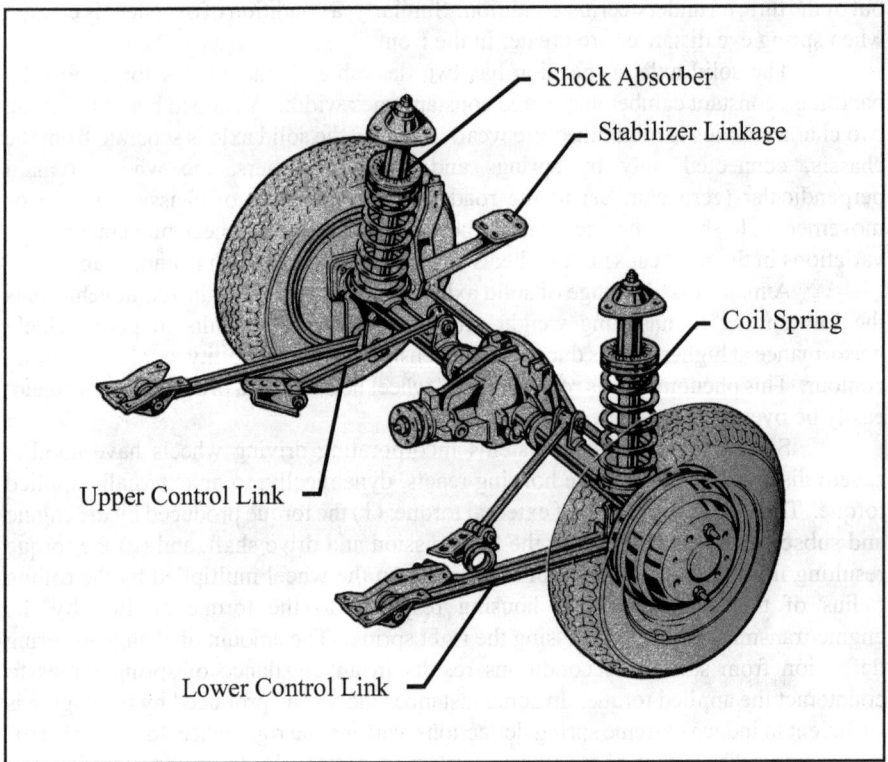

Figure (1.5): FOUR LINK SOLID AXLE REAR SUSPENSION WITH DRIVING WHEELS {Fiat 131}

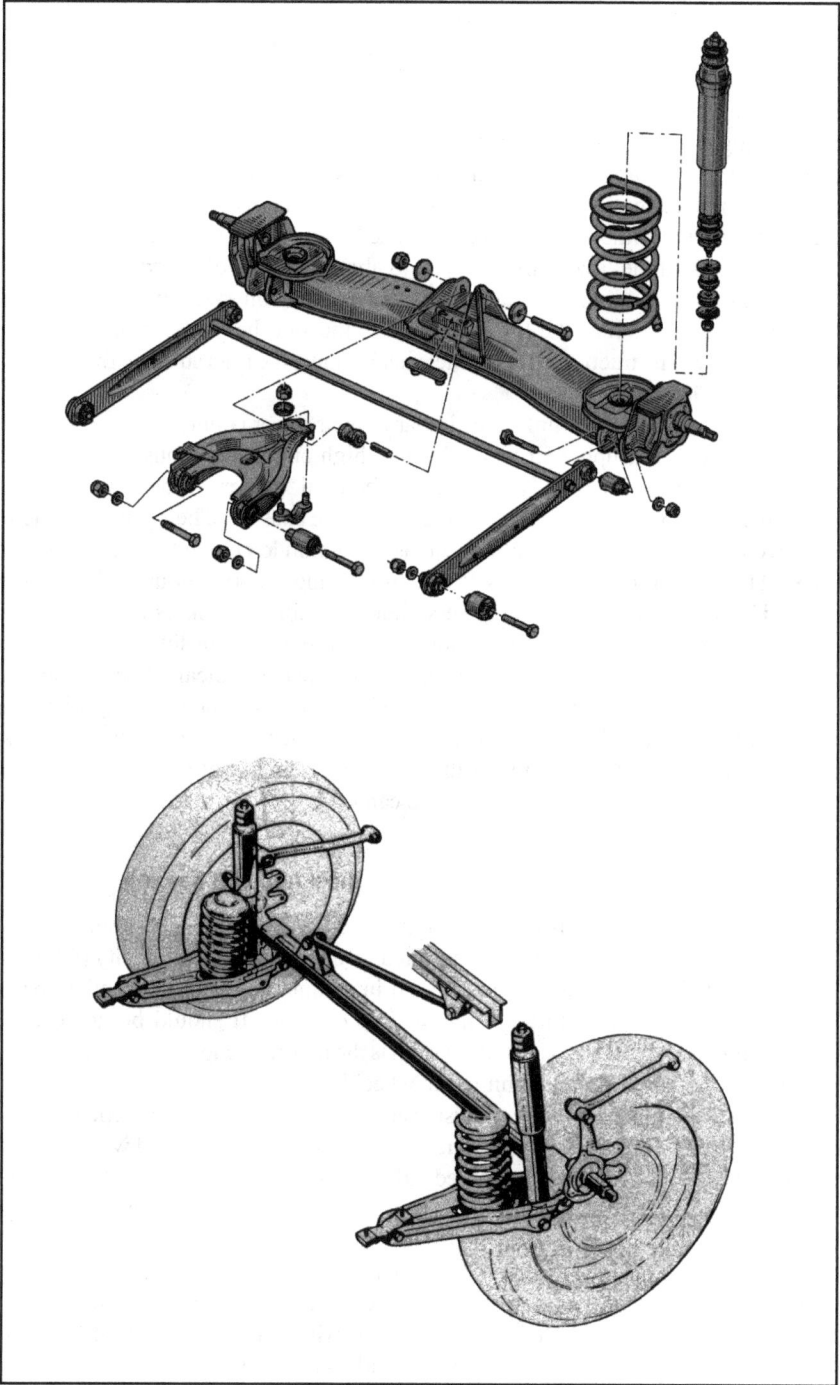

Figure (1.6): FOUR LINK SOLID REAR SUSPENSION WITHOUT DRIVING WHEELS {top: Renault 18i, bottom: Saab}

Chapter 1: INTRODUCTION

chassis. One end of the link is connected to a mounting bracket fastened to the chassis; the other end is secured to a bracket on the axle housing. During the fabrication process, the bracket is welded to the housing below the axle centerline.

The geometrical constraints of the four link axle prevent the wheels from moving straight up and down when traveling over a bump. Instead, the wheel must move in an arc. The arc movement means there is some change in wheelbase which results from bounce motion of the suspension. Designing long lower struts minimizes wheelbase variation, although there is an increased unsprung weight penalty. Generally, the design length of the lower strut is equal to about one-half of the track width. The longer length also minimizes, to a certain degree, the roll steer experienced while cornering. The function of the lower strut is to transmit the load, resulting from the force produced by the tractive effort of the wheel against the road surface, from the axle housing to the chassis.

The upper struts, or links are similar to their lower counter parts, except they are much shorter. The upper struts are located high on the axle housing towards the middle, or in some instances above the differential carrier. These links extend diagonally about 45° from the center of the axle to the chassis. Their primary function is to prevent axle housing rotation induced as a reaction to engine torque and braking torque. They also provide lateral stability in locating the axle housing beneath the chassis. However, an additional transverse link, or stabilizer link, is used quite often to absorb lateral loads and limit lateral movement of the axle housing.

Figure (1.5) illustrates a rear suspension that is typical of most four link designs. Two variations of the four link suspension for axles without driving wheels are shown in Figure (1.6). The performance characteristics of the four link solid axle suspension are essentially the same as those of the foregoing solid axle system. Only slight improvements in vehicle performance can be realized.

(1.2.3) SWING AXLE

The simplest of all independent suspension systems is probably the swing axle. The rear suspension system, as shown in Figure (1.7), consists of a wheel/hub assembly mounted on a driving axle, which is connected through a constant velocity (CV) joint to the differential unit. A radius arm is used in conjunction with the driving axle to transmit the wheel tractive force to the vehicle chassis. It should be noted that a distinguishing feature of the swing axle system is the non-telescoping design of the half-shaft connecting the differential unit to the wheel hub.

The design of the swing axle suspension, as well as other independent rear suspension systems, eliminates the torque reaction problems associated with solid axle systems and has the benefit of improved ride comfort. However, the inherent poor geometry of the swing axle in controlling wheel movement during cornering discourages incorporating this design on all but the most modest of vehicles.

The major disadvantage of the swing axle is a phenomenon called "jacking", which inevitably occurs to some degree while executing a cornering maneuver. The elastic behavior of a pneumatic tire is such that when the tire is steered a twisting deformation occurs between the rim and the road contact patch. The energy absorbed by the continuous distortion of the tire produces an effect equivalent to a slight amount of vehicle braking. The braking action causes a transfer of vehicle sprung weight towards the front of the car which, in turn, results in lifting the rear of the vehicle. Since the final drive unit is rigidly mounted to the vehicle chassis, the drive unit raises up with the rear of the vehicle, forcing both rear wheels into a positive camber attitude. The high roll center of the swing axle, together with the inertia effects of cornering, further complicates matters because a significant portion of the sprung weight is placed on the outside wheel. Consequently, there is not enough weight remaining on the inside tire to bring the rear of the vehicle back down.

Another problem with the swing axle geometry is the large amount of wheel camber change that occurs during suspension deflection, either from single wheel bumps

Figure (1.7): SWING AXLE REAR SUSPENSION {Corvair}

or from body roll. The suspension geometry constrains the rear wheel to move in an arc about an axis which runs through the radius arm pivot at the chassis, and the center of the CV joint. Since the wheel pivot axis is not parallel to the fore/aft centerline of the vehicle, the arc movement of the wheel produces changes in camber angle and wheel toe. The effect of these wheel alignment variations is most noticeable while cornering, causing changes in the vehicle yaw. This effect is referred to as roll steer.

(1.2.4) TRAILING ARM

The trailing arm suspension is used primarily for rear suspension applications, and can be used on vehicles with either front wheel drive or rear wheel drive. It is a simple suspension system with relatively low unsprung weight. The most popular modern day application is a rear suspension for vehicles which have front wheel drive. The main structural component of the trailing arm suspension is a rather stout radius

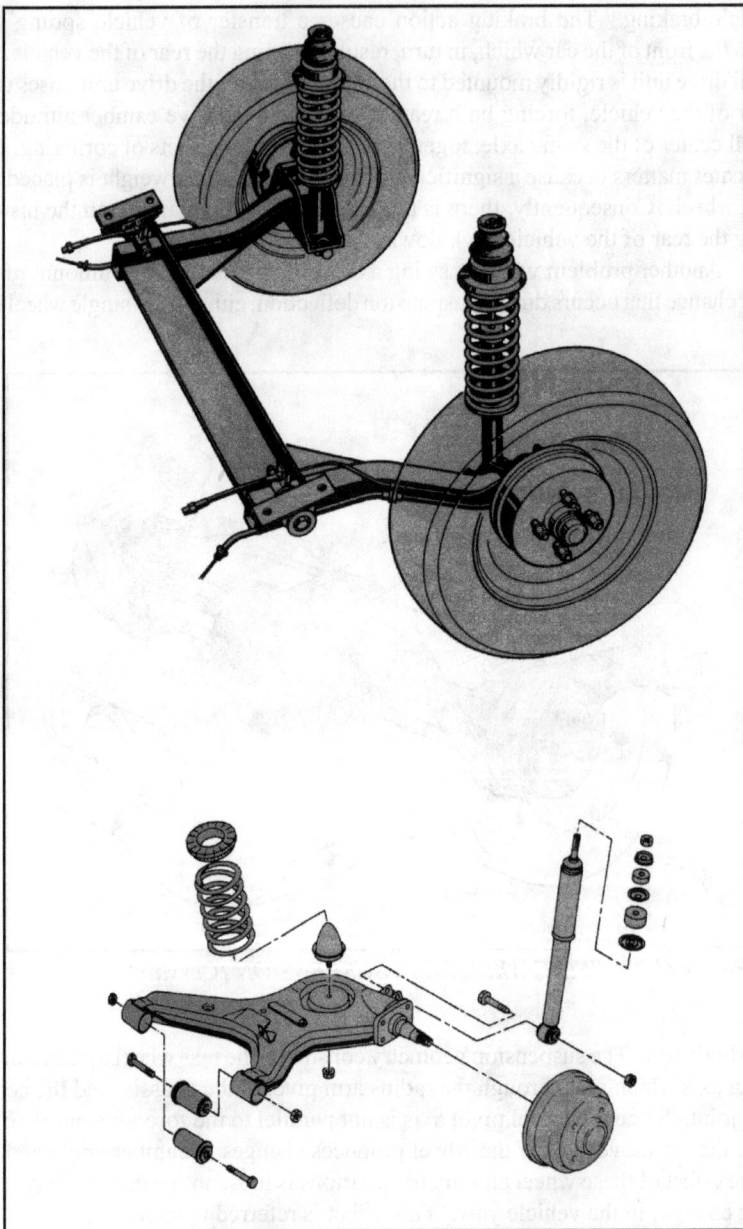

Figure (1.8): TRAILING ARM REAR SUSPENSION {top: Volkswagen Jetta, bottom: Nissan F10}

arm. The front of the radius arm pivots on a short horizontal shaft attached to the chassis. The wheel hub is mounted on a stub axle, permanently secured to the rear of the radius arm. The motion of the suspension is controlled by coil springs or a torsion spring, either of which can be easily adapted to the radius arm. For applications in which the rear wheels drive, universal joints with splined slip shafts are used to transmit power from the final drive differential to the wheels. Two typical rear suspension designs are shown in Figure (1.8).

Although the trailing arm suspension is a definite improvement over the swing axle suspension, there are two distinct disadvantages for performance cars. The motion of the radius arm about the pivot on the chassis confines the wheel to move in the vertical direction only, with respect to the chassis. Except for a slight amount of deflection in the mechanical components, there is no change in wheel camber relative to the chassis, with vertical movement of the tire. As a result, when the vehicle rolls in a corner, the camber relative to the road surface changes an amount equivalent to the body roll angle. Other problems associated with the motion of the radius arm are large caster angle changes, and changes in wheelbase dimension. However, these can be minimized to some degree by limiting body roll and restricting suspension travel.

(1.2.5) SEMI-TRAILING ARM

The semi-trailing arm suspension is a variation of the trailing arm rear suspension, and is used exclusively for rear suspension applications. The mechanical components of this suspension are essentially the same as the trailing arm; however, the geometry is quite different. The control arm of the semi-trailing arm suspension pivots about an axis rotated approximately 10 to 40 degrees from the usual trailing arm transverse axis. This suspension has been popular with European automobile manufacturers such as BMW, Porsche, Peugeot, and Mercedes-Benz. The semi-trailing arm rear suspension developed by BMW for the 528e is shown in Figure (1.9).

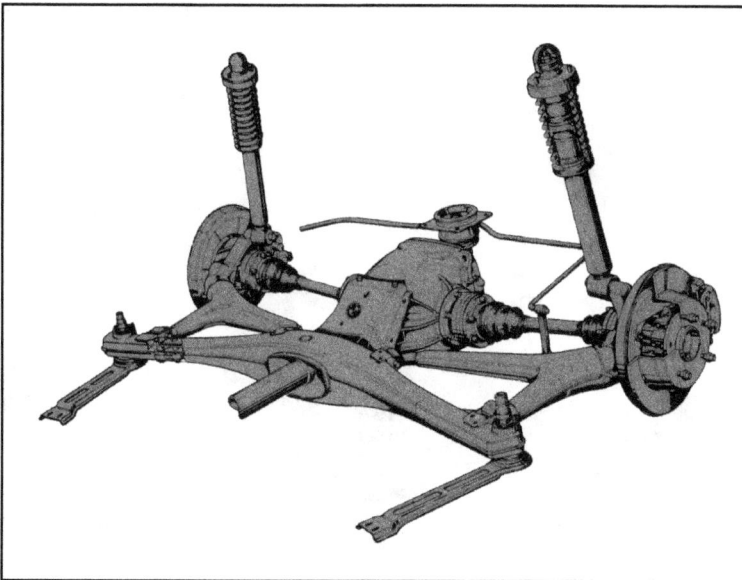

Figure (1.9): SEMI-TRAILING ARM REAR SUSPENSION {BMW 528e}

The variation in the pivot axis orientation of the semi-trailing arm allows for a small amount of wheel camber change with respect to body roll. This is an improvement to the trailing arm design because cornering performance is enhanced. In addition, the roll center characteristics of this suspension can be adjusted by raising or

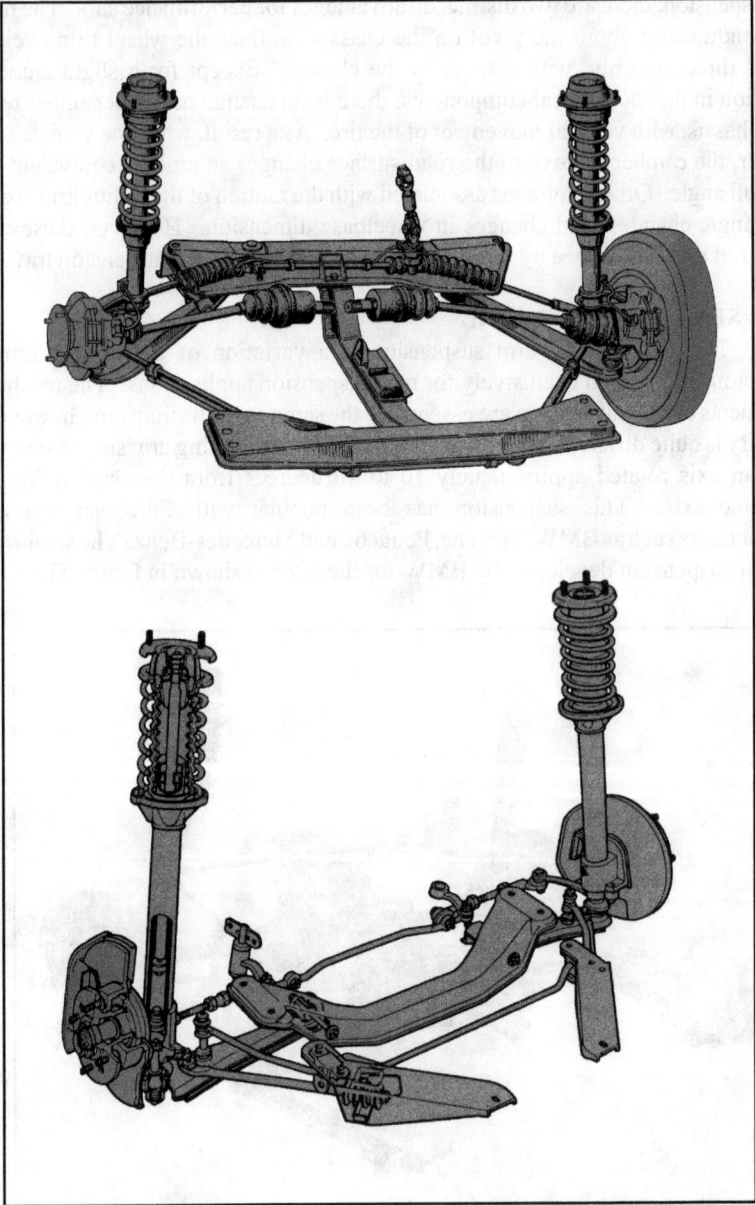

Figure (1.10): MacPHERSON STRUT SUSPENSION {top: Honda, bottom: Datsun}

lowering the pivot point on the inside hinge of the control arm. The disadvantage of the semi-trailing arm suspension is the unavoidable change in wheel toe resulting from suspension movement. The variation in wheel toe manifests itself in an effect called bump steer and roll steer, an unwanted steering effect hindering handling performance.

(1.2.6) MacPHERSON/CHAPMAN STRUT SUSPENSION

Earle S. MacPherson first developed the strut suspension for General Motors in the 1940's. However, it has not been until recent years that the suspension has become popular with automobile manufacturers. The main components of the MacPherson suspension are a lower control arm, a wheel knuckle, and a near vertical telescoping member referred to as a strut. The top of the strut is secured to the chassis, and the lower end of the strut fits within a sleeve formed into the wheel knuckle. The design of the sleeve allows partial rotation of the wheel knuckle, enabling the wheel to be steered. Integrated into the construction of the strut are the internal workings of a shock absorber and an external coil spring. The lower control arm is hinged with a chassis cross-member riveted or bolted to the vehicle frame. The opposite end fits into a lower ball joint connected to the wheel knuckle. A horizontal stabilizer link mounted between the frame and the control arm is generally used to control fore and aft deflection of the suspension. Roll stiffness of the suspension can be increased by adding a torsion bar (spring) between the two adjacent lower control arms and the chassis.

Figure (1.11): CHAPMAN STRUT SUSPENSION {Datsun 280Z}

The geometry aspects of the strut suspension, i.e., tilt of the strut and placement of the control arm, can be designed to produce good handling characteristics. A disadvantage of the suspension is that it is relatively tall compared to other types of suspension systems. This somewhat compromises the aerodynamic design of the vehicle. Figure (1.10) illustrates strut designs developed by Honda and Datsun. These designs are representative of strut suspensions currently in production.

The Chapman strut suspension concept is essentially the same as the MacPherson strut system. It uses the same components and has the same handling characteristics. The only difference between the two is that the Chapman strut suspension is used for rear suspension applications, e.g., the Datsun 280Z rear strut suspension shown in Figure (1.11).

(1.2.7) DOUBLE A-ARM INDEPENDENT SUSPENSION

The double A-arm suspension is the most versatile of all independent suspension systems, and excels in handling performance. It is also the most complicated. The suspension geometry mimics a mechanical four bar linkage. As depicted in the example schematic of Figure (1.12), the chassis and wheel knuckle are two of the links, and the upper and lower control arms, referred to as *A-arms* or *wishbones* owing to their physical appearance, are the remaining links.

Figure (1.12): DOUBLE A-ARM SUSPENSION

The ideal performance goal of the suspension is to keep all of the wheels nearly perpendicular to the road surface regardless of vehicle attitude; however, actually achieving this goal is not possible and a compromise is required. The lengths of the A-arms are designed so that the wheel cambers in with jounce. This allows the vehicle to roll in a corner without changing the outside wheel camber relative to the road surface. The inside wheel generally cambers in more than ideally desired with this design. Additional advantages of the double A-arm suspension are that it can be designed to provide anti-dive and anti-squat characteristics. These are important considerations in vehicle acceleration and braking.

Chapter 2:
TIRES

Tires, as mentioned in the previous chapter, are an integral design component of the vehicle suspension design process. But, it's actually the tire characteristics that govern vehicle performance. In particular, their purpose is to (1) provide a cushion between the vehicle and the road surface, and (2) grip the road surface adequately to provide traction and directional control. The suspension designer has the responsibility of developing a suspension system which will utilize all of the performance advantages of the tire.

The first task is to select an appropriate tire which will match the intended use of the vehicle. The selection process includes consideration of such items as tire size (in terms of diameter and section width), tread design, rim size, and for racing applications, rubber compound. For example, the outside diameter of the tire influences several aspects of vehicle design: (1) it determines, in part, the ride height of the vehicle, (2) the overall ratio between engine speed and vehicle ground speed, which must be taken into account when specifying the gear ratios for the transmission and differential, and (3) the load capacity of the vehicle. The single most important consideration influencing the dynamic behavior of a vehicle is, however, the interaction of the tire contact patch with the road surface. This is governed by the tractive (braking) and cornering performance characteristics of the prospective tire. Clearly, gaining a basic understanding of the mechanics of pneumatic tires and tire characteristics is essential to the successful design of a vehicle suspension system.

(2.1) TIRE CONSTRUCTION

The concept of the pneumatic tire was developed around 1877, primarily for the fledgling bicycle industry; but, it was J.B. Dunlop who made the first practical pneumatic tire in 1888 [1,2]. Since that time there has been a considerable evolution in tire technology. The most notable improvement of recent years is tire adhesion or grip. This is mainly due to changes in tire construction and development of new rubber compounds.

The bias-ply tire is the oldest established method of tire construction. This tire was almost exclusively used in U.S. car and truck applications until the late 1960's. At that time radial-ply tires which had been developed and proven very successful in Europe by Michelin and Pirelli [3], were introduced into the American market. Initially, U.S. tire manufacturers were not eager to spend the large sums of money required to re-tool their factories for radial tires. Instead, they began to produce bias-belted tires, which had some of the performance features of radial tires. It was a short lived tire that was quickly replaced by radial tires, which have now become the standard for automobile suspension systems. Although bias-ply tires and radial tires have about an equal share of truck applications.

(2.1.1) THE CARCASS

The major components involved in constructing a tire are shown in Figure (2.1). The tire carcass refers to the combination of tread and sidewall parts of the tire.

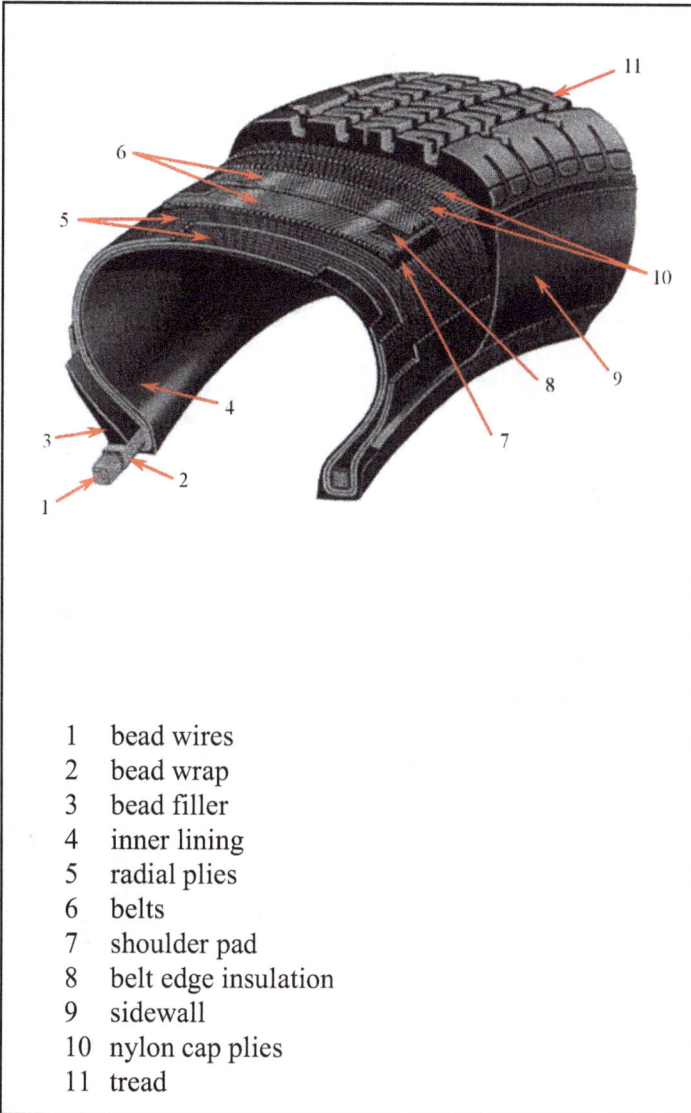

1	bead wires
2	bead wrap
3	bead filler
4	inner lining
5	radial plies
6	belts
7	shoulder pad
8	belt edge insulation
9	sidewall
10	nylon cap plies
11	tread

Figure (2.1): TIRE CONSTRUCTION AND COMPONENTS

It can be thought of as a rubber structural element reinforced with a matrix of tread bracing belts and plies. It is the design of the tire carcass that largely determines the performance characteristics of the tire.

The construction of the tire carcass, which is also used to categorized the tire, is identified as either bias-ply, radial-ply, or bias-belted. The basic construction features of these tires are shown in Figure (2.2). The carcass plies are composed of chords spun from rayon, nylon, aramid fiber polyester or fiberglass materials [3]. The crown angle, defined as the angle between the chord and the tire circumferential centerline, specification significantly affects the behavior exhibited by the tire. Generally speaking, crown angle values near zero provide good directional stability and good cornering performance, but the resulting ride is harsh. At the upper limit of crown angle, i.e., approaching 90 degrees, the tire provides very good ride characteristics, but the cornering performance is poor.

In an effort to achieve a compromise between ride and cornering performance, traditional bias-ply tires are constructed with crown angles of 32 to 40 degrees. In this design, two or more plies are used in the carcass. The chord layers of alternate plies have an opposite direction; thus, when the chords are overlapped they create the chord patterns as depicted in Figure (2.2). The bias-ply tire does have good directional stability due to its design; but, the relatively large lateral stiffness of the tire detracts from its cornering performance by allowing the inside edge of the tread to be picked up off the road surface. This affect decreases the contact area of the tire with the road.

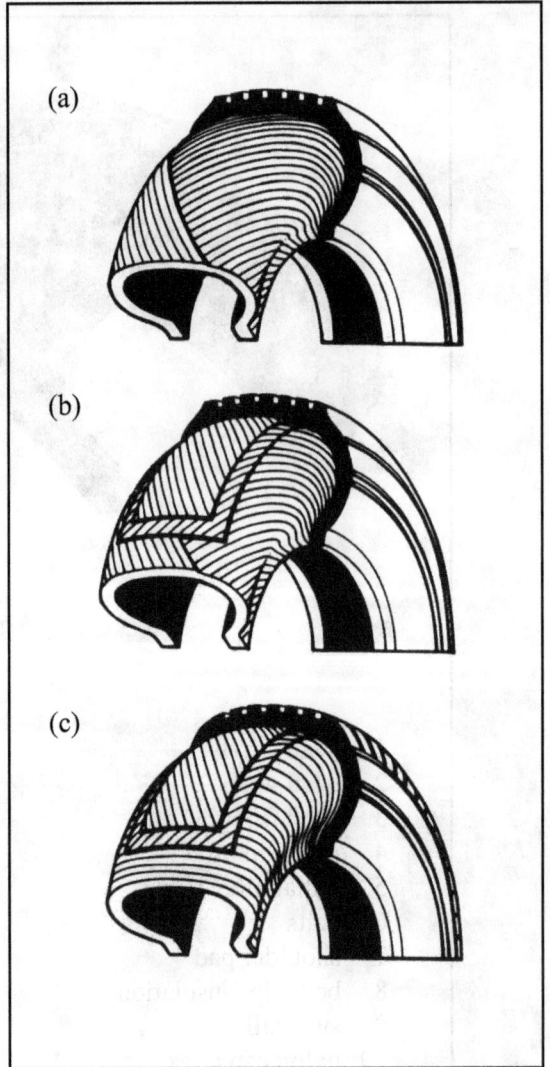

Figure (2.2): CARCASS CONSTRUCTION FEATURES OF (a) BIAS-PLY TIRE, (b) BIAS-BELTED TIRE, and (c) RADIAL-PLY TIRE

The crown angle of chords in a radial-ply carcass are near $90°$. They are adjusted by about $2°$ to facilitate wrapping the chords around the bead wire. Looking back at Figure (2.2), several tread stabilizer belts, usually referred to as just "belts", are secured to the radial plies underneath the tread. The belts are an essential component of the radial tire. The crown angle of the belts, which is in the range of 18 to $20°$, is set to provide a high degree of directional stability. To a large extent, the design and construction of the radial tire prevents sidewall deflection from influencing the performance of the tread. Consequently, the tread maintains better contact with the road surface during cornering. We can summarize radial tire performance by saying it has very good ride qualities and good cornering characteristics, but does not have quite the directional stability as the bias-ply tire.

A bias-belted tire is fabricated by adding several belts to a conventional bias-ply tire. The performance characteristics of a bias-belted tire are roughly half way in between those of a bias-ply tire and a radial-ply tire. Historically, the bias-belted tire was an intermediate step in changing from bias-ply to radial-ply tire production, and is now rarely used.

(2.1.2) TIRE MATERIALS

A tire contains a large number of individual components, and may be constructed from as many as 100 separate materials [4,5]. The "rubber" part of the tire is actually a blend of natural rubber, synthetic rubber(s), and processing agents which have been vulcanized. Vulcanization is a manufacturing process in which the starting rubber blend is heated in a steel mold to about $150°C$. During the vulcanization process, sulfur molecules in the processing agent form cross-links between double carbon bonds on adjacent polymer chains [4]. The material strength of a tire is a direct result of this process.

Different blends, or rubber compounds, are used in fabricating the tread, sidewall and linear components of a tire. A few of the traditional synthetic elastomers used in tire production are given in Table (2.1). The processing agent found in the rubber compound usually consists of : (1) a vulcanizing agent, which is basically sulphur, (2) curatives, (3) stabilizers, including antioxidants and antiozanants, (4) reinforcing agents, (5) extenders, and (6) fillers. Carbon black is the main reinforcing agent. It not only gives the tire its characteristic color, but allows the tire to be heat treated, which significantly adds strength and abrasion resistance. Carbon black also acts as an ultra-violet filter. Oil is used as an extender or plasticizer. It facilitates the incorporation of carbon black into the tire manufacturing process, and makes the finished tire tougher and greatly improves wet-grip performance. Fillers are used to cheapen tires and to some extent modify material properties. Materials used as fillers include silica and silicates.

It is interesting to note that today's radial tires are constructed using two tread layers of different rubber compounds. The tread base is designed for low rolling resistance and durability; whereas the outer layer is designed for optimal traction and long wear. The liner of tubeless tires is a halogenated rubber blended mainly with natural rubber and carbon black. The linear acts as a moisture barrier and minimizes air permeation through the carcass.

Table (2.1): LISTING OF COMMON SYNTHETIC RUBBERS USED IN TIRE CONSTRUCTION

CHEMICAL NAME	ABBREVIATION	COMMENTS
polychloroprene {trade name: Neoprene}	CR	- resembles natural rubber - high tensile strength - stands up well to aging - poor splice or bond with tire carcass fabric
styrene butadiene rubber {trade name: Buna}	SBR	- strong bond with carcass materials - excellent resistance to abrasion - high degree of hysteresis - must be compounded with carbon black - inferior tear resistance - high-styrene SBR blended with natural rubber {racing tire compounds}
oil-extended SBR	OESBR	- addition of oil results in improved wet road grip
polybutadiene	BR	- stable rubber properties with respect to temperature changes - mixed with SBR degrades wet road holding but improves wear - great resistance to tear propagation and cutting
polyisoprene	PI	- wear properties better than natural rubber - low sensitivity to build up of heat - blended with natural rubber and BR for improved abrasion resistance
ethylene-propylene	EPM	- experimental synthetic rubber
butyl	IIR	- low permeability to gases {used for lining of tubeless tires}

(2.1.3) TIRE SIZE DESIGNATION AND MAXIMUM LOAD

Two of the most important tire properties vehicle suspension designers must be cognizant of is tire size and maximum load. Most modern tires are sized using the "P-Metric" system, e.g., P215/60R14 80H. The tire size label starts out with a letter **P** or letters **LT** which indicate a passenger car or light truck tire respectively. Next, the section width, in millimeters, is specified followed by a forward slash and the aspect ratio. The tire aspect ratio is the percent of the sidewall height relative to the section width. A letter representing the tire construction code, **R** for radial-ply, **B** for bias-belted, and **D** for bias-ply tires, is given after the aspect ratio. Finally, the rim diameter is specified in units of inches.

At the end of the tire size/construction label is the *service description*, which identifies the tire's load index and speed rating. The load index ranges from 70 to 126; a few examples of which are:

$$80 = 992 \text{ lb } (450 \text{ kg}) \quad 104 = 1984 \text{ lb } (900 \text{ kg}) \quad 120 = 3086 \text{ lb } (1400 \text{ kg})$$

The speed rating is a code that specifies the maximum vehicle speed which can be safely sustained for a particular tire. The following code has been established by the Tire Industry Safety Council and the Tire and Rim Association:

P = 94 MPH	S = 112	H = 130	Y = 186
Q = 100	T = 118	V = 149	Z = over 149
R = 106	U = 124	W = 168	

In addition to the tire size, the Department of Transportation (D.O.T.) Has a "Uniform Quality Tire Grade" label embossed on the sidewall. The grading scheme established by D.O.T. is a method to rate tread wear, traction, and temperature resistance under controlled test conditions. The tread wear rating is usually in the range of 100 to 300, with the softer rubber compounds closer to 100. The traction grade is designated by the letter grades **AA**, **A**, **B**, and **C**. The grade represents the tire's decelerating performance on wet pavement. The last tire quality is temperature resistance, also designated by **A**, **B**, and **C**. The temperature resistance indicated the tire's ability to control the rise in tire operating temperature at highway speeds caused by energy dissipation due to rolling resistance.

Since the D.O.T. quality tire grade is based on laboratory conditions it does not indicate how well the tire will perform under actual conditions, nor can the grades be used in suspension design. As we will learn, suspension design requires an entirely different, and more extensive, set of performance data.

(2.2) TRACTIVE EFFORT AND BRAKING FORCE

All tires, regardless of their size and construction, exhibit similar performance characteristics and behavior. For instance all tires exhibit elastic deformation over the road surface. At any instant the deformation creates a flat spot in the tire, which when viewed from underneath may book similar to the "footprint" shown in Figure (2.3). The contact patch is the interface between the footprint and the ground; it defines region where the interaction of forces takes place between the tire and the road surface. In simple terms, the contact patch is the area where the tire exerts a grip on the road, and vice versa.

A theory that completely describes grip, is not available, but conventional thinking is that grip results from the combination of two different phenomena; adhesion and hysteresis. Adhesion resembles Coulomb friction and is the larger of the two components. Although, due to the complexities of tire construction, adhesion does not precisely follow Coulomb's dry friction model, except near the limit of adhesion when the tire is completely sliding through the contact patch. Hysteresis describes the tire's ability to fold over irregularities in the road surface. It is dependent on rubber compound, contact patch area, and temperature.

In our quest to understand the mechanics of pneumatic tires it is convenient for us to first study the longitudinal forces (i.e., tractive and braking forces) that exist within the contact patch. We will then move on to lateral forces (or cornering forces), and then finally the combination of longitudinal and lateral forces.

(2.2.1) TRACTIVE EFFORT AND LONGITUDINAL SLIP

Longitudinal forces, F_x

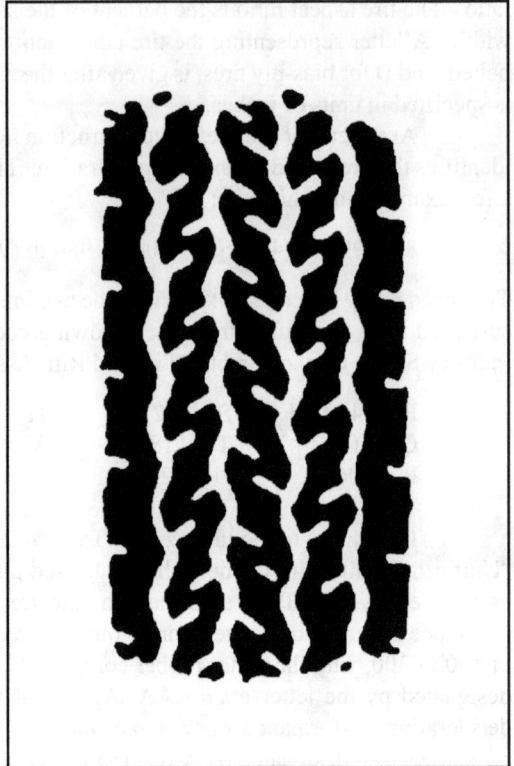

Figure (2.3): EXAMPLE OF TIRE FOOTPRINT

as shown in Figure (2.4), are developed in the contact patch when either a driving torque or a braking torque is applied to the wheel. These forces are generally referred to as either tractive effort or braking force, depending on the application of wheel torque. At the same time the longitudinal forces are being developed, elastic deformations within the tire carcass are also taking place. The tire deformation under an applied driving torque results in a small region of tread compression in front of the contact patch. Compression in the tread causes a small decrease in the distance between the spin axis and the contact patch. Consequently, the rolling radius of a tire in a free-rolling condition, r, is larger than that of the same tire experiencing a driving torque. The latter rolling radius is called the effective rolling radius, r_e. The ratio of the effective rolling radius to the free rolling radius provides the basis for the concept of deformation slip, or longitudinal slip:

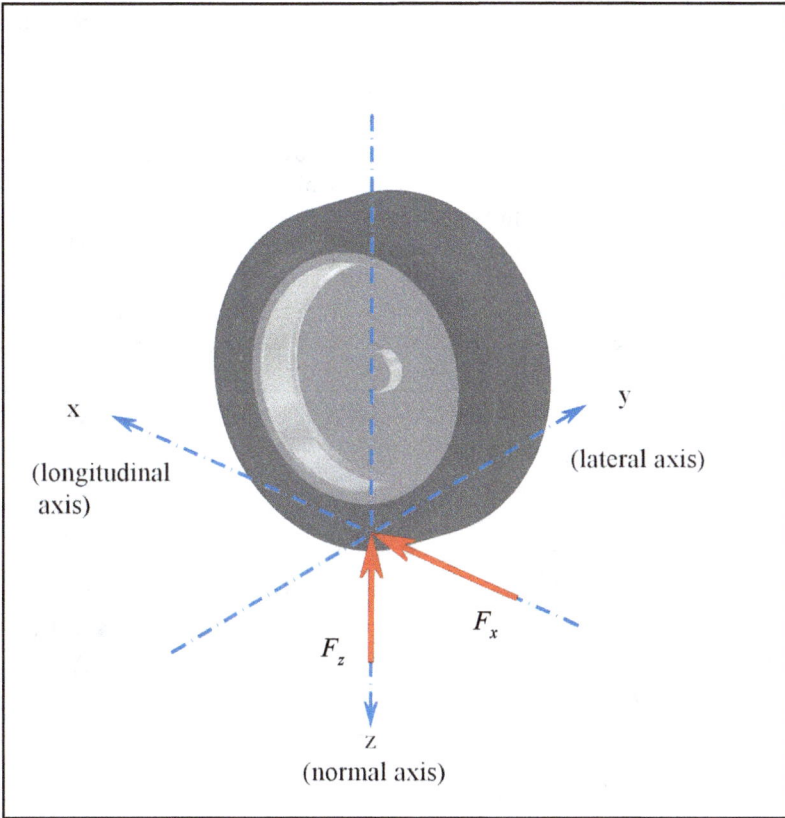

Figure (2.4): TIRE COORDINATE SYSTEM WITH NORMAL AND LONGITUDINAL FORCES

$$\kappa = 1 - \frac{r_e}{r} \qquad (2.1a)$$

We can derive an alternate expression for the longitudinal slip by noting that the forward velocity of the tire at the spin axis is related to the effective rolling radius by $V = r_e \omega$, then Eq. (2.1a) can be rewritten as:

$$\kappa = 1 - \frac{V}{r\omega} \qquad (2.1b)$$

From our earlier discussion we conclude that while the tire is under the influence of a driving torque $V < r\omega$, which means $\kappa > 0$. The upper limit on longitudinal slip is obtained when the tire is in a pure sliding condition without forward progress. In this case $\kappa = 1$. The range of longitudinal slip is therefore $0 < \kappa \le 1$ for a tire with an applied driving torque, and $\kappa = 0$ for a free rolling tire.

Experimental studies of tire performance characteristics reveal that there is a relationship between the tractive effort being developed by the tire and the longitudinal slip. As indicated by Figure (2.5) initially the tractive effort increases almost linearly with longitudinal slip. The linear variation which is initially observed can be attributed to the elastic deformation of the tire tread. As the applied driving torque is increase, a portion of the tread begins to slide at the trailing edge of the contact patch. This results in a non-linear increase in tractive effort until the maximum effort is achieved, somewhere between 12 and 20 percent longitudinal slip. At this point the tire is at the limit of adhesion. Increasing the driving torque any further results in an unstable

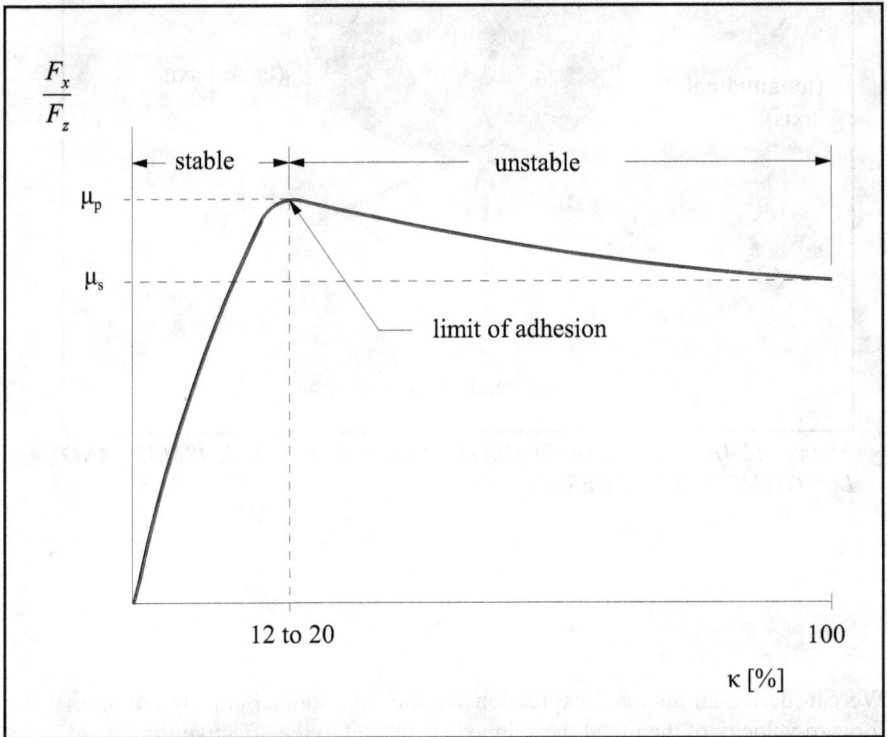

Figure (2.5): VARIATION OF TRACTIVE EFFORT WITH LONGITUDINAL SLIP

condition in which the tractive effort quickly drops from its peak value to a lower value representing complete sliding.

A tractive effort coefficient, defined as $\quad \mu = \dfrac{F_x}{F_z}\quad$, is sometimes used in

quantifying the traction characteristics of tires at their peak performance, μ_p and at their pure sliding condition, μ_s. Some fairly representative values of μ_p and μ_s, although somewhat dated (1957) by modern standards, are given in Table (2.2). It should be realized, however, that actual values are dependent on a number of factors, including tread geometric design, tread material (i.e., rubber compound) carcass construction, tire surface temperature, and road surface conditions. In addition, inflation pressure, operating speed and vertical load also influence the tractive effort coefficients. It has been found [5] that a 10% increase in vertical load will reduce both μ_p and μ_s by about 0.01. Thus, over the entire load range the peak tractive effort coefficient, for example, may drop from 0.9 at light loads to 0.8 at heavy loads. Increasing the vehicle speed will also reduce the peak and sliding tractive effort coefficients. The inflation pressure does not have a drastic affect on the coefficients when the road surface is dry; however, significant improvement in both coefficients is realized on wet surfaces.

Table (2.2): REPRESENTATIVE VALUES OF TRACTIVE EFFORT COEFFICIENT

Surface	Peak Value, μ_p	Sliding Value, μ_s
Asphalt/Concrete (dry)	0.8 - 1.0	0.65 - 0.8
Asphalt (wet)	0.5 - 0.7	0.45 - 0.6
Concrete (wet)	0.8	0.7
Gravel	0.6	0.55
Dirt road (dry)	0.68	0.65
Dirt road (wet)	0.55	0.4 - 0.5
Snow (packed)	0.2	0.15
Ice	0.1	0.07

The information supplied in this table is primarily based on work done by J.J. Taborek, "Mechanics of Vehicles," MACHINE DESIGN, May 30 - December 26, 1957.

(2.2.2) JULIEN'S ANALYTICAL TIRE MODEL

A comprehensive tire model which describes all of the tire performance behavior previously discussed, does not exist. However, a simple analytical tire model which relates tractive effort to longitudinal slip has been developed by Julien [6]. The basis of his model is the premise that the tire tread can be modeled as an elastic band. As shown in Figure (2.6), when a driving torque is applied to a tire it produces a region of tread compression in front of the contact patch. The tread contraction, e_o, within this region can be related to an average longitudinal strain, ε, by the equation

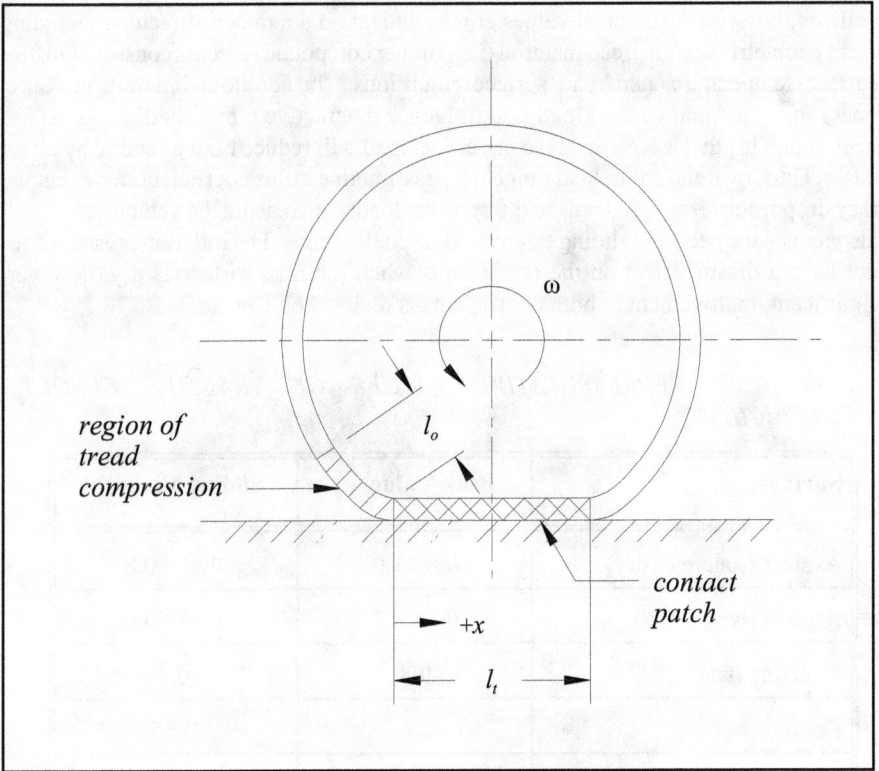

Figure (2.6): JULIEN'S ELASTIC BAND TIRE MODEL

$$e_o = \frac{\varepsilon}{l_o} \tag{2.2}$$

where l_o represents the circumferential distance of the compression region. If it is assumed that the average longitudinal strain remains constant throughout the tread compression region and the contact patch, then the tread contraction in the contact patch can be expressed as

$$e = e_o + \varepsilon x$$
$$= \left(l_o + x \right) \varepsilon \tag{2.3}$$

For a moment, consider that no sliding of the tread takes place within the contact patch. In this case the fundamental relationship between tractive effort and contraction is

$$\frac{dF_x}{dx} = k_t e$$

$$= k_t\left(l_o + x\right)\varepsilon$$

$$(2.4)$$

where k_t is the elastic tangential stiffness of the tread, a material property. The tractive force at any location inside the contact patch is found by integrating Eq. (2.4):

$$F_x = \int_0^x k_t\left(l_o + x\right)\varepsilon\, dx$$

$$= k_t\varepsilon\left(l_o x + \frac{x^2}{2}\right)$$

$$(2.5)$$

The upper limit of x is fixed by either the length of the contact patch or the location where sliding occurs, which ever is less.

Under ideal conditions of uniform normal force distribution over the entire contact patch the local maximum tractive force is limited by Coulomb static friction, i.e.,

$$\frac{dF_{x,\,max}}{dx} = \mu_p\frac{F_z}{l_t}$$

$$(2.6)$$

If l_c is used to denote the location which defines the onset of sliding, then Eq. (2.4) indicates

$$\frac{dF_{x,\,max}}{dx} = k_t\left(l_o + l_c\right)\varepsilon$$

$$(2.7)$$

Equating Eqs. (2.6) and (2.7) results in

$$l_c = \frac{\mu_p F_z}{k_t l_t \varepsilon} - l_o$$

$$(2.8)$$

When $l_c \geq l_t$ the tractive effort created by the entire contact patch can be evaluated from Eq. (2.5) as:

$$F_x = k_t \left(l_o l_t + \frac{l_t^2}{2} \right) \varepsilon \qquad (2.9)$$

In a physical sense the longitudinal slip is a measure of contraction. Then from Eq. (2.9) we conclude that the tractive force is a linear function of longitudinal slip until the critical tread length becomes equal to the tread contact length. At this point the local tractive force reaches impending sliding at the trailing edge of the contact patch.

A further increase in longitudinal slip causes l_c to decrease, which reduces the no slip or adhesion region while increasing the impending sliding region of the contact patch. In this case the tractive force developed by the adhesion region is

$$F_{xa} = k_t \left(l_o l_c + \frac{l_c^2}{2} \right) \varepsilon$$

or, using Eq. (2.8)

$$F_{xa} = \frac{k_t \varepsilon}{2} \left[\left(\frac{\mu_p F_z}{k_t l_t \varepsilon} \right)^2 - l_o^2 \right] \qquad (2.10)$$

The tractive force developed by the impending sliding region follows Coulomb's friction model:

$$\begin{aligned} F_{xs} &= \mu_p F_z \left(1 - \frac{l_c}{l_t} \right) \\ &= \mu_p F_z \left(1 + \frac{l_o}{l_t} \right) - \left(\frac{\mu_p F_z}{l_t} \right)^2 \frac{1}{k_t \varepsilon} \end{aligned} \qquad (2.11)$$

The total tractive force, which is developed by the entire contact patch, is the sum of force components as described by Eqs. (2.10) and (2.11):

$$\begin{aligned} F_x &= F_{xa} + F_{xs} \\ &= \mu_p F_z - \frac{\left(\mu_p F_z - k_t l_t l_o \varepsilon \right)^2}{2 k_t l_t \varepsilon} \end{aligned} \qquad (2.12)$$

Eq. (2.12) obviously suggests that a nonlinear relationship exists between the tractive force and longitudinal slip when impending sliding occurs in the contact patch. With the aid of Eqs. (2.9) and (2.12), a graphical interpretation, Figure (2.7), of tractive force versus longitudinal slip can be made. The portion of the plot originating at **A** and going to **B** is governed by Eq. (2.9), and indicates the tire behavior while the contact patch experiences adhesion conditions. The nonlinear behavior shown from **B** to **C** results from an increasing impending sliding region, which grows from the trailing edge forward until the region encompasses the entire contact patch at point **C**. An increase is slip beyond this point produces an unstable situation in which the tractive force rapidly falls from its peak value of $\mu_p F_z$ to a pure sliding value of $\mu_s F_z$.

There are definitely a multitude of assumptions made in Julien's tire model, but nevertheless it does provide us with better insight into mechanics of pneumatic tires.

(2.2.3) BRAKING FORCE AND SKID

The discussion presented thus far has concerned itself mainly with tire behavior and tractive effort resulting from the application of a driving torque. Aside from the obvious directional differences, similar tire behavior and performance is observed under braking conditions. The fundamental difference between the two is the tread behavior occurring before and after the contact patch. When a braking torque is applied to a tire a region of tread elongation is created in front of the contact patch, followed by tread compression behind the patch. In this situation the tread elongation causes an increase in the tire rolling radius. Since this is opposite to the effect created by a driving torque, an adjustment in the deformation slip definition is necessary. Under braking conditions the deformation slip has been historically referred to as skid number[1], and is defined as

$$SN = 1 - \frac{r}{r_e}$$

$$= 1 - \frac{r\omega}{V}$$

$$(2.13)$$

Analogous to the longitudinal slip, the range of skid number is $0 \le SN \le 1$. In this case, however, the upper limit of $SN = 100\%$ represents a "locked wheel" condition, i.e., forward motion ($V > 0$) without wheel rotation ($\omega = 0$).

The performance of a braking tire is essentially the same as that shown in Figure (2.5) for a tire under the influence of a driving torque; but, the nomenclature is slightly different. Tire braking performance is expressed in terms of a braking force coefficient, defined as $\frac{F_x}{F_z}$, and skid number instead of longitudinal slip. Then when a braking torque is applied to the tire we would say that the braking force coefficient

---

[1] *Newer studies in vehicle dynamics and tire modeling treat the deformation slip under braking conditions as negative longitudinal slip.*

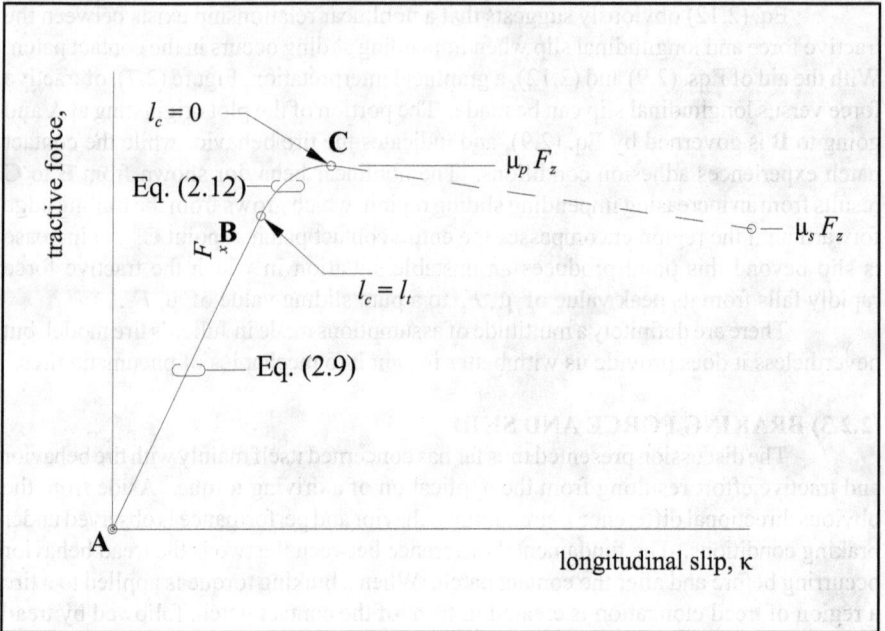

Figure (2.7): TRACTIVE EFFORT RESULTS OF JULIEN'S THEORY

increases with skid, first as a linear function, then becoming non-linear until the skid reaches a value of 12 to 20%. At this point the braking force coefficient achieves a maximum value equivalent to μ_p. A further increase in skid produces an unstable situation where the braking force coefficient quickly drops to its sliding value μ_s. The same operating parameters, e.g., road conditions, which affect peak and sliding tractive effort coefficients also affect the braking force coefficients. Although, the affect is not necessarily symmetric. For example, due to tire construction, the affect of vertical load on the peak and sliding tractive effort coefficients may be slightly different than its affect under braking conditions.

(2.2.4) ROLLING RESISTANCE

The characteristics of pneumatic tires which make them indispensable in vehicle suspensions are also responsible for a significant portion of the energy losses associated with vehicle motion. For example consider a tire rolling on a hard surface. The sidewall deflection and tread compression/elongation near the contact patch cause a certain amount of energy to be dissipated due to the visco-elastic properties of the tire. This is referred to as internal hysteresis, and accounts for approximately 90% of the tire energy losses. The remaining losses can be attributed to friction between the tire and ground, and aerodynamic drag on the tire. The affect of the tire energy losses is a non-uniform distribution of normal pressure within the contact patch. As illustrated in Figure (2.8), higher normal pressures prevail in the leading half of the contact patch, which results in the normal force shifting forward of the spin axis. The shift in normal force produces a moment, called a rolling resistance moment, about the tire spin axis.

If we consider a free rolling tire, i.e., one without the benefit of an applied wheel torque, a horizontal force, F_r in Figure (2.8), acting at the contact patch must exist in order for equilibrium to be maintained. This force is referred to as the rolling resistance.

There are a number of tire properties and operating parameters that affect the magnitude of rolling resistance. These include tire construction and materials, road surface conditions, vehicle speed, tire inflation pressure and temperature. It is customary when correlating these affects to use the *coefficient of rolling resistance*, f_r, defined as the ratio of rolling resistance to vertical load, i.e., $F_r = f_r F_z$.

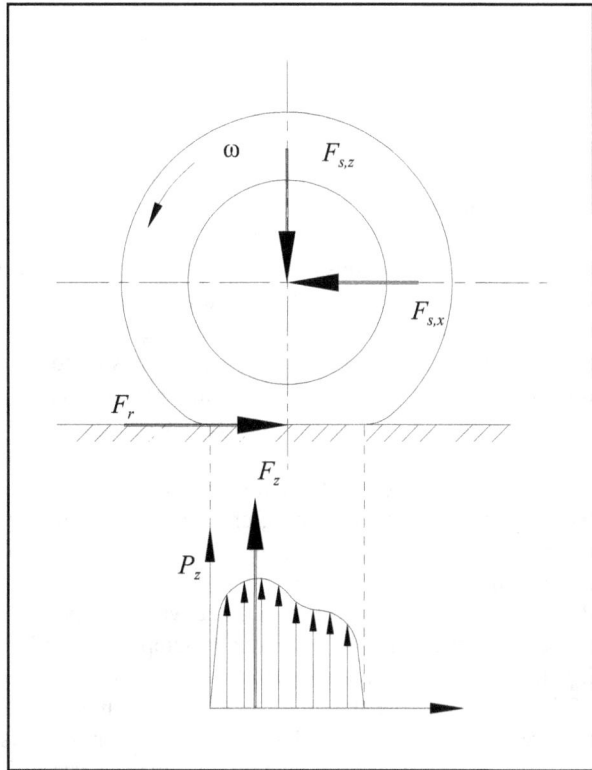

Figure (2.8): FORCE DIAGRAM OF TIRE ROLLING RESISTANCE

To great extent the flexibility and stiffness of a tire is determined by the inflation pressure. However, it is the combination of inflation pressure and road (or off-road) surface conditions that affect the rolling resistance. In fact these are the two most important parameters influencing the rolling resistance. On concrete pavement or other hard surfaces the rolling resistance decreases slightly with increasing inflation pressure. This is due to a decrease in sidewall distortion and a reduction in contact patch area. These effects tend to reduce the internal hysteresis and minimize the frictional loss between the tire and road.

An entirely different result is obtained for soft road surfaces. To understand this phenomenon we will introduce the concept of contact pressure, defined as the wheel vertical load divided by the contact patch area. Any ground surface, whether it be clay, loam, sand, or asphalt, can only provide a certain amount of support without deforming. If the contact pressure should exceed the support capacity of the ground, the surface will yield and the vehicle will "sink". As the tire sinks into the ground the contact patch area increases, thereby decreasing the contact pressure, until equilibrium is achieved. Of course if we have a very soft surface, such as mud, it is possible that equilibrium is unattainable, and our vehicle will sink right up to the belly pan. When the tire penetrates into the ground surface, a significant amount of energy is required to "push" the ground out of the way as the tire rolls. This energy penalty appears as a dramatic increase in the rolling resistance. A similar situation exists when the roads are covered with snow. Decreasing the inflation pressure in this case will increase the contact patch area, which causes a corresponding reduction in the contact pressure. This improves the floatation characteristics of the tire. The implication is less earthen material to displace in front of the tire, which in turn reduces the rolling resistance.

As the vehicle speed increases so does the work involved in deforming the tire carcass. The energy consumed in the deformation process appears as an increase in rolling resistance. In addition, tire vibration becomes a problem at higher speeds. We have learned earlier in this section that a region of tread compression exists in front of the contact patch, as well as a region of expansion following the patch. The tire tread does not respond instantaneously with the transition from the contact patch to the expansion region   At higher vehicle speeds the time lag in tread response allows compressed tread elements to enter the expansion region. As the tread elements attempt to reach equilibrium a standing wave is created. We can best describe a standing wave by imagining the tread as being unwrapped from the tire carcass. Starting near the beginning of the expansion region we would notice a large vertical displacement in the tread (radial deflection if the tread was still attached). From that point the tread displacement would appear to oscillate with an exponential decay in amplitude, much the same as the response of an under-damped, second order, linear system to a step input. Under severe conditions the standing wave will extend over the entire periphery of the tire. Once the standing wave has been initiated, energy losses due to internal hysteresis greatly increase. This energy loss manifests itself as a dramatic increase in tire temperature, which will eventually cause catastrophic failure. Consequently, the standing wave phenomenon places an upper limit on the safe, continuous operating speed of the tire. The actual speed at which a standing wave is formed depends on inflation pressure, vertical load, and the finer details of the tire construction. However, tire manufacturers are required by law to display a maximum speed code, see Section (2.1.3), on the tire sidewall.

The tire aspect ratio also has an impact on rolling resistance. Tires with a relatively low aspect ratio tend to be stiffer. This results in a shorter contact patch length and less sidewall distortion when compared to tires of similar diameter. Low aspect ratios, therefore, provide a relative decrease in rolling resistance.

The interdependence of tire construction, performance characteristics and operating conditions makes it virtually impossible to develop an analytical expression describing rolling resistance. However, there is one relationship which has been used for many years to estimate the coefficient of rolling resistance for passenger car tires on concrete pavement [7]:

$$ f_r = f_o + f_s \left(\frac{V}{100} \right)^{2.5} \qquad ; \ V \ [km/hr] \qquad (2.14a) $$

or

$$f_r = f_o + f_s \left(\frac{1.6093 \, V}{100} \right)^{2.5} \qquad ; \; V \, [\text{MPH}] \qquad (2.14b)$$

The coefficients f_o and f_s can be determined from Eq. (2.15):

$$\left. \begin{array}{c} f_o = 0.008 + \dfrac{0.012}{\exp\left(0.18 \, P^{0.7}\right)} \\[4mm] f_s = \dfrac{0.020}{\exp\left(0.145 \, P^{0.7}\right)} \end{array} \right\} \qquad P \, [\text{psi}] \qquad (2.15)$$

The previous equations provide a reasonable estimate of rolling resistance suitable for vehicle performance simulation. However, exact values of rolling resistance can only be obtain through experimental means.

(2.3) CORNERING FORCES AND MOMENTS

Tire behavior observed during cornering is quite a bit different from that for driving or braking. Cornering requires lateral forces, or cornering forces, to be developed at the contact patch instead of longitudinal forces. In order for equilibrium to be maintained, an equal but opposite reaction force must exist along the spin axis. Under these conditions the elastic nature of the sidewall allows the contact patch to distort and deflect away from the wheel longitudinal center plane. Although Figure (2.9) is an exaggerated depiction of the contact patch distortion, it does give us some idea of the contact patch movement and sidewall distortion that occurs during a cornering maneuver for a free rolling tire.

Figure (2.9) also portrays the tire cornering phenomenon referred to as sideslip. From the illustration we notice that the direction the tire is actually rolling in, i.e., heading, is different from the direction indicated by the tire longitudinal axis. The angular deviation between these two directions is called the slip angle, α. Under stable cornering conditions the front portion of the contact patch has a firm grip on the road surface, and the front of the contact patch is aligned with the heading. However, the rear of the contact patch tends to bend towards the longitudinal axis. As result of this bending behavior, a certain amount of sliding occurs at the rear of the contact patch. For small slip angles, the contact patch is essentially aligned with the direction of motion with only a small amount of sliding. As the slip angle is increased, the contact patch bending moves forward with a subsequently larger area of sliding, until the slip angle reaches a value of 12° to 15°. At this point the entire contact patch area is sliding and the lateral force has reached its maximum.

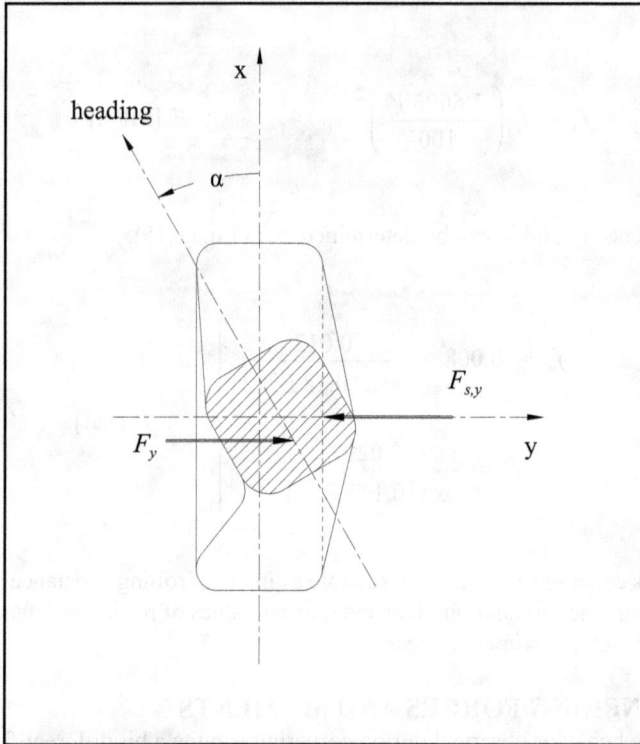

Figure (2.9): CONTACT PATCH GEOMETRY AND CORNERING FORCES

(2.3.1) LATERAL FORCE AND SLIP ANGLE

There are two explanations that are used to describe the interaction between lateral force and slip angle: (1) the application of a lateral force to a pneumatic tire generates a slip angle, and (2) when a tire is steered it creates a slip angle which produces a lateral force. We will find both explanations useful when we study the cornering behavior of vehicles later on. In any event, it is clear that a relationship exists between the lateral force and the tire slip angle. A general plot of lateral force coefficient, defined as $\dfrac{F_y}{F_z}$, versus slip angle is shown in Figure (2.10). The lateral force is a fairly linear function of slip angle up to about four degrees for common road tires. Beyond four degrees the sliding within the contact patch becomes significant, and begins to limit the rate of increase in lateral force.

There are a number of factors that influence the limit of adhesion, and consequently the cornering characteristics of a particular tire. The list of tire parameters and operating conditions are basically the same as those encountered in driving and braking performance, i.e., tread compound and carcass construction, tread surface temperature, road conditions, inflation pressure, and vertical load. In addition, two more factors must be considered: rim width and camber. Figures (2.11) through (2.13) qualitatively illustrate the effects of vertical load, inflation pressure, and tread temperature, respectively, on the lateral limit of adhesion. With respect to inflation

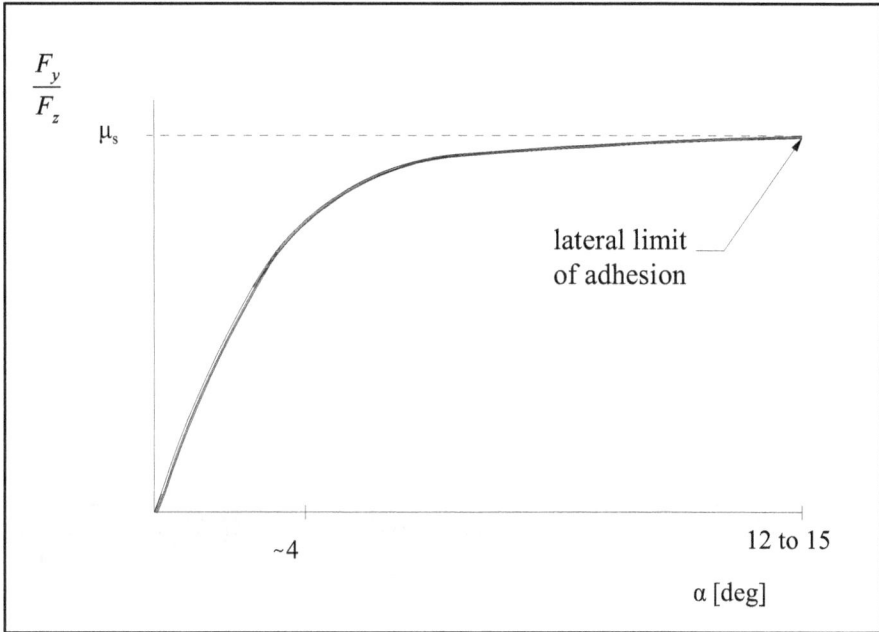

$$\frac{F_y}{F_z}$$

μ_s

lateral limit
of adhesion

~4

12 to 15

α [deg]

Figure (2.10): TYPICAL VARIATION OF LATERAL FORCE COEFFICIENT WITH SLIP ANGLE

pressure and tread temperature we see that there is an optimum point which extends the lateral limit of adhesion to its maximum. The improvement in the adhesion limit manifests itself as a relative increase in lateral force at lower slip angles. For a particular carcass and tread compound the optimum operating pressure and temperature is interrelated and varies depending on the road surface and ambient conditions. Generally speaking, maintaining the proper inflation pressure on conventional car and truck tires is sufficient, as the temperature does not have a great effect on these tires. On the other hand, cornering characteristics attributed to rubber compounds used in racing tires are greatly affected by temperature. Finding the optimum operating point for racing tires is usually accomplished by trial-and-error means.

Rim width also has an impact on cornering performance. To maximize cornering performance it is desirable to use a rim that produces a minimal amount of contact patch lateral distortion. A rim width that is too narrow forces the tread into a convex shape, while a rim too wide creates a concave curvature in the tread. In either case, the result is a decrease in the contact patch area and a decrease in the uniformity of vertical force distribution. The tire manufacturer generally specifies the rim width required for a given tire, and it is advisable to follow these recommendations. If for some reason this information is unavailable, we can use the following "rule of thumb"

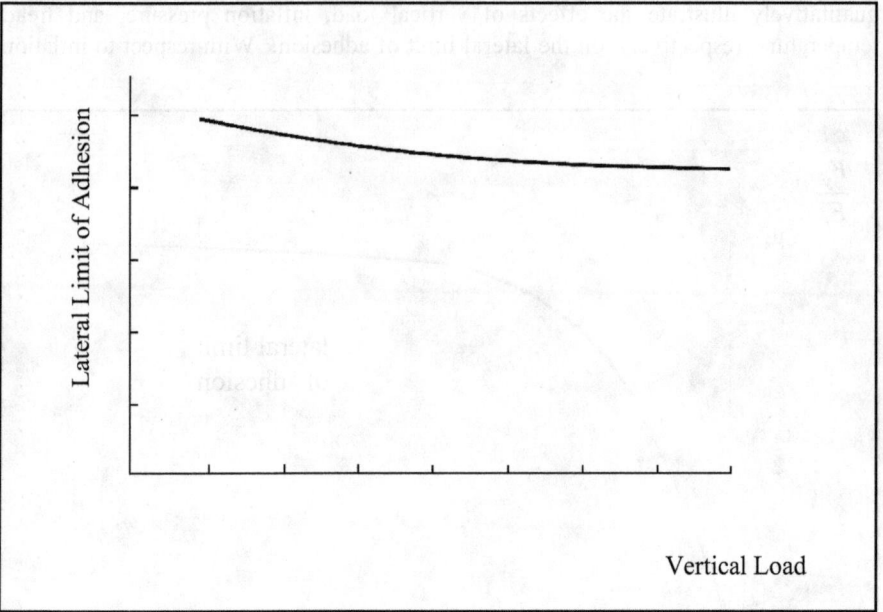

Figure (2.11): EFFECT OF VERTICAL LOAD ON LATERAL LIMIT OF ADHESION

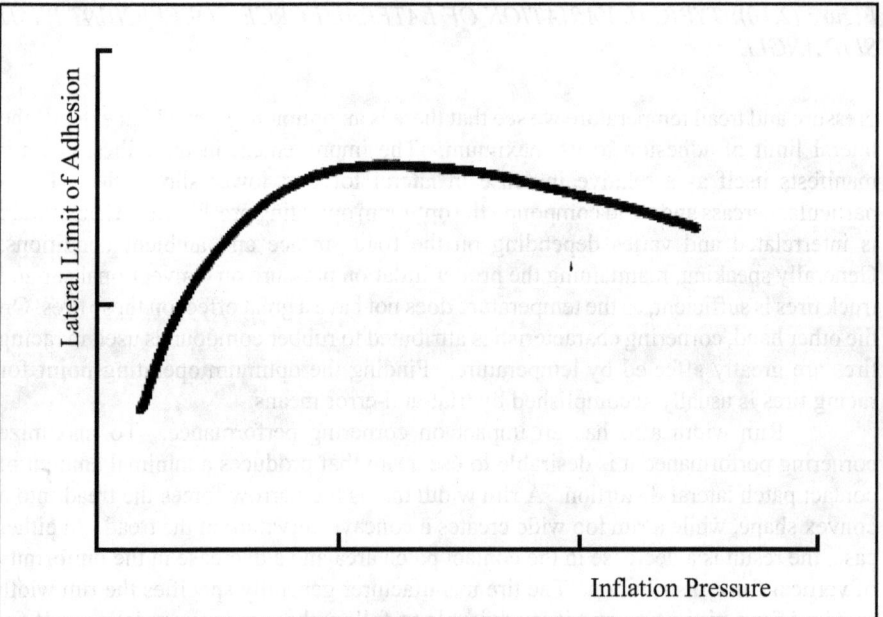

Figure (2.12): EFFECT OF TIRE INFLATION PRESSURE ON LATERAL LIMIT OF ADHESION

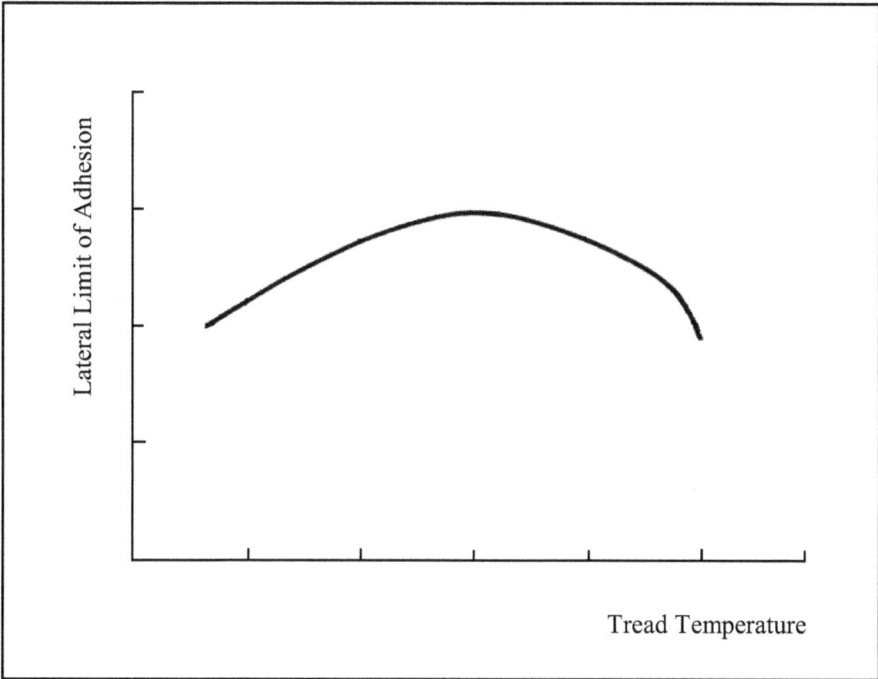

Figure (2.13): EFFECT OF TIRE TREAD TEMPERATURE ON LATERAL LIMIT OF ADHESION

for sizing rims: the rim width should not be less than 70% of the tire section width, nor greater than 85%.

(2.3.1A) Camber and Camber Thrust

Camber, γ, is the angle between the wheel-centered vertical axis and the wheel centerline when viewed from either the front or the rear of the vehicle. Figure (2.14) indicated that camber is either positive or negative, depending on the direction the wheel centerline is tilted in. A positive value is assigned to camber angles when the top of the tire is leaning away from the chassis, while negative camber angles indicate that the tire is leaning in.

Originally, camber was a design parameter of early suspension systems. The camber angle was fixed such that a majority of the wheel load was transferred to the inner wheel bearing of front suspensions. Due to advances in metallurgy and improvements in bearing design, this is no longer the case. Camber also has an impact on vehicle cornering performance. We will study this topic in greater detail in Chapter 5.

The longitudinal centerline of a contact patch is distorted into a semi-elliptical shape whenever the tire is cambered in or out. For the moment we will consider this to be the natural condition, or unstressed state, of the contact patch. However, if this cambered tire was restricted to roll in a straight line (as would be the case if the tire was

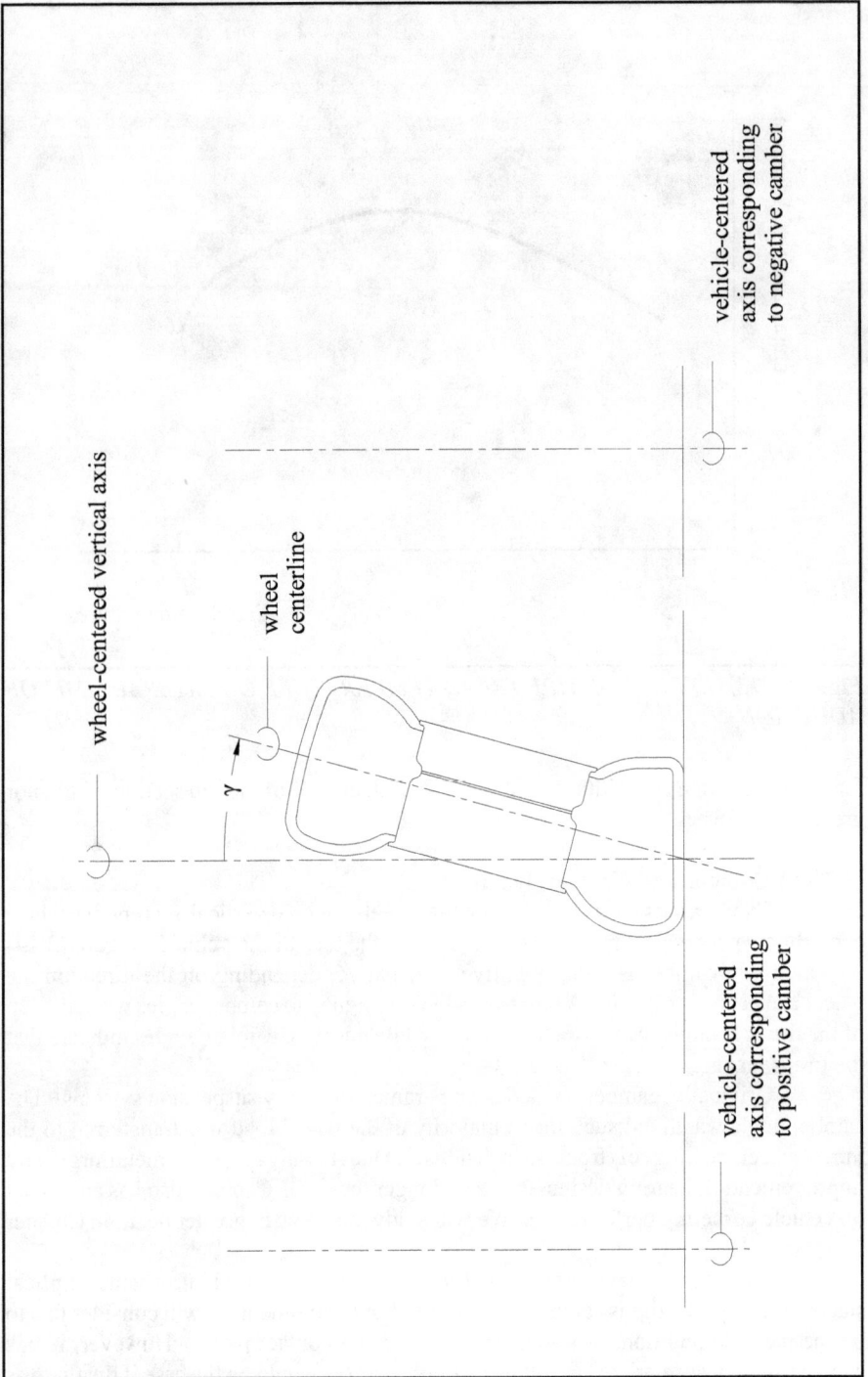

Figure (2.14): CAMBER GEOMETRIC DEFINITION

attached to a vehicle suspension), then the contact patch centerline would be forced into a straight line parallel to the direction of travel. The linear shape of the centerline for the rolling tire is an unnatural condition. This suggests that lateral compressive stresses must exist within the contact patch that "undistort" the centerline. The integration of the compressive stresses over the length of the contact patch yields a lateral force. Which is referred to as camber thrust. The reaction to the camber trust is a lateral ground force which acts in the same direction as the tire is tilted, e.g., the ground force points away from the vehicle centerline for a tire with positive camber.

Camber thrust also exists for tires with non-zero values of slip. Under conditions of tire slip, the total lateral force is occasionally expressed as a sum of the lateral forces due to slip and camber, i.e.,

$$F_{y,\,total} \;=\; F_{y,\,slip} \;+\; F_{y,\,camber}$$

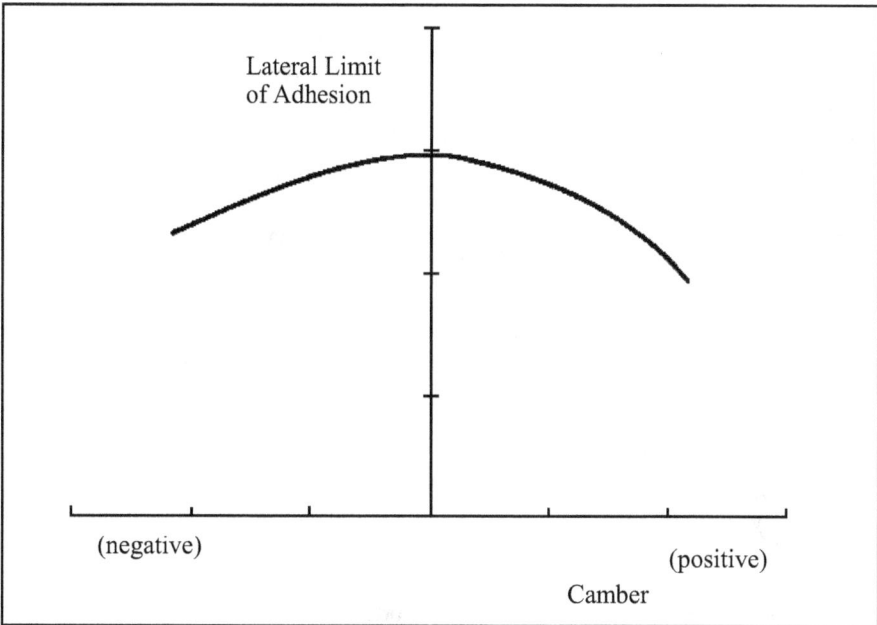

Figure (2.15): EFFECT OF CAMBER ON LATERAL LIMIT OF ADHESION

At first glance it appears from the previous equation that the combined lateral force is simply the slip lateral force linearly offset by the camber thrust. This, as you might expect, is not the case. The compressive stresses introduced into the tread as a result of camber have an influence on the lateral force being developed through tire slip. The effect is most noticeable at high angles of slip. Since it is not possible to describe the influence of camber in simple terms, it is most convenient to express the total lateral force as a single nonlinear function of slip angle and camber. In this regard Figure (2.15) indicates that slightly negative values of camber will extend the lateral limit of adhesion. The use of too much negative camber can be detrimental to tire performance;

as it accelerates tire wear and in some case, especially for wider tires, causes the outside portion of the tread to lift while driving or braking.

(2.3.2) SELF-ALIGNING TORQUE AND PNEUMATIC TRAIL

Up to this point our discussion of lateral forces has been without any reference to the location at which the lateral force acts on the contact patch. The friction behavior of the tread inside the contact patch causes a non-uniform distribution of lateral force intensity, $\dfrac{dF_y}{dx}$ or F_y'. It is convenient for us when describing the cornering characteristics of tires, either analytically or experimentally, to use just a single lateral force. Mathematically, the lateral force results from an integration of the lateral force intensity over the longitudinal length of the contact patch. However, in order to obtain a true representation of the contact patch behavior, the point of application for the lateral force must be at the area centroid of the lateral force intensity distribution.

The two examples provided in Figure (2.16) show that the lateral force acts behind the center of the contact patch. The distance between the center of the contact patch and the point of application of the lateral force is called the pneumatic trail, e. The fact that the lateral force is not co-linear with the reaction force along the spin axis, shown in Figure (2.17), gives rise to a force couple or moment equivalent to $F_y\,e$. This moment is referred to as the self-aligning torque. Two different graphs showing typical characteristics of self-aligning torque are given in Figure (2.17). The graphs demonstrate that the self-aligning torque achieves a maximum before the lateral limit of adhesion is reached. The decay in self-aligning torque is caused by a decrease in pneumatic trail at higher slip angles. The self-aligning torque does not significantly influence vehicle cornering performance or stability, but it does play a role in the design of the steering system.

(2.3.3) ELASTIC BEAM ANALYTICAL TIRE MODEL

The mechanics of pneumatic tire behavior during a cornering maneuver is considerably more complex than that of driving or braking. Consequently the analytical tire models for studying cornering mechanics are more involved as well, but not impossible to understand.

One of the classic analytical tire models used in the study of cornering behavior is the elastic beam model. In this model the tire tread is at first considered to be separate from the rest of the carcass. Furthermore, the portion of the tread that defines the contact patch is broken up into an infinite number of individual lateralcolumns, Figure (2.18), each supporting a fraction of the total lateral force. The relationship between stress and strain for the tread column is $\sigma_y = \varepsilon_y E$, or

$$dF_y = E\,\frac{t}{w}\,\delta\,dx \qquad (2.15)$$

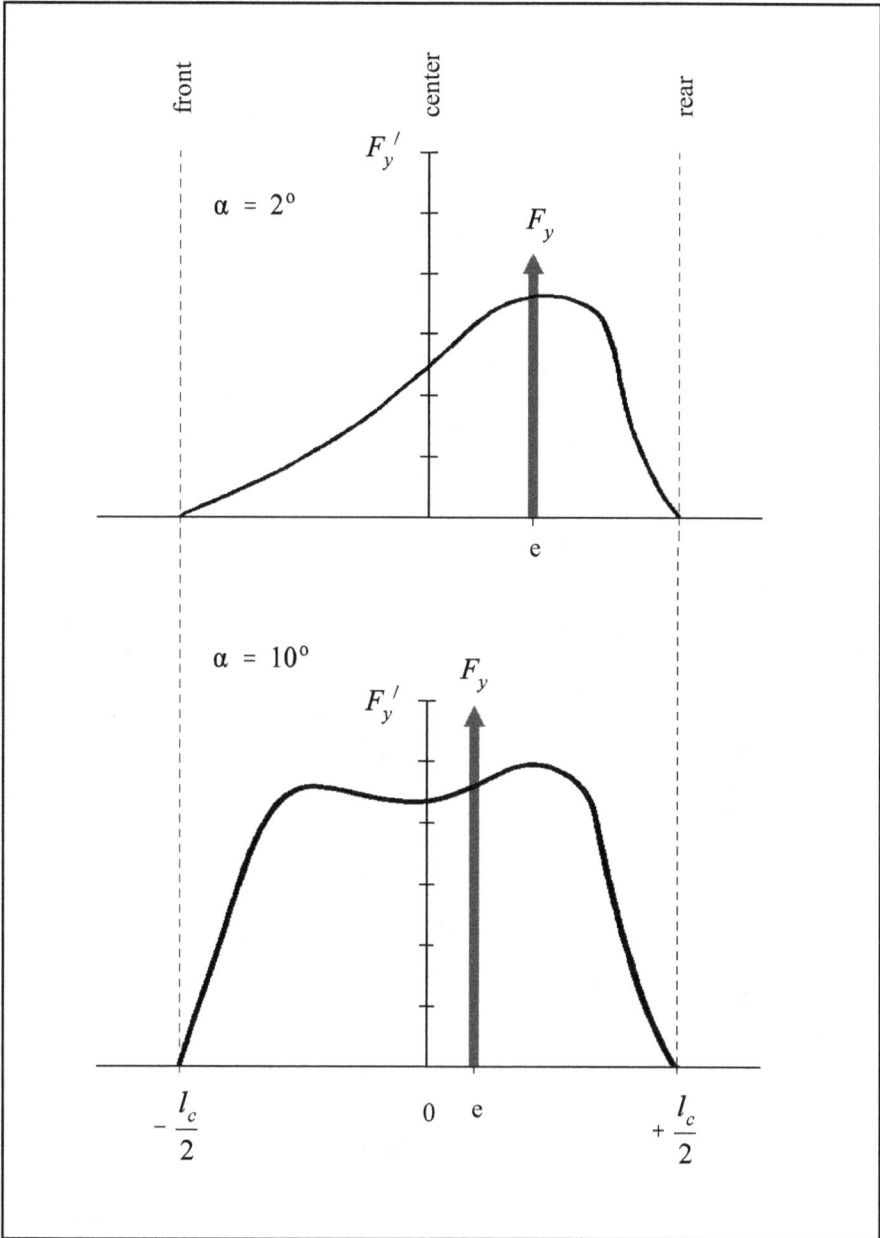

Figure (2.16): EXAMPLE LATERAL FORCE INTENSITY DISTRIBUTION INSIDE CONTACT PATCH

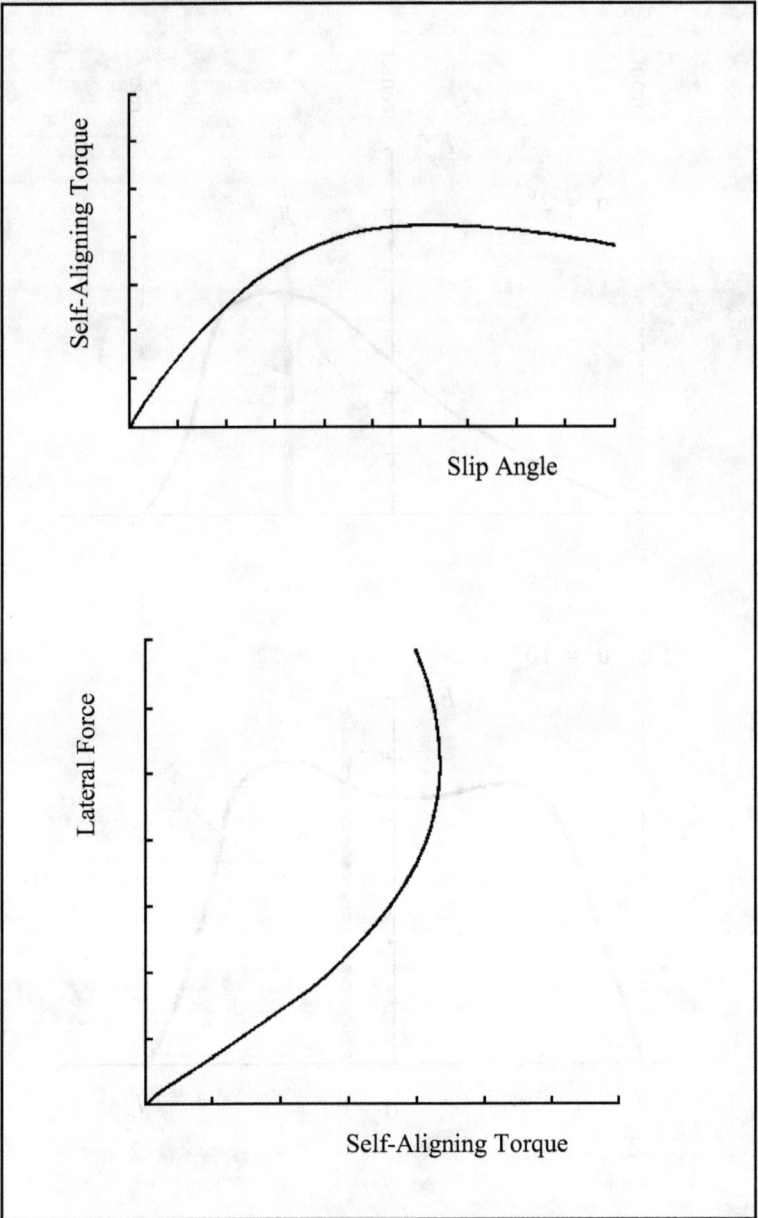

Figure (2.17): SELF-ALIGNING TORQUE CHARACTERISTICS

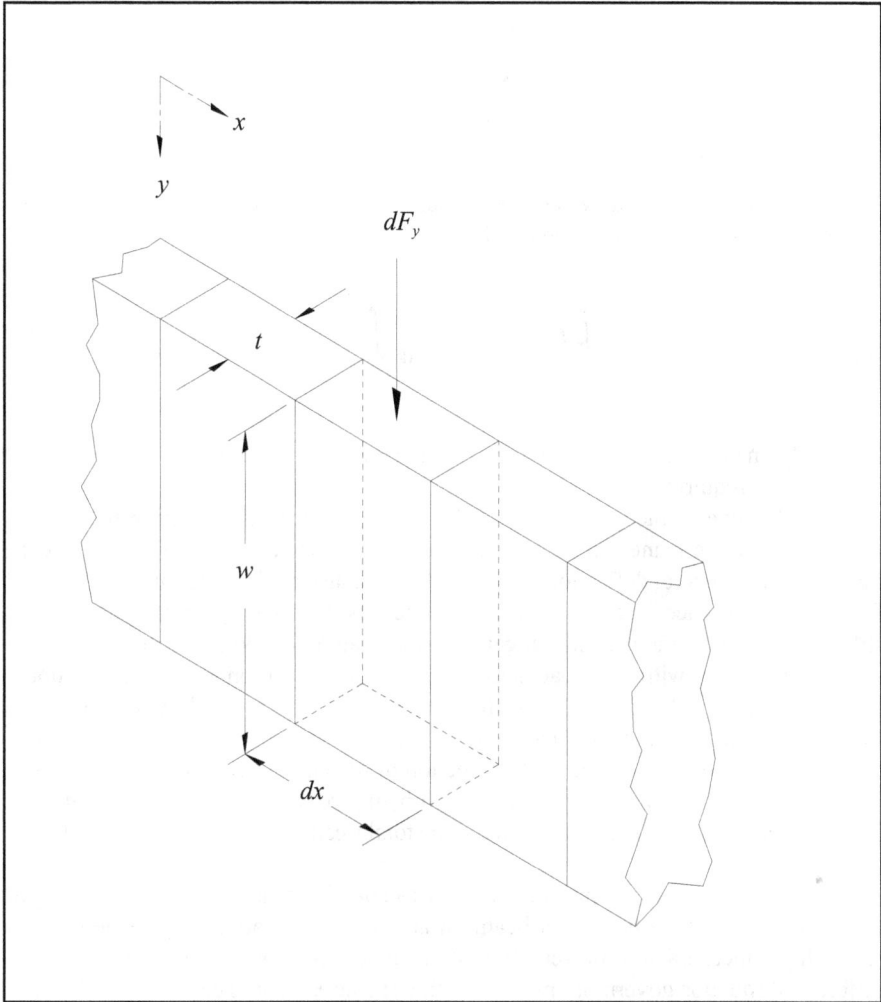

Figure (2.18): TREAD ELEMENTS FOR ELASTIC BEAM MODEL

where

$E \equiv$ tread modulus of elasticity
$t \equiv$ tread thickness
$w \equiv$ tread width, and
$\delta = \delta(x)$, tread elongation.

Eq. (2.15) can also be expressed in terms of the lateral force intensity, F_y' :

$$F_y' = \frac{dF_y}{dx} = E\frac{t}{w}\delta(x) \qquad (2.16)$$

The lateral force supported by the tread, which is the reaction to the force exerted by the road surface on the bottom of the tread, is found from

$$F_y = \int_0^{l_c} F_y' \, dx = E\frac{t}{w}\int_0^{l_c} \delta(x) \, dx \qquad (2.17)$$

Obviously, in order to evaluate the previous integral a relationship describing the tread elongation is required.

Relative to the rim centerline, the total lateral tread deflection that occurs during a cornering maneuver can be thought of as the result of two different effects. First, there is a general deflection in the tread band cause by the application of a lateral force. In this instance the displacement of the tread is resisted only through the sidewall stiffness. Secondly, there is the elongation in the tread rubber below the tread band due to the interaction with the road surface. The superposition of the tread rubber elongation on top of the tread band displacement defines the total deflection of the tread. It should be pointed out that the total tread deflection (remember this is relative to the rim centerline) must match the direction of travel, until relative sliding starts in the contact patch. Returning to our earlier problem, we can state that the elongation in the tread rubber is the difference between the total tread deflection and the tread band displacement.

An approximate expression for the tread band displacement can be developed by modeling the tread band as a beam on an elastic foundation as provided by the sidewall. A mechanical representation of this model is shown in Figure (2.19). The differential equation governing small deflections in the equatorial centerline of the tread band is:

$$EI\frac{d^4v}{dx^4} = -k_l v + F_y'(x) \qquad (2.18)$$

where

$v(x)$ ≡ tread band displacement relative to rim centerline,

EI ≡ tread bending modulus, and

k_l ≡ lateral stiffness {also referred to as a foundation modulus}.

An exact representation of the equitorial displacement is difficult to obtain. However, we can develop an estimate of the displacement by assuming: (1) the tread band is infinity long, and (2) a single external lateral force, F_y, acts at $x = 0$. Under these conditions the solution to Eq. (2.18) is [8]:

$$v = \frac{n F_y}{2 k_l} e^{-nx} \left[\cos(nx) + \sin(nx) \right]$$ (2.19)

where

$$n = \left(\frac{k_l}{4 E I} \right)^{\frac{1}{4}}$$

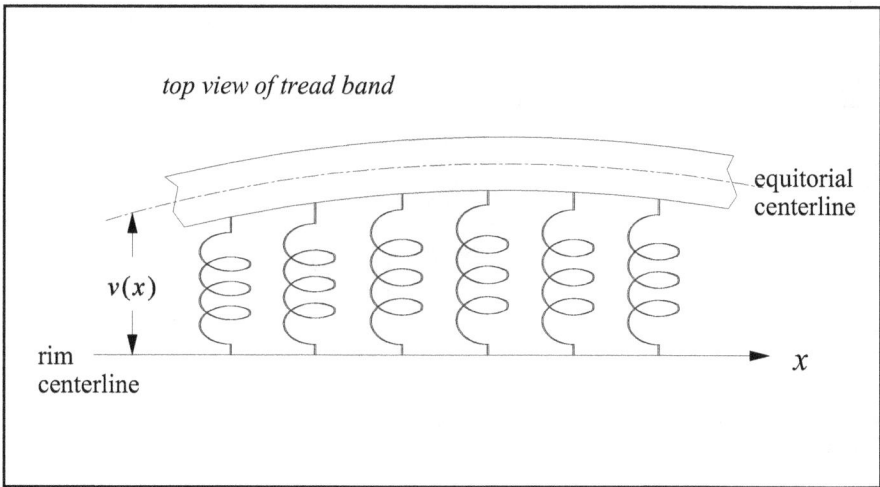

Figure (2.19): MECHANICAL SCHEMATIC OF ELASTIC BEAM TIRE MODEL

A variation of Eq. (2.19) for small values of nx is found by replacing the exponential and trigonometric function with their power series equivalents, i.e.,

$$e^{-nx} = 1 - nx + \frac{(nx)^2}{2!} - \cdots$$

$$\approx 1 - nx$$ (2.20)

and

$$\sin(nx) = nx - \frac{(nx)^3}{3!} + \frac{(nx)^5}{5!} - \cdots$$

$$\approx nx$$ (2.22)

Substituting Eqs. (2.20), (2.21), and (2.22) into Eq. (2.19) results in

$$v \approx \frac{n F_y}{2 k_l} \left[1 - (nx)^2 \right]$$

(2.23)

Eq. (2.23) indicates that the equitorial displacement within the contact patch is approximately parabolic in shape, with a maximum deflection of

$$v_{max} = \frac{n F_y}{2 k_l}$$

(2.24)

We know from practical experience, however, that the tread band is finite in length and has no deflection at the front or rear of the contact patch. Using the parabolic shape suggested by Eq. (2.23) and the maximum deflection defined by Eq. (2.24), we are finally able to approximate the equitorial displacement over the length of the contact patch as

$$v = 4 v_{max} \frac{x}{l_c} \left(1 - \frac{x}{l_c} \right)$$

or

$$v(x) = \frac{2 n F_y}{k_l} \frac{x}{l_c} \left(1 - \frac{x}{l_c} \right) \qquad ; 0 \le x \le l_c$$

(2.25)

As previously mentioned, the tread centerline must match the direction of travel until sliding occurs in the latter part of the contact patch. A general diagram of the tread centerline is given in Figure (2.20). The tread centerline displacement for the adhesion region of the contact patch is determined from the slip angle, i.e.,

$$u(x) = x \tan \alpha \qquad ; 0 \le x \le l_o$$

(2.26)

The sliding region relationship is not quite as simple. The lateral force that is developed as a result of sliding is governed by

$$F_y' = \mu_s F_z'$$

(2.27)

where F_z' is the vertical load intensity. If we assume that the vertical load distribution is parabolic in nature, then the vertical load intensity can be expressed as

$$F_z' = 6 F_z \frac{x}{l_c^2} \left(1 - \frac{x}{l_c} \right) \tag{2.28}$$

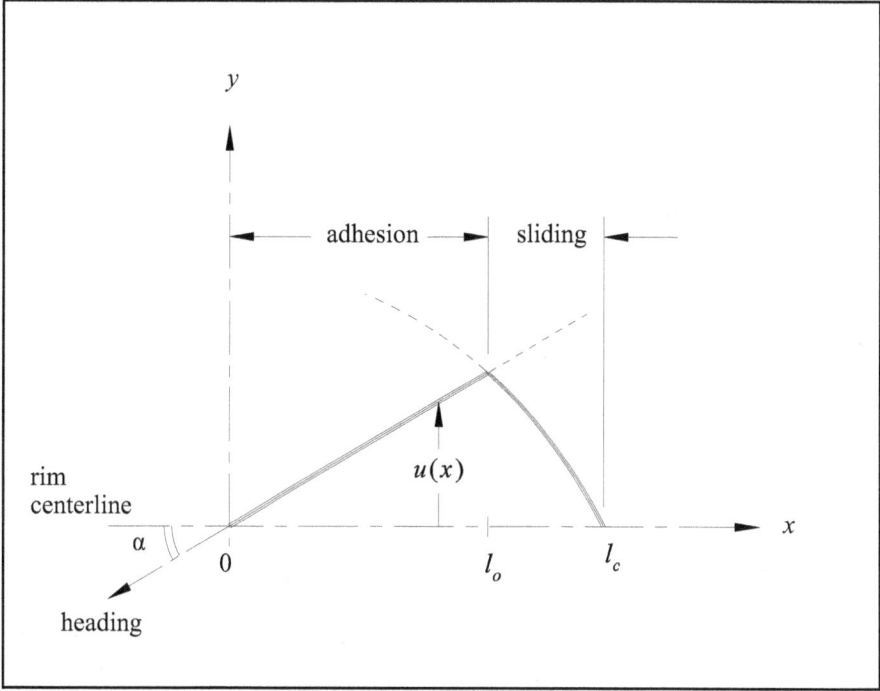

Figure (2.21): TREAD CENTERLINE DISPLACEMENT DIAGRAM

Based on Eqs. (2.27) and (2.28), the lateral load intensity for the sliding region is

$$F_y' = 6 \mu_s F_z \frac{x}{l_c^2} \left(1 - \frac{x}{l_c} \right) \qquad ; l_o \le x \le l_c \tag{2.29}$$

The elongation of the tread rubber within the sliding region of the contact patch is found by substituting Eq. (2.29) into (2.16):

$$\delta(x) = 6 \frac{\mu_s w}{E t} F_z \frac{x}{l_c^2} \left(1 - \frac{x}{l_c} \right) \qquad ; l_o \le x \le l_c \tag{2.30}$$

The tread centerline displacement for the sliding region is

$$u(x) = v(x) + \delta(x)$$

$$= \left(\frac{2nF_y}{k_l} + \frac{6\mu w F_z}{Etl_c} \right) \frac{x}{l_c} \left(1 - \frac{x}{l_c} \right) \qquad ; l_o \le x \le l_c$$

(2.31)

The location l_o which defines the point at which the adhesion region meets the sliding region is determined by the intersection of Eqs. (2.26) and (2.31):

$$u(l_o) = l_o \tan \alpha = \left(\frac{2nF_y}{k_l} + \frac{6\mu_s w F_z}{Etl_c} \right) \left(\frac{l_o}{l_c} \right) \left(1 - \frac{l_o}{l_c} \right)$$

$$\therefore l_o = l_c \left[1 - \frac{l_c \tan \alpha}{\dfrac{2nF_y}{k_l} + \dfrac{6\mu_s w F_z}{Etl_c}} \right]$$

(2.32)

We should pause here for a moment to contemplate the significance of the previous result. A discontinuity in the tread centerline displacement cannot physically exist. This means the growth of the adhesion region is limited by the sliding region, which is always present to some extent at the rear of the contact patch. Furthermore, the local lateral force intensity, $\mu_s F_z'$, of the sliding region must become the upper limit for the adhesion region. Consequently, the maximum lateral force developed under cornering conditions is limited by the sliding friction coefficient, rather than the peak coefficient as in driving or braking.

Now that we have defined the displacements of the tread band and the tread centerline, we are ready to complete the integration specified by Eq. (2.17), i.e.,

$$F_y = \int_0^{l_c} F_y' \, dx$$

$$= \frac{Et}{w} \int_0^{l_o} \left[u(x) - v(x) \right] dx + \int_{l_o}^{l_c} F_y' \, dx$$

$$= \frac{Et}{w} \left[\int_0^{l_o} x \tan\alpha \, dx - \frac{2nF_y}{k_l} \int_0^{l_o} \frac{x}{l_c} \left(1 - \frac{x}{l_c} \right) dx \right]$$

$$+ 6\mu_s F_z \int_{l_o}^{l_c} \frac{x}{l_c^2} \left(1 - \frac{x}{l_c} \right) dx$$

With a little bit of work, the previous integrals reduce to

$$F_y = \frac{Et}{w} \left[\frac{l_o^2}{2} \tan\alpha - \frac{nF_y}{k_l} \frac{l_o^2}{l_c} \left(1 - \frac{2}{3} \frac{l_o}{l_c} \right) \right]$$

$$+ \mu_s F_z \left[1 - 3 \frac{l_o^2}{l_c^2} \left(1 - \frac{2}{3} \frac{l_o}{l_c} \right) \right] \tag{2.33}$$

We can rewrite Eq. (2.32) as

$$\frac{nF_y}{k_l} = \frac{l_c^2 \tan\alpha}{2 \left(l_c - l_o \right)} - \frac{3\mu_s w F_z}{Et l_c} \tag{2.34}$$

Substituting Eq. (2.34) into Eq. (2.33) produces

$$F_y = \mu_s F_z - \frac{Et l_o^3}{6w \left(l_c - l_o \right)} \tan\alpha \tag{2.35}$$

Our tire model now consists of two unknowns, l_o and F_y, in two equations, Eq. (2.32) and Eq. (2.35). If Eq. (2.35) is back-substituted into Eq. (2.32), then the following algebraic equation describing l_o can be obtained:

$$l_o^3 + a\, l_o + b = 0 \qquad (2.36)$$

where

$$a = \left(\frac{6\, \mu_s\, w\, F_z}{E\, t\, \tan \alpha} \right) \left(1 + \frac{3\, w\, k_l}{n\, E\, t\, l_c} \right) \qquad (2.37)$$

and

$$b = \frac{3\, w\, k_l\, l_c^2}{n\, E\, t} - \left(\frac{6\, \mu_s\, w\, F_z\, l_c}{E\, t\, \tan \alpha} \right) \left(1 + \frac{3\, w\, k_l}{n\, E\, t\, l_c} \right) \qquad (2.38)$$

The real solution to Eq. (2.36) is given by

$$l_o = A_1 + A_2 \qquad (2.39)$$

where

$$A_{1,2} = \left(-\frac{b}{2} \pm \sqrt{ \frac{b^2}{4} + \frac{a^3}{27} } \right)^{\frac{1}{3}} \qquad (2.40)$$

After l_o is calculated from Eqs. (2.37) through (2.40), the lateral force, F_y, can be evaluated from Eq. (2.35).

As an example, consider a tire which has the following geometric and mechanical properties:

$$w = 6 \text{ in} \qquad\qquad k_l = 27 \text{ lb/in/in}$$
$$t = 1.5 \text{ in} \qquad\qquad E = 562 \text{ psi}$$
$$l_c = 10 \text{ in}$$

In addition, assume that the sliding friction coefficient is 0.8 and that the vertical load is 800 lb. Under these conditions, l_o and F_y can be evaluated as functions of the tire slip angle, α. The results for this example are shown in Figure (2.21).

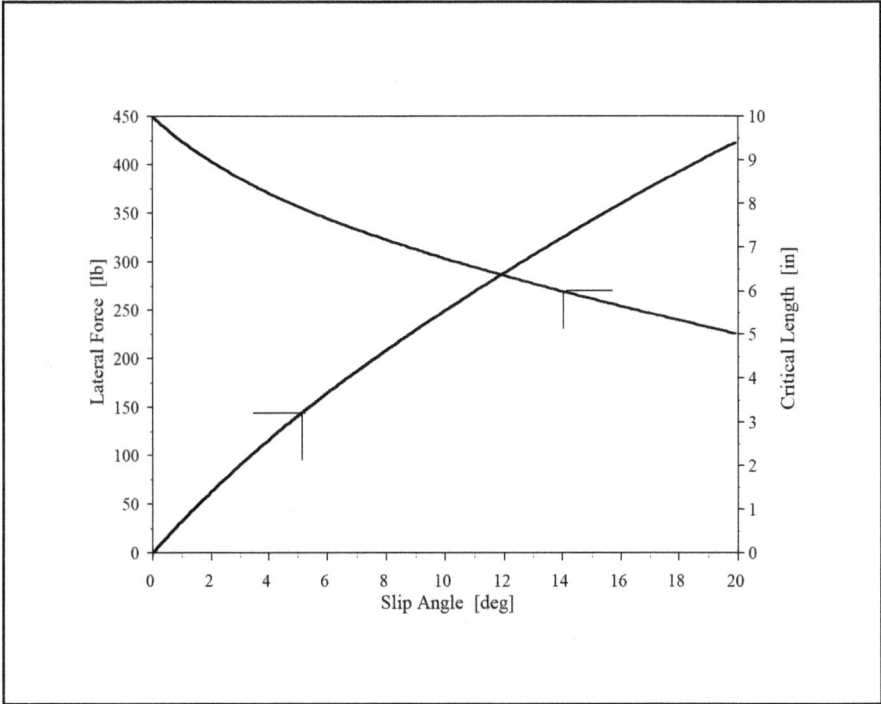

Figure (2.21): LATERAL FORCE RESULTS OF ELASTIC BEAM MODEL

The self-aligning torque is determined from the integral

$$M_z = \int_0^{l_c} F_y' \left(x - \frac{l_c}{2} \right) dx$$

or

$$M_z = \int_0^{l_o} F_y' \left(x - \frac{l_c}{2} \right) dx + \int_{l_o}^{l_c} F_y' \left(x - \frac{l_c}{2} \right) dx \qquad (2.41)$$

Substituting the appropriate relationships for F_y' into the previous equation results in

$$M_z = \frac{Et}{w}\left[\tan\alpha \int_0^{l_o} x\left(x - \frac{l_c}{2}\right)dx - \frac{2nF_y}{k_l}\int_0^{l_o} \frac{x}{l_c}\left(1 - \frac{x}{l_c}\right)\left(x - \frac{l_c}{2}\right)dx\right]$$

$$+ 6\mu_s F_z \int_{l_o}^{l_c} \frac{x}{l_c^2}\left(1 - \frac{x}{l_c}\right)\left(x - \frac{l_c}{2}\right)dx$$

Again with a little bit of work, the foregoing integrals reduce to

$$M_z = \frac{Et}{w}\left[\tan\alpha\left(\frac{l_o^3}{3} - \frac{l_o^2 l_c}{4}\right) - \frac{2nF_y}{k_l l_c}\left(\frac{l_o^3}{2} - \frac{l_o^2 l_c}{4} - \frac{l_o^4}{4 l_c}\right)\right]$$

$$- \frac{6\mu_s F_z}{l_c^2}\left(\frac{l_o^3}{2} - \frac{l_o^2 l_c}{4} - \frac{l_o^4}{4 l_c}\right)$$

or, with the help of Eq. (2.32)

$$M_z = \frac{Et l_o^3}{12 w}\tan\alpha \tag{2.42}$$

Continuing our earlier example, the result for self-aligning torque is shown in Figure (2.22). The pneumatic trail, calculated from $e = \dfrac{M_z}{F_y}$, is also shown in this figure.

In closing this section we should realize that even though we are able to predict general trends for tire cornering behavior, the numeric results for lateral force and self-aligning torque are not particularly accurate. Real tires, as we have learned, contain a complex matrix of chords and belts within the sidewalls and tread. The cornering performance for these tires is significantly different from the homogenious rubber tire used in our analytical model.

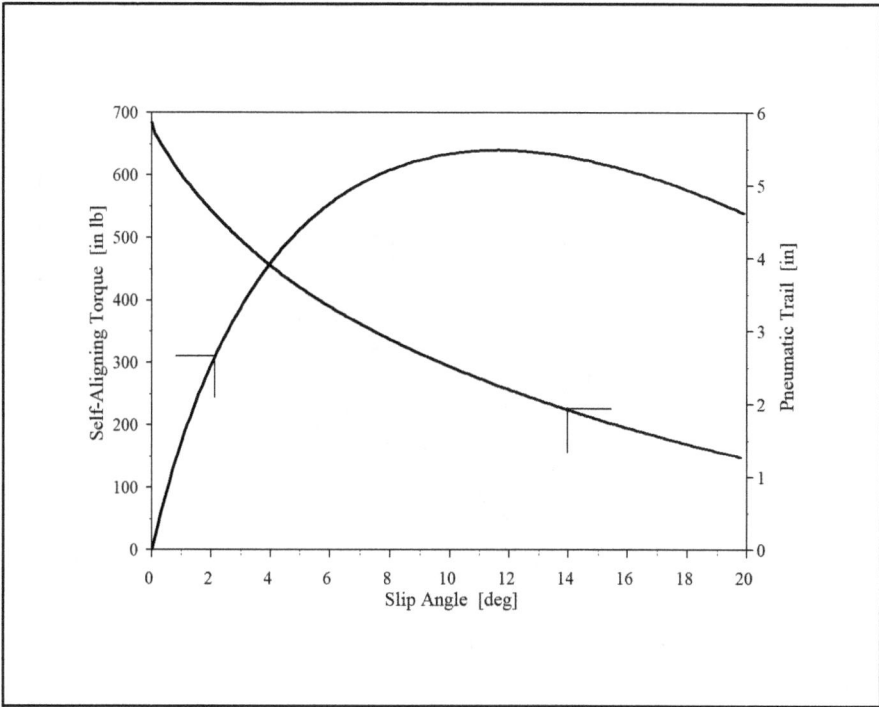

Figure (2.22): SELF-ALIGNING TORQUE RESULTS OF ELASTIC BEAM TIRE MODEL

(2.4) EMPIRICALLY BASED TIRE MODELS

Considering all of the different tire fabrication methods and all of the operating factors influencing tire performance, it is not surprising that a definitive treatise on tire behavior has not been developed. At present, describing tire performance through experimental correlations is the best we can do. Tire performance experiments are usually conducted using either a stationary tire fixture on top of a spinning drum, or a moving tire fixture mounted to a test trailer. In both cases slip angle, longitudinal skid, and camber can be varied. Measurements generally include lateral force, brake force (opposite of tractive effort), and self-aligning torque as defined in Figure (2.23)

The amount of tabular data generated by tire tests can be enormous. Two possible ways to reduce the tabular data into a more compact and useful format are: (1) performance graphs, and (2) empirical correlations. The lateral force coefficient[2] graph for a racing tire shown in Figure (2.24) is an example of a modern performance graph. Graphical representations are easy to generate from tabular data, but they are cumbersome to use in suspension design calculations, and impossible to use in computer

---

[2] *Refer to Section (2.3.1).*

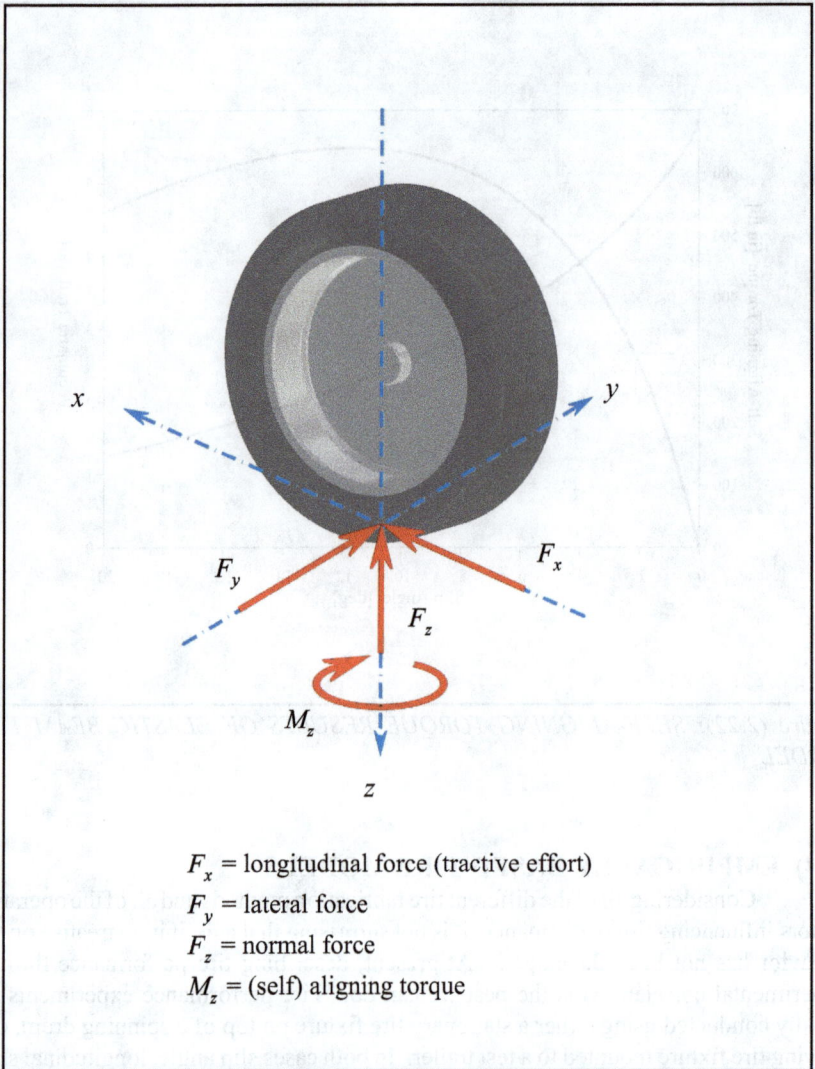

$$F_x = \text{longitudinal force (tractive effort)}$$
$$F_y = \text{lateral force}$$
$$F_z = \text{normal force}$$
$$M_z = \text{(self) aligning torque}$$

Figure (2.23): DEFINITION OF TIRE FORCES AND MOMENTS

simulations. Empirical correlations are more difficult to generate but have the advantage of being easily incorporated into vehicle simulation programs.

An alternate approach of correlating tire performance data is an empirically based tire model. These models usually start with a simple function which imitates the general shape of the desired performance curve. For example, looking at the shape of the curve in Figure (2.24), a lateral force relationship may take the form of:

$$F_y = F_{y,\,\text{max}} \sin\left(B\,\alpha\right) \tag{2.43}$$

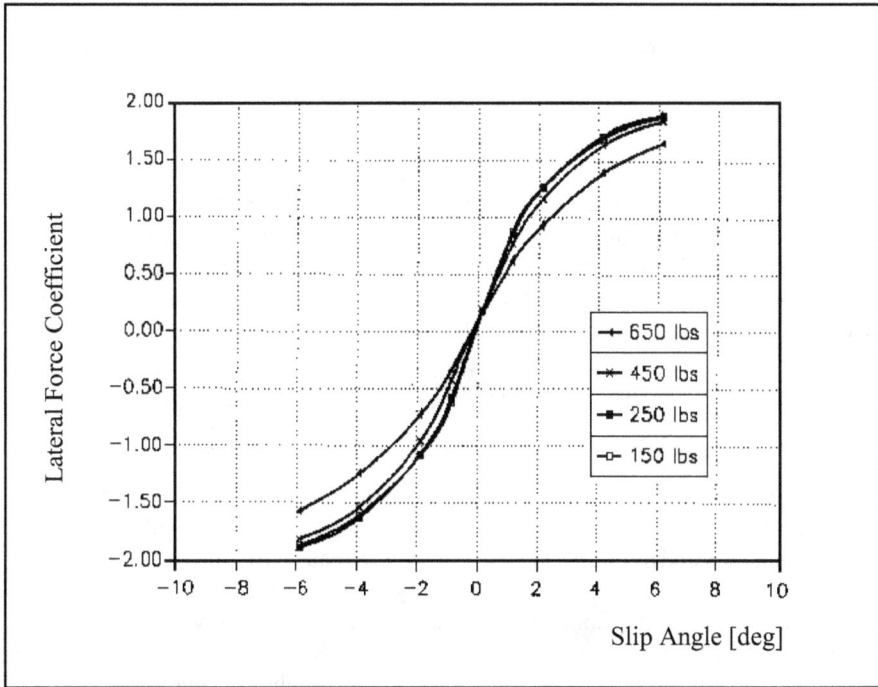

Figure (2.24): EXAMPLE LATERAL FORCE PERFORMANCE GRAPH

This tire model become empirically based by requiring the correlation constant "*B*" to be deduced from experimental measurements. This is accomplished by first noting that

$$\frac{\partial F_y}{\partial \alpha} = B F_{y,\,\text{max}} \cos(B\alpha) \tag{2.44}$$

Evaluating the previous equation at zero slip results in

$$\left.\frac{\partial F_y}{\partial \alpha}\right|_{\alpha=0} = B F_{y,\,\text{max}} \tag{2.45}$$

The cornering stiffness, C_α, of a particular tire is defined as the initial slope of the lateral force versus slip angle performance curve, i.e., $C_\alpha \equiv \left.\dfrac{\partial F_y}{\partial \alpha}\right|_{\alpha=0}$. The correlation constant "*B*" is evaluated from $B = \dfrac{C_\alpha}{F_{y,\,\text{max}}}$ once the cornering stiffness and the peak lateral force have been determined by experimental means. As a practical matter, however, the cornering stiffness and peak lateral force are dependent on the vertical load. To obtain a better performance predictor it is necessary to modify the simple empirical tire model to account for the influence of other tire performance parameters.

A very comprehensive lateral force empirical tire model has been introduced by Bakker, et. al. [9] and has become known as the "magic tyre formula." In an attempt to achieve a good fit between the tire model and actual tire performance curves, the basic sine function was modified using an arctangent function in the sine argument along with four performance trend factors. These factors include: (1) a stiffness factor, B; (2) a dimensionless shape factor, C; (3) a peak force factor, D; and (4) a curvature factor, E. Using these factors, the lateral force equation takes on the general form of:

$$F_y(\Phi) = D \sin\left[C \arctan(B \Phi)\right] + S_v \qquad (2.46)$$

where $\Phi = fn(\alpha, S_h)$.

The terms S_h and S_v refer to horizontal shift and vertical shift factors, respectfully. These factors are included in the tire model are to account for non-ideal effects of conicity and ply steer. These effects tend to cause real tires to develop a small amount of cornering force at zero slip. Conicity refers to the conical shape in the tread relative to the spin axis. Conicity occurs as a result from slightly skewing the belts off-center during the manufacturing process. Conicity behavior is analogous to rolling a cone on a flat surface. The cone will not roll in a straight line, but rather in an arc. Ply steer is directly attributed to the angle of the plies. The tire construction techniques that we studied earlier taught us that alternating layers of plies have opposite angles, i.e., a mirror image. This construction technique attempts to compensate for ply steer, but it is nearly impossible to achieve a balance in the tire design in which it will be completely eliminated.

The application of the magic tyre model has led to several successful tire models. Two of the more popular models, namely the Pacejka '94 model and the MF5.2 model, are discussed next.

(2.4.1) PACEJKA '94 TIRE MODEL

Accuracy improvements in modeling tire behavior in the original magic tyre formula, Eq. (2.46), have led to several improved tire models. Most notably the Pacejka '89 model [10] and a short time later the Pacejka '94 model [11]. The Pacejka '94 tire model for lateral force under conditions of pure slip is summarized in Figure (2.25). This tire model does a good job of predicting tire performance from a fairly limited amount of experimental data, but it is at the expense of a large number of correlation constants, which are referred to as Pacejka coefficients. An example set of coefficients representative of a high performance tire, along with their definitions, are given in Table (2.3). Figure (2.26) shows a plot of lateral force as a function of slip angle using these sample values.

$$F_y = D \sin\left[C \arctan\left(B \, \Phi \right) \right] + S_v$$

with

$$D = \mu_y \, F_z$$

$$BCD = a_3 \sin\left[2 \arctan\left(\frac{F_z}{a_4} \right) \right] \left(1 - a_5 \, |\gamma| \right)$$

$$B = \frac{BCD}{CD}$$

$$\Phi = \left(1 - E \right)\left(\alpha + S_h \right) + \frac{E}{B} \arctan\left[B\left(\alpha + S_h \right) \right]$$

where

$$C = a_0$$

$$\mu_y = a_1 \, F_z + a_2$$

$$E = a_6 \, F_z + a_7$$

$$S_h = a_8 \, \gamma + a_9 \, F_z + a_{10}$$

$$S_v = a_{11} \, F_z + a_{12} + \left(a_{13} \, F_z + a_{14} \right) \gamma \, F_z$$

Figure (2.25): PACEJKA '94 LATERAL FORCE TIRE MODEL UNDER CONDITIONS OF PURE SLIP

The magic tyre formula has been successfully applied to model not only lateral force, but longitudinal force[3] and aligning torque[4] as well. These tire models are given in Figures (2.27) and (2.28) along with their corresponding empirical coefficients in Tables (2.5) and (2.6). In addition, example tire output results for longitudinal force

---

[3] *Longitudinal force as represented in tire modeling is calculated as either positive, indicating a tractive effort, or negative indicating a braking action.*

[4] *Current nomenclature uses "aligning torque" to replace the traditional wording of self-aligning torque and/or self-aligning moment.*

and aligning torque under conditions of pure slip are provided in Figures (2.29) and (2.30).

Table (2.3): LATERAL FORCE SAMPLE COEFFICIENT VALUES FOR PACEJKA '94 TIRE MODEL UNDER CONDITIONS OF PURE SLIP

Coefficient	Sample Value	Definition
a_0 [--]	1.6929	shape factor
a_1 [1/kN]	-0.0552084	load influence on lateral friction coefficient
a_2 [--]	1.27128	lateral friction coefficient
a_3 [N/deg]	1601.8	change in stiffness with slip
a_4 [kN]	6.4946	change in stiffness progress with load
a_5 [1/deg]	0.0047966	camber influence on stiffness
a_6 [1/kN]	-0.3875	curvature change per load
a_7 [--]	1	curvature factor
a_8 [--]	-0.045399	camber influence on horizontal shift
a_9 [deg/kN]	0.0042832	load influence on horizontal shift
a_{10} [deg]	0.086536	horizontal shift at zero load
a_{11} [N/kN]	7.668	load influence on vertical shift at zero camber
a_{12} [N]	45.8764	vertical shift at zero load
a_{13} [N/deg/MN]	-7.973	load dependent camber influence on vertical shift
a_{14} [N/deg/kN]	-0.2231	camber influence on vertical shift

The required input variables for the Pacejka 94 tire model are the normal load, F_z [kN], camber angle, γ [deg] and slip anlge, α [deg] (positive value indicates a right-hand turn). The corresponding output is lateral force, F_y [N].

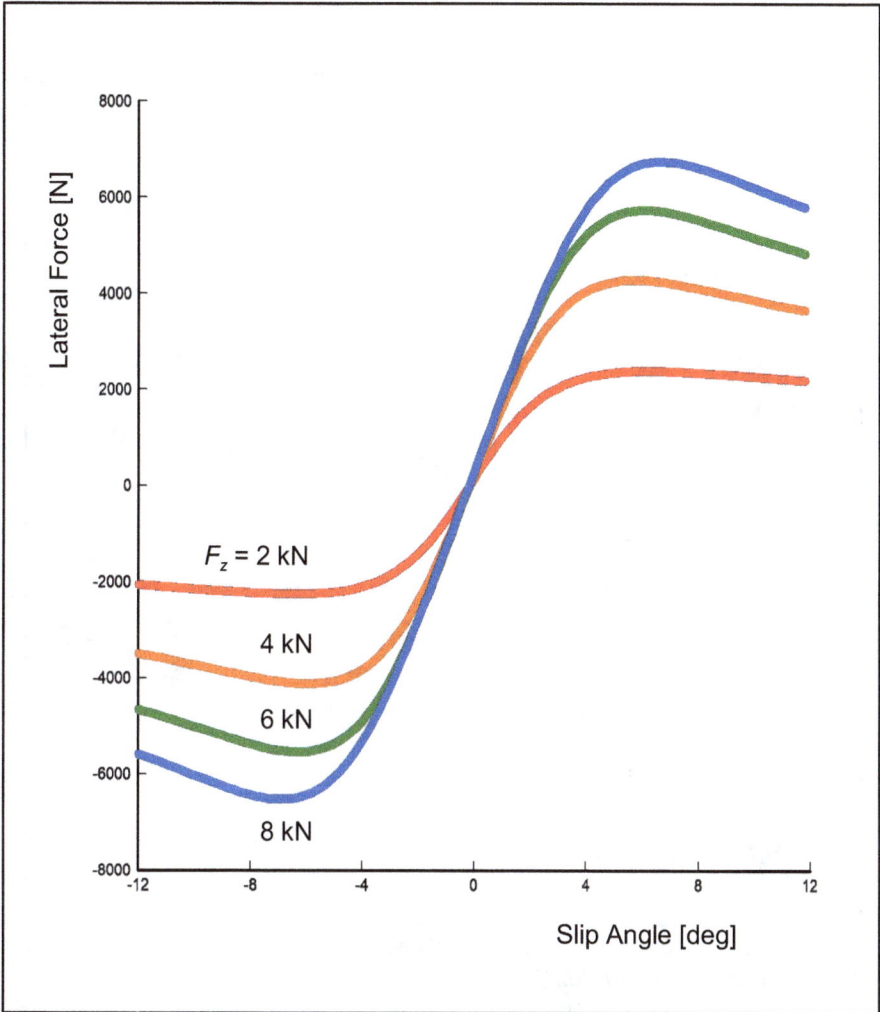

Figure (2.26): EXAMPLE LATERAL FORCE RESULTS BASED ON PACEJKA '94 TIRE MODEL

$$F_x = D \sin\left[C \arctan\left(B \Phi \right) \right] + S_v$$

with

$$D = \mu_x F_z$$

$$BCD = \left(b_3 F_z^2 + b_4 F_z \right) e^{-b_5 F_z}$$

$$B = \frac{BCD}{CD}$$

$$\Phi = \left(1 - E \right)\left(\kappa + S_h \right) + \frac{E}{B} \arctan\left[B \left(\kappa + S_h \right) \right]$$

where

$$C = b_0$$

$$\mu_x = b_1 F_z + b_2$$

$$E = b_6 F_z^2 + b_7 F_z + b_8$$

$$S_h = b_9 F_z + b_{10}$$

$$S_v = b_{11} F_z + b_{12}$$

Figure (2.27): PACEJKA '94 LONGITUDINAL FORCE TIRE MODEL UNDER CONDITIONS OF PURE SLIP

Table (2.4): SAMPLE COEFFICIENT VALUES FOR PACEJKA '94 LONGITUDINAL FORCE TIRE MODEL UNDER CONDITIONS OF PURE SLIP

Coefficient	Sample Value	Definition
b_0 [--]	1.65	shape factor
b_1 [1/kN]	-0.076118	load influence on longitudinal friction coefficient
b_2 [--]	1.1226	longitudinal friction coefficient
b_3 [1/%/MN]	-7.36	change in stiffness per load squared
b_4 [N/%/kN]	144.82	change in stiffness with load
b_5 [1/kN]	-0.076614	change in stiffness progress with load
b_6 [1/kN^2]	-0.00386	curvature change per load squared
b_7 [1/kN]	0.085055	curvature change with load
b_8 [--]	0.075719	curvature factor
b_9 [%/kN]	0.023655	load influence on horizontal shift
b_{10} [%]	0.023655	horizontal shift at zero load
b_{11} [N/kN]	0	load influence on vertical shift at zero camber
b_{12} [N]	0	vertical shift at zero load

The required input variables for the Pacejka '94 tire model are the normal load, F_z [kN], camber angle, γ [deg] and slip ration, κ [%] (positive value indicates tractive effort, negative value used for braking). The corresponding output is longitudinal force, F_x [N].

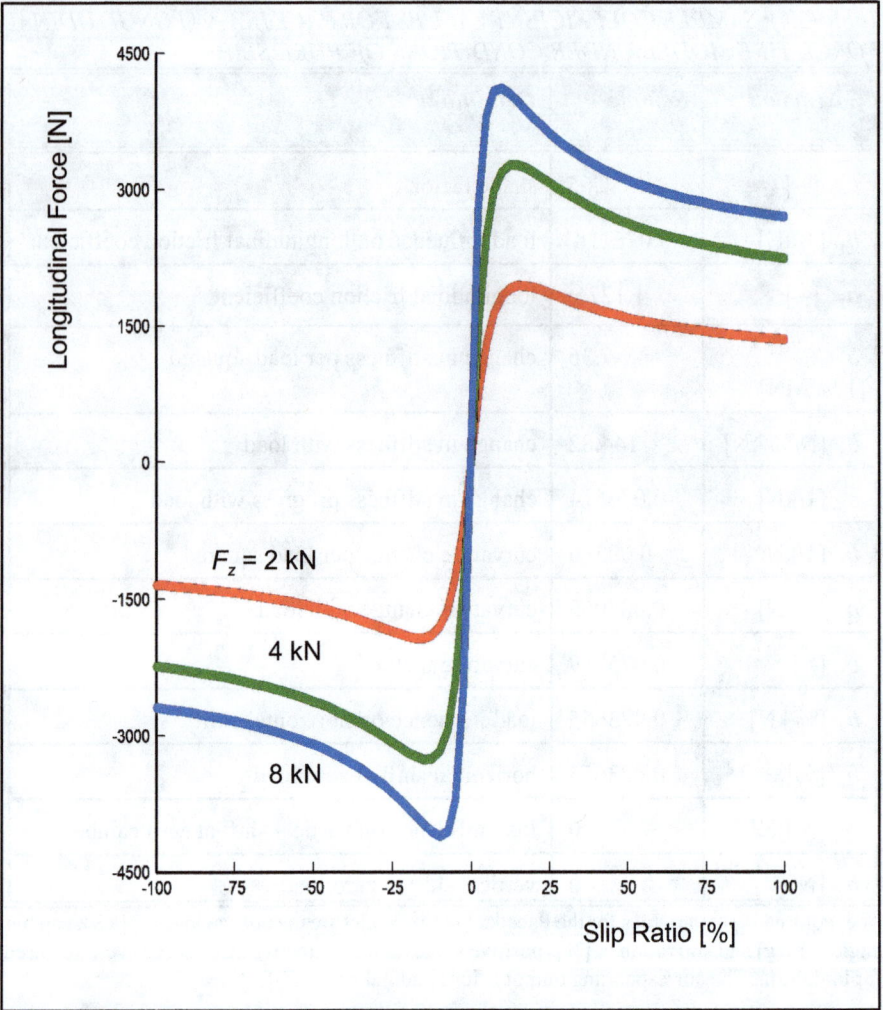

Figure (2.28): EXAMPLE LONGITUDINAL FORCE RESULTS BASED ON PACEJKA '94 TIRE MODEL

$$M_z = D \sin\left[C \arctan\left(B \Phi \right) \right] + S_v$$

with

$$D = c_1 F_z^2 + c_2 F_z$$

$$BCD = \left(c_3 F_z^2 + c_4 F_z \right)\left(1 - c_6 |\gamma| \right) e^{-c_5 F_z}$$

$$B = \frac{BCD}{CD}$$

$$\Phi = \left(1 - E \right)\left(\alpha + S_h \right) + \frac{E}{B} \arctan\left[B\left(\alpha + S_h \right) \right]$$

where

$$C = c_0$$

$$E = \left(c_7 F_z^2 + c_8 F_z + c_9 \right)\left(1 - c_{10} |\gamma| \right)$$

$$S_h = c_{11} \gamma + c_{12} F_z + c_{13}$$

$$S_v = \left(c_{14} F_z^2 + c_{15} F_z \right)\gamma + c_{16} F_z + c_{17}$$

Figure (2.29): PACEJKA '94 ALIGNING TORQUE TIRE MODEL UNDER CONDITIONS OF PURE SLIP

Table (2.5): ALIGNING TORQUE SAMPLE COEFFICIENT VALUES FOR PACEJKA '94 TIRE MODEL UNDER CONDITIONS OF PURE SLIP

Coefficient	Sample Value	Definition
c_0 [--]	2.2264	shape factor
c_1 [m/MN]	-3.0428	change in peak torque per load squared
c_2 [mm]	-9.2284	change in peak torque with load
c_3 [m/MN/deg]	0.500088	change in stiffness per load squared
c_4 [mm/deg]	-5.56696	change in stiffness with load
c_5 [1/kN]	-0.25964	change in stiffness progress with load
c_6 [1/deg]	-0.00129724	camber influence on stiffness
c_7 [1/kN^2]	-0.358348	curvature change per load squared
c_8 [1/kN]	3.74476	curvature change with load
c_9 [--]	-15.1566	curvature factor
c_{10} [1/deg]	0.0021156	camber influence on curvature
c_{11} [--]	0.000346	camber influence on horizontal shift
c_{12} [deg/kN]	0.0	load influence on horizontal shift
c_{13} [deg]	0.0	horizontal shift at zero load
c_{14} [m/MN/deg]	0.100695	load squared dependent camber influence on vertical shift
c_{15} [mm/deg]	-1.398	load dependent camber influence on vertical shift
c_{16} [mm]	0.0	load influence on vertical shift at zero camber
c_{17} [Nm]	0.0	vertical shift at zero load

The required input variables for the Pacejka 94 tire model are the normal load, F_z [kN], camber angle, γ [deg] and slip anlge, α [deg] (positive value indicates a right-hand turn). The corresponding output is aligning torque, M_z [Nm].

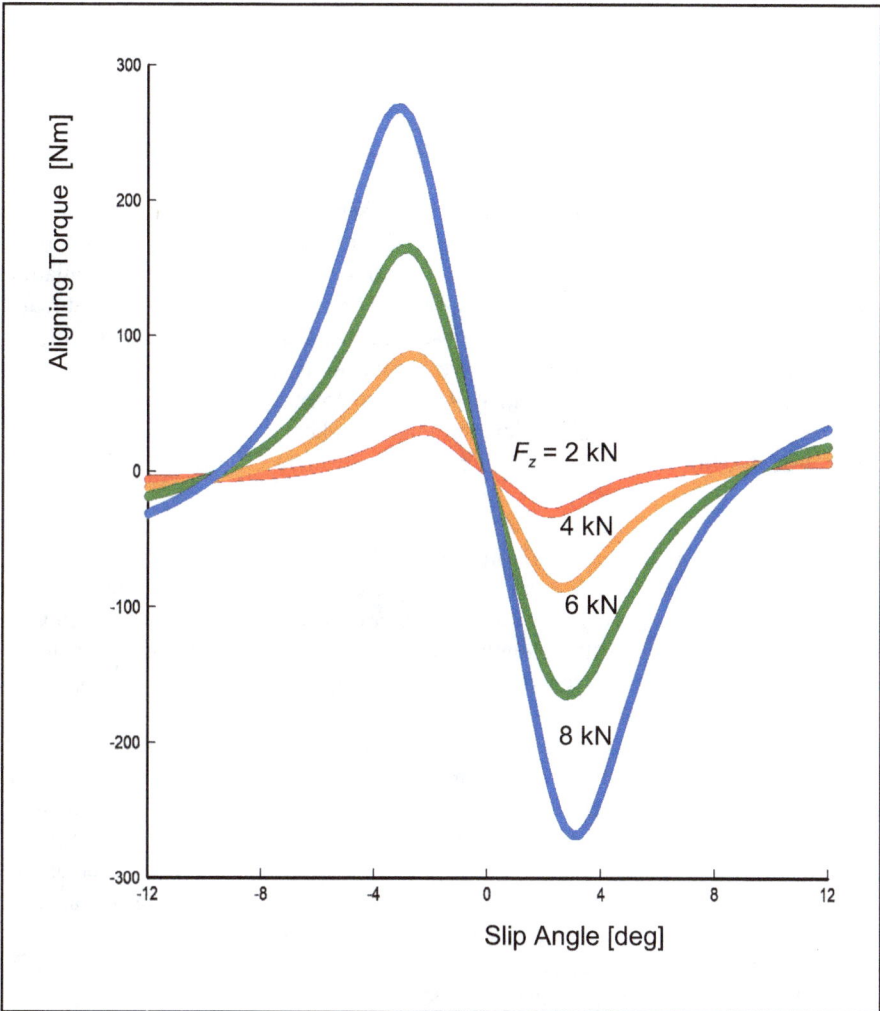

Figure (2.30): EXAMPLE ALIGNING TORQUE RESULTS BASED ON PACEJKA '94 TIRE MODEL

(2.4.2) MF 5.2 TIRE MODEL

While the Pacejka '94 tire model has enjoyed quite a bit of early success, especially among racers, the quest has remained for a more robust model to predict behavior of actual tire performance in vehicle dynamics simulation. One of the more recent models introduced is the Pacejka 2002 model [12] which was slightly reworked into the MF 5.2 tire model. The latter model uses the concept of "scaling factors", in addition to an extended set of Pacejka coefficients, to allow the user to tweak the modeling equations to more closely match the experimental behavior observed under test conditions. This also allows better extrapolated prediction of tire behavior outside the original test conditions of the tire.

The complete set of MF 5.2 equations for longitudinal force, lateral force and aligning torque are given in Figures (2.31), (2.32) and (2.33), respectively. The definitions of the coefficients and scaling factors for the equation sets are given in Tables (2.6), (2.7) and (2.8).

When using the MF 5.2 tire model, two observations should be made. The first is that a parameter, ε, which is usually given a value on the order of 0.001, is used in the model. This parameter is used to prevent "divide-by-zero" errors in the simulation which can occur at low forward vehicle speeds. The second observation is that many of the lateral force components are used in calculating the aligning torque. Consequently, the lateral force must be evaluated prior to the aligning torque.

Sample tire data representative of a high performance racing tire are provided in Table (2.9). Graphical results of longitudinal force, lateral force and aligning torque obtain from the MF 5.2 tire models using this data are shown in Figures (2.34), (2.35) and (2.36).

(2.4.2A) Combined Longitudinal and Lateral Forces

The empirical tire correlations as described by either of the foregoing tire models are applicable to only conditions of either pure lateral slip (singular cornering force) or pure longitudinal slip (singular tractive effort). In actual vehicle maneuvers however, cornering forces seldom exist without braking or tractive forces. The presence of a longitudinal force affects the amount of lateral force developed by the tire at a given slip angle. For instance, a small amount of tractive effort results in a decrease in the lateral force. This is particularly noticeable in bias-ply tires and is caused by a reduction in the cornering stiffness associated with the tire. Braking, on the other hand, causes a slight increase in cornering stiffness. Further increases in either tractive or braking force diminishes the cornering force relative to a pure cornering maneuver. In this instance the tractive (brake) force is utilizing a significant portion of the local adhesion available between the tire and the road surface. In order to accurately predict tire behavior during cornering, we must develop a revised tire model which accounts for the combined effects of longitudinal and lateral forces.

In our earlier discussion of tire behavior we have talked about longitudinal slip, κ, (or longitudinal skid, κ_s) and slip angle, α. Both of these quantities are used to describe, to a certain degree, the tire deformation which results from the application of longitudinal and lateral forces, respectively. We can extend this concept by introducing a new slip quantity, σ, which applies for any combination of longitudinal and lateral forces. The combined slip quantity σ is based on the slip velocities shown in Figure (2.37). Specifically

$$\sigma = \sqrt{\sigma_x^2 + \sigma_y^2} \qquad (2.47)$$

$$F_{x0} = D_x \sin\left[C_x \arctan\left(B_x \Phi_x \right) \right] + S_{Vx}$$

with

$$df_z = \frac{F_z - F_{z0} \lambda_{F_{z0}}}{F_{z0} \lambda_{F_{z0}}}$$

$$C_x = p_{Cx1} \lambda_{Cx} \qquad\qquad D_x = \mu_x F_z$$

$$\mu_x = \left(p_{Dx1} + p_{Dx2}\, df_z \right)\left(1 - p_{Dx3}\, \gamma_x^2 \right) \lambda_{\mu_x}$$

$$B_x = \begin{cases} \dfrac{K_x}{C_x D_x} & : C_x D_x \geq \varepsilon \\[3mm] \dfrac{K_x}{C_x D_x + \varepsilon} & : C_x D_x < \varepsilon \end{cases}$$

$$K_x = F_z \left(p_{Kx1} + p_{Kx2}\, df_z \right) \exp\left(p_{Kx3}\, df_z \right) \lambda_{Kx}$$

$$\Phi_x = \kappa_x \left(1 - E_x \right) + \frac{E_x}{B_x} \arctan\left(B_x \kappa_x \right)$$

where

$$E_x = \left(p_{Ex1} + p_{Ex2}\, df_z + p_{Ex3}\, df_z^2 \right)\left[1 - p_{Ex4}\, \mathrm{sgn}(\kappa_x) \right] \lambda_{Ex}$$

$$\kappa_x = \kappa + S_{Hx} \qquad\qquad \gamma_x = \gamma \lambda_{\gamma x}$$

$$S_{Hx} = \left(p_{Hx1} + p_{Hx2}\, df_z \right) \lambda_{Hx}$$

$$S_{Vx} = F_z \left(p_{Vx1} + p_{Vx2}\, df_z \right) \lambda_{Vx} \lambda_{\mu_x}$$

Figure (2.31): MF 5.2 LONGITUDINAL FORCE TIRE MODEL UNDER PURE LONGITUDINAL SLIP CONDITIONS
Note: If the value of E_x is calculated to be greater than 1.0, then it should be replaced with 1.0.

Chapter 2: TIRES

$$F_{y0} = D_y \sin\left[C_y \arctan\left(B_y \, \Phi_y \right) \right] + S_{Vy}$$

with

$$df_z = \frac{F_z - F_{z0} \, \lambda_{F_{z0}}}{F_{z0} \, \lambda_{F_{z0}}}$$

$$C_y = p_{Cy1} \lambda_{C_y} \qquad\qquad D_y = \mu_y F_z$$

$$\mu_y = \left(p_{Dy1} + p_{Dy2} \, df_z \right)\left(1 - p_{Dy3} \, \gamma_y^2 \right) \lambda_{\mu_y}$$

$$B_y = \begin{cases} \dfrac{K_y}{C_y D_y} & : C_y D_y \geq \varepsilon \\[3mm] \dfrac{K_y}{C_y D_y + \varepsilon} & : C_y D_y < \varepsilon \end{cases}$$

$$K_y = p_{Ky1} F_{z0} \sin\left[2 \arctan\left(\frac{F_z}{p_{Ky2} F_{z0}} \right) \right]\left(1 - p_{Ky3} \, |\gamma_y| \right) \lambda_{F_{z0}} \lambda_{Ky}$$

where

$$\Phi_y = \alpha_y \left(1 - E_y \right) + \frac{E_y}{B_y} \arctan\left(B_y \, \alpha_y \right)$$

$$E_y = \left(p_{Ey1} + p_{Ey2} \, df_z \right)\left[1 - \left(p_{Ey3} + p_{Ey4} \, \gamma_y \right) \mathrm{sgn}(\alpha_y) \right] \lambda_{Ey}$$

$$\alpha_y = \alpha + S_{Hy} \qquad\qquad \gamma_y = \gamma \, \lambda_{\gamma y}$$

$$S_{Hy} = \left(p_{Hy1} + p_{Hy2} \, df_z \right) \lambda_{Hy} + p_{Hy3} \, \gamma_y$$

Figure (2.32): MF 5.2 LATERAL FORCE TIRE MODEL UNDER PURE LATERAL SLIP CONDITIONS
Note: If the value of E_y is calculated to be greater than 1.0, then it should be replaced with 1.0.

Chapter 2: TIRES   69

$$M_{z0} = -t_0 F_{y0} + D_r \cos\left[\arctan\left(B_r \alpha_r\right)\right] \cos\alpha$$

with

$$D_r = F_z R_0\left[\left(q_{Dz6} + q_{Dz7}\, df_z\right)\lambda_r + \left(q_{Dz8} + q_{Dz9}\, df_z\right)\gamma_z\right]\lambda_{\mu_y}$$

$$B_t = \left(q_{Bz1} + q_{Bz2}\, df_z + q_{Bz3}\, df_z^2\right)\left(1 + q_{Bz4}\gamma_z + q_{Bz5}\left|\gamma_y\right|\right)\frac{\lambda_{Ky}}{\lambda_{\mu_y}}$$

$$B_r = q_{Bz9}\frac{\lambda_{Ky}}{\lambda_{\mu_y}} + q_{Bz10}\, B_y C_y$$

$$C_t = q_{Cz1}$$

$$\Phi_t = \alpha_t\left(1 - E_t\right) + \frac{E_t}{B_t}\arctan\left(B_t \alpha_t\right)$$

$$t_0 = D_t \cos\left[C_t \arctan\left(B_t \Phi_t\right)\right]\cos\alpha$$

$$E_t = \left(q_{Ez1} + q_{Ez2}\, df_z + q_{Ez3}\, df_z^2\right)\left[1 + \frac{2}{\pi}\left(q_{Ez4} + q_{Ez5}\gamma_z\right)\arctan\left(B_t C_t \alpha_t\right)\right]$$

where

$$\alpha_t = \alpha + S_{Ht} \qquad \alpha_r = \alpha + S_{Hf} \qquad \gamma_z = \gamma\lambda_{\gamma z}$$

$$S_{Ht} = q_{Hz1} + q_{Hz2}\, df_z + \left(q_{Hz3} + q_{Hz4}\, df_z\right)\gamma_z$$

$$S_{Hf} = S_{Hy} + \frac{S_{Vy}}{K_y}$$

Figure (2.33): MF 5.2 ALIGNING TORQUE TIRE MODEL UNDER PURE LATERAL SLIP CONDITIONS
Note: (1) if the value of E_t is calculated to be greater than 1.0, then it should be replaced with 1.0, and (2) if the absolute value of K_y is less than ε then one of two adjustments must be made: for positive values $K_y = \varepsilon$ and for negative values $K_y = -\varepsilon$.

Table (2.6): MF 5.2 LONGITUDINAL FORCE TIRE MODEL COEFFICIENTS AND SCALING FACTORS FOR PURE LONGITUDINAL SLIP

Empirical Coefficients	Scaling Factors[*]	Description
F_{z0}	$\lambda_{F_{z0}}$	nominal rated load
p_{Dx1}, p_{Dx2}, p_{Dx3}	λ_{μ_x}	peak friction
p_{Cx1}	λ_{Cx}	shape factor
p_{Kx1}, p_{Kx2}, p_{Kx3}	λ_{Kx}	slip stiffness
p_{Ex1}, p_{Ex2}, p_{Ex3}, p_{Ex4}	λ_{Ex}	curvature factor
p_{Hx1}, p_{Hx2}	λ_{Hx}	horizontal shift
p_{Vx1}, p_{Vx2}	λ_{Vx}	vertical shift
	$\lambda_{\gamma x}$	camber factor

* These are user defined values that should normally be set to a value of one.

Table (2.7): MF 5.2 LATERAL FORCE TIRE MODEL COEFFICIENTS AND SCALING FACTORS FOR PURE LATERAL SLIP

Empirical Coefficients	Scaling Factors[*]	Description
F_{z0}	$\lambda_{F_{z0}}$	nominal rated load
p_{Dy1}, p_{Dy2}, p_{Dy3}	λ_{μ_y}	peak friction
p_{Cy1}	λ_{Cy}	shape factor
p_{Ky1}, p_{Ky2}, p_{Ky3}	λ_{Ky}	slip stiffness
p_{Ey1}, p_{Ey2}, p_{Ey3}, p_{Ey4}	λ_{Ey}	curvature factor
p_{Hy1}, p_{Hy2}, p_{Hy3}	λ_{Hy}	horizontal shift
p_{Vy1}, p_{Vy2}, p_{Vy3}, p_{Vy4}	λ_{Vy}	vertical shift
	$\lambda_{\gamma y}$	camber factor

* These are user defined values that should normally be set to a value of one.

Table (2.8): MF 5.2 ALIGNING TORQUE TIRE MODEL COEFFICIENTS AND SCALING FACTORS FOR PURE LATERAL SLIP

Empirical Coefficients	Scaling Factors*	Description
F_{z0}		nominal rated load
R_0		unloaded tire radius
q_{Dz1}, q_{Dz2}, q_{Dz3}, q_{Dz4}	λ_t	peak pneumatic trail
q_{Dz6}, q_{Dz7}	λ_r	peak residual torque
q_{Dz8}, q_{Dz9}	$\lambda_{\gamma z}$	peak camber torque
q_{Cz1}		shape factor
q_{Bz1}, q_{Bz2}, q_{Bz3}, q_{Bz4}, q_{Bz5} q_{Bz9}, q_{Bz10}		camber force stiffness
q_{Ez1}, q_{Ez2}, q_{Ez3}, q_{Ez4}, q_{Ez5}		curvature factor
q_{Hz1}, q_{Hz2}, q_{Hz3}, q_{Hz4}		horizontal shift

* These are user defined values that should normally be set to a value of one.

where

$$\sigma_x = \frac{V_{s,x}}{V_x} = \frac{-\kappa_s}{1 + \kappa_s} \tag{2.48}$$

and

$$\sigma_y = \frac{V_{s,y}}{V_x} = \frac{-\tan\alpha}{1 + \kappa_s} \tag{2.49}$$

Theoretically, at least, we should be able to predict the brake and cornering forces from

$$F_x = -\frac{\sigma_x}{\sigma} F(\alpha) \tag{2.50}$$

$$F_y = -\frac{\sigma_y}{\sigma} F(\alpha) \tag{2.51}$$

where $F(\alpha)$ represents the resultant tire force as a function of theoretical slip. However, this theory only applies to a tire with isotropic properties; that is to say, a tire which has the same performance characteristics in both the longitudinal and lateral

Chapter 2: TIRES

Table (2.9): MF 5.2 EXAMPLE TIRE DATA

Tire Characteristics			
F_{z0} [N]	659.47	R_0 [m]	0.2650

Longitudinal Force		Lateral Force				Aligning Torque	
p_{Cx1}	1.3000	p_{Cy1}	1.5307	r_{By1}	-41.2712	q_{Bz1}	3.9317
p_{Dx1}	2.4602	p_{Dy1}	1.9404	r_{By2}	35.3659	q_{Bz2}	-1.3409
p_{Dx2}	-0.2498	p_{Dy2}	-0.2296	r_{By3}	-0.02275	q_{Bz3}	0.5890
p_{Dx3}	0.0000	p_{Dy3}	-7.3961	r_{Cy1}	1.0020	q_{Bz4}	-12.4526
p_{Ex1}	-2.0000	p_{Ey1}	-0.5613	r_{Ey1}	1.0000	q_{Bz5}	16.6470
p_{Ex2}	0.0000	p_{Ey2}	-0.1752	r_{Ey2}	0.0000	q_{Bz9}	0.0000
p_{Ex3}	0.0000	p_{Ey3}	-0.2298	r_{Hy1}	0.005493	q_{Bz10}	0.0000
p_{Ex4}	-0.1250	p_{Ey4}	8.1420	r_{Hy2}	0.0000	q_{Cz1}	3.7679
p_{Kx1}	67.8218	p_{Ky1}	-98.3897	r_{Vy1}	-0.05100	q_{Dz1}	0.03506
p_{Kx2}	-1.8E-05	p_{Ky2}	-3.1264	r_{Vy2}	0.03462	q_{Dz2}	.006438
p_{Kx3}	-0.1830	p_{Ky3}	0.8756	r_{Vy3}	0.04912	q_{Dz3}	-0.033175
p_{Hx1}	0.00588	p_{Hy1}	-0.001112	r_{Vy4}	12.0000	q_{Dz4}	-0.002324
p_{Hx2}	-0.00589	p_{Hy2}	-0.001773	r_{Vy5}	6.0000	q_{Dz6}	-0.005938
p_{Vx1}	0.00632	p_{Hy3}	0.001753	r_{Vy6}	-2.0000	q_{Dz7}	0.009806
p_{Vx2}	0.00426	p_{Vy1}	0.03305			q_{Dz8}	-0.99811
r_{Bx1}	21.3280	p_{Vy2}	-0.02343			q_{Dz9}	-0.27981
r_{Bx2}	44.8568	p_{Vy3}	-3.1561			q_{Ez1}	-3.1336
r_{Cx1}	1.1304	p_{Vy4}	-0.4298			q_{Ez2}	1.1869
r_{Ex1}	1.0000					q_{Ez3}	-0.8540
r_{Ex2}	0.0000					q_{Ez4}	0.4186
r_{Hx1}	0.03008					q_{Ez5}	0.7286
						q_{Hz1}	-0.01162
						q_{Hz2}	-0.01520
						q_{Hz3}	0.2460
						q_{Hz4}	0.1291

Note: The sample coefficient values have units consistent with the model input variables of (1) slip ratio [--], i.e., a nondimensional decimal value, (2) slip angle [rad], (3) camber angle [rad], and (4) normal force [N]. This MF 5.2 model will yield results of longitudinal force [N], lateral force [N], and aligning torque [Nm].

directions. To compensate for all of the non-isotropic tire properties, a much more elaborate scheme is required to model real tire behavior. Essentially the theory must be adjusted by expressing the resultant force as a function of previously established pure longitudinal force characteristics together with pure lateral force characteristics.

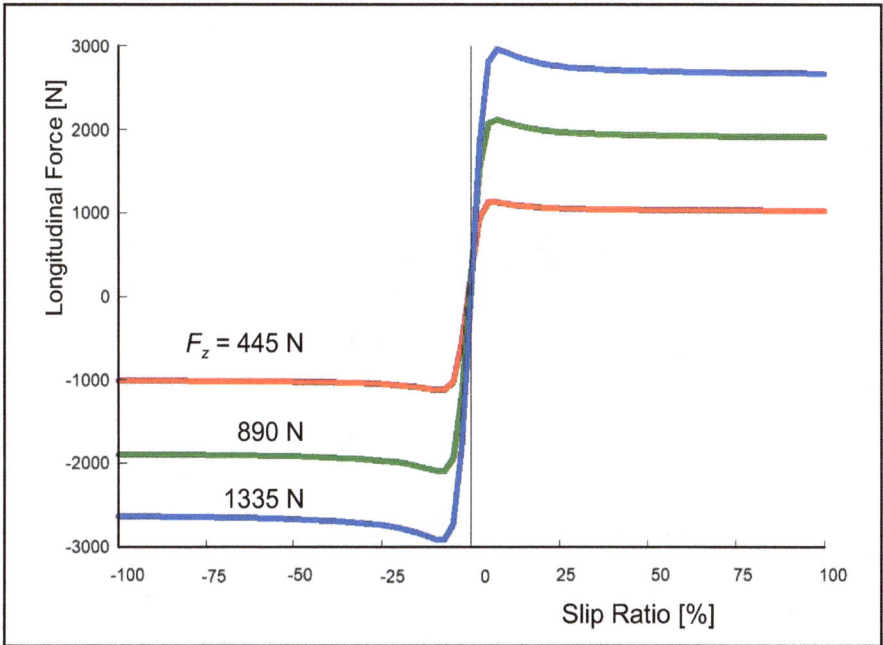

Figure (2.34): EXAMPLE LONGITUDINAL FORCE RESULTS BASED ON MF 5.2 TIRE MODEL

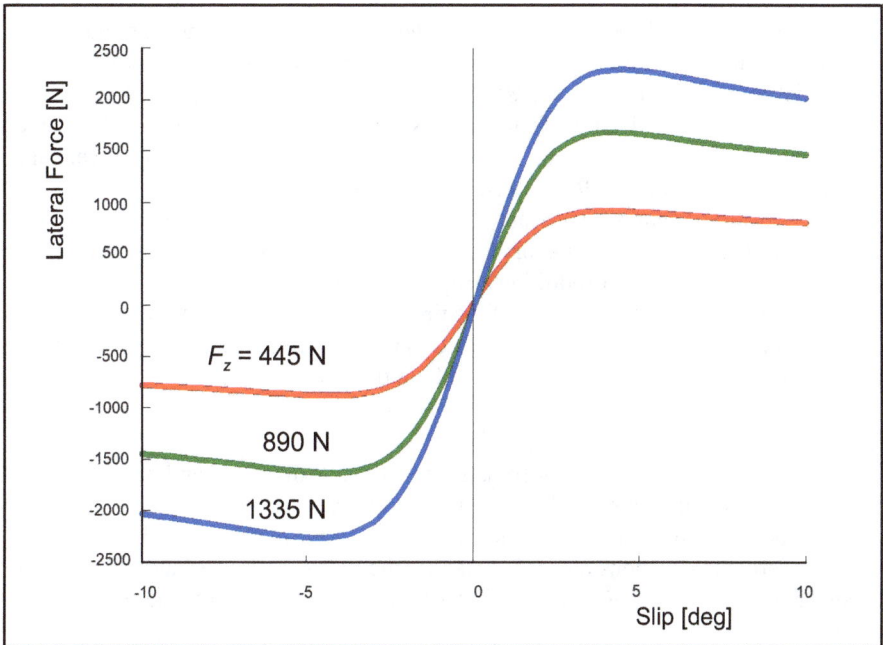

Figure (2.35): EXAMPLE LATERAL FORCE RESULTS BASED ON MF 5.2 TIRE MODEL

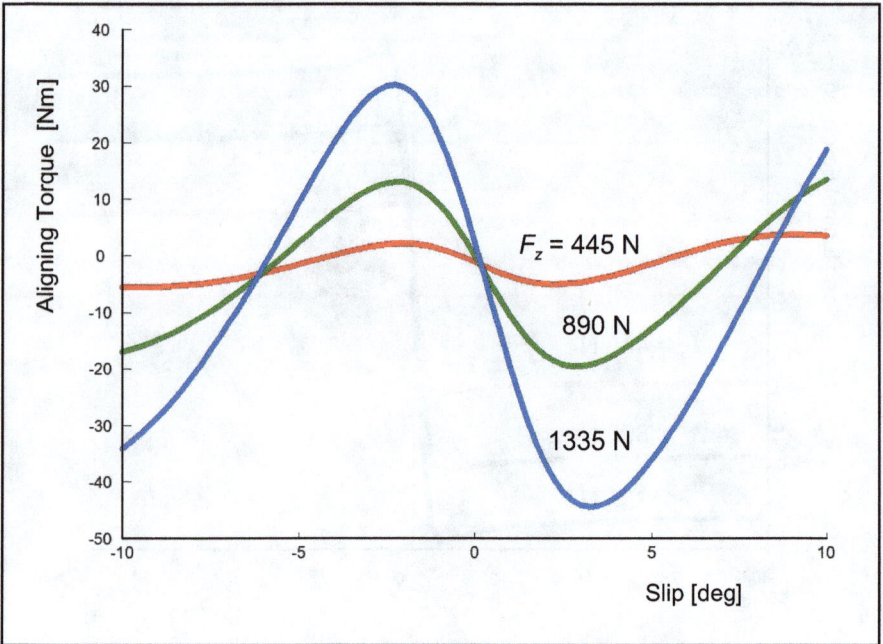

Figure (2.36): EXAMPLE ALIGNING TORQUE RESULTS BASED ON MF 5.2 TIRE MODEL

A summary of the equations and methodology for calculating longitudinal and lateral forces under conditions of combined longitudinal and lateral slip for the MF 5.2 tire model is given in Figures (2.38) and (2.39). As might be expected, additional correlation constants and scaling factors are also required. These are provided in Tables (2.10) and (2.11). A graph of the effect of longitudinal slip on lateral force, using previous data from Table (2.9), is illustrated in Figure (2.40).

A word of caution is warranted. Considering all of the available tire models, it is imperative that the designer apply the experiment data (coefficients) corresponding to the intended tire model. In addition, to appropriately apply the tire models to vehicle dynamic studies, the test conditions of the tire, e.g., tire pressure, must also be known. No doubt new tire models will be developed as research continues, mostly in adjusting for the affects of tire pressure and rim width on tire performance.

(2.5) RIDE PROPERTIES OF TIRES

The characteristics for pneumatic tires we have discussed and studied thus far has centered around traction and cornering performance. Other important qualities of a tire are its ride properties. Tires obviously have desirable road holding characteristics, but tires are also responsible for isolating the vehicle chassis and driver from irregularities in the road surface. The latter function is referred to as the ride property of tires.

In the study of vehicle ride behavior it is convenient to use a vertical (normal) coordinate system originating at the road surface, but translates at a speed equivalent

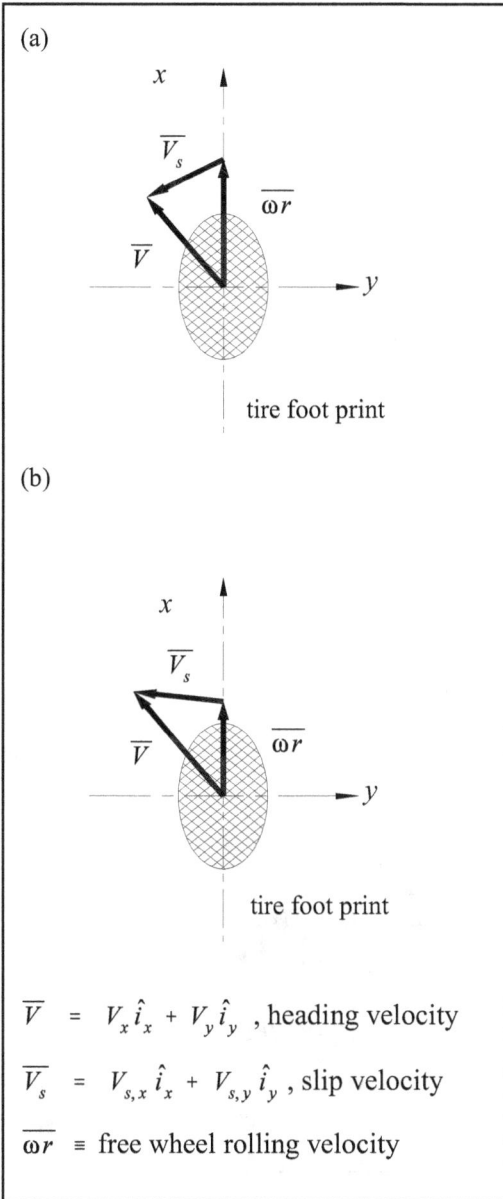

(a)

x

$\overline{V}_s$

$\overline{\omega r}$

$\overline{V}$

y

tire foot print

(b)

x

$\overline{V}_s$

$\overline{V}$

$\overline{\omega r}$

y

tire foot print

$$\overline{V} = V_x \hat{i}_x + V_y \hat{i}_y \text{ , heading velocity}$$

$$\overline{V}_s = V_{s,x} \hat{i}_x + V_{s,y} \hat{i}_y \text{ , slip velocity}$$

$$\overline{\omega r} \equiv \text{free wheel rolling velocity}$$

Figure (2.37): TIRE VELOCITY DIAGRAM OF COMBINED LONGITUDINAL AND LATERAL SLIP FOR (a) TRACTION and (b) BRAKING

to the forward velocity of the vehicle. Relative to this coordinate system the tires would appear to move only in an up and down (bounce) motion. There are two tire ride models that are used in conjunction with this coordinate system: the linear spring and damper model and the viscoelastic model [13]. A mechanical schematic of each model is shown in Figure (2.41). The linear model consists of the tire and rim weight supported by a linear spring, with spring constant k, in parallel with a damper with damping rate b. The viscoelastic model recognizes that the rubber portion of the tire has both spring and damping qualities. Consequently, single damper found in the linear model is replaced with a series arrangement of a linear spring k_2 and damper b_2. In both models, the spring and damping rates are determined by experimental means and, thus, are referred to as semi-empirical constants.

In terms of mathematical manipulation, the linear tire model is easier to use than the viscoelastic model; but, the viscoelastic model tends to yield slightly more realistic results. However, neither model fully describes the non-linear behavior actually observed in pneumatic tires.

There are various experimental methods used to evaluate tire ride properties. A static (non-rolling) tire test can be used to determine the spring rate. In this test a known vertical load is placed on the tire and rim combination at the spin axis. The vertical deflection of the spin axis is then measured. A force-deflection plot is generated by repeating this process for different

$$F_x = D_{xa} \cos\left[C_{xa} \arctan\left\{B_{xa}\,\alpha_s - E_{xa}\left[B_{xa}\,\alpha_s - \arctan\left(B_{xa}\,\alpha_s\right)\right]\right\}\right]$$

with

$$B_{xa} = r_{Bx1} \cos\left[\arctan\left(r_{Bx2}\,\kappa\right)\right]\lambda_{xa}$$

$$C_{xa} = r_{Cx1}$$

$$E_{xa} = r_{Ex1} + r_{Ex2}\,df_z$$

$$G_{xa0} = \cos\left[C_{xa} \arctan\left\{B_{xa}S_{Hxa} - E_{xa}\left[B_{xa}S_{Hxa} - \arctan\left(B_{xa}S_{Hxa}\right)\right]\right\}\right]$$

$$D_{xa} = \begin{cases} \dfrac{F_{x0}}{G_{xa0}} & : G_{xa0} > \varepsilon \\[2ex] 0 & : G_{xa0} \le \varepsilon \end{cases}$$

where

$$\alpha_s = \alpha + S_{Hxa}$$
$$S_{Hxa} = r_{Hx1}$$

Figure (2.38): MF 5.2 LONGITUDINAL FORCE TIRE MODEL UNDER CONDITIONS OF COMBINED LATERAL AND LONGITUDINAL SLIP

Table (2.10): MF 5.2 LONGITUDINAL FORCE TIRE MODEL COEFFICIENTS AND SCALING FACTOR FOR COMBINED LATERAL AND LONGITUDINAL SLIP

Empirical Coefficients	Scaling Factor*	Description
r_{Bx1}, r_{Bx2}	λ_{xa}	combined slip ratio multipliers
p_{Cx1}		combined shape factor
r_{Ex1}, r_{Ex2}		combined curvature multipliers
r_{Hx1}		combined horizontal shift

* This is a user defined value that should normally be set to a value of one.

$$F_y = D_{y\kappa} \cos\left[C_{y\kappa} \arctan\left\{B_{y\kappa}\,\kappa_s - E_{y\kappa}\left[B_{y\kappa}\,\kappa_s - \arctan\left(B_{y\kappa}\,\kappa_s\right)\right]\right\}\right] + S_{Vy\kappa}$$

with

$$B_{y\kappa} = r_{By1} \cos\left\{\arctan\left[r_{By2}\left(\alpha - r_{By3}\right)\right]\right\} \lambda_{y\kappa\alpha}$$

$$C_{y\kappa} = r_{Cy1}$$

$$E_{y\kappa} = r_{Ey1} + r_{Ey2}\,df_z$$

$$G_{y\kappa 0} = \cos\left[C_{y\kappa} \arctan\left\{B_{y\kappa}\,S_{Hy\kappa} - E_{y\kappa}\left[B_{y\kappa}\,S_{Hy\kappa} - \arctan\left(B_{y\kappa}\,S_{Hy\kappa}\right)\right]\right\}\right]$$

$$D_{y\kappa} = \frac{F_{y0}}{G_{y\kappa 0}}$$

where

$$\kappa_s = \kappa + S_{Hy\kappa}$$

$$D_{Vy\kappa} = \mu_y\,F_z\left(r_{Vy1} + r_{Vy2}\,df_z + r_{Vy3}\,\gamma\right)\cos\left[\arctan\left(r_{Vy4}\,\alpha\right)\right]$$

$$S_{Vy\kappa} = D_{Vy\kappa} \sin\left[r_{Vy5} \arctan\left(r_{Vy6}\,\kappa\right)\right] \lambda_{Vy\kappa\alpha}$$

Figure (2.39): MF 5.2 LATERAL FORCE TIRE MODEL UNDER CONDITIONS OF COMBINED LATERAL AND LONGITUDINAL SLIP

Table (2.11): MF 5.2 LATERAL FORCE TIRE MODEL COEFFICIENTS AND SCALING FACTORS FOR COMBINED LATERAL AND LONGITUDINAL SLIP

Empirical Coefficients	Scaling Factors*	Description
r_{By1}, r_{By2}, r_{By3}	$\lambda_{y\kappa\alpha}$	combined slip multipliers
r_{Cy1}		combined shape factor
r_{Ey1}, r_{Ey2}		combined curvature multiplier
r_{Hy1}, r_{Hy2}		combined horizontal shift multipliers
r_{Vy1}, r_{Vy2}, r_{Vy3} r_{Vy4}, r_{Vy5}, r_{Vy6}	$\lambda_{Vy\kappa\alpha}$	combined vertical shift multipliers

* These are user defined values that should normally be set to a value of one.

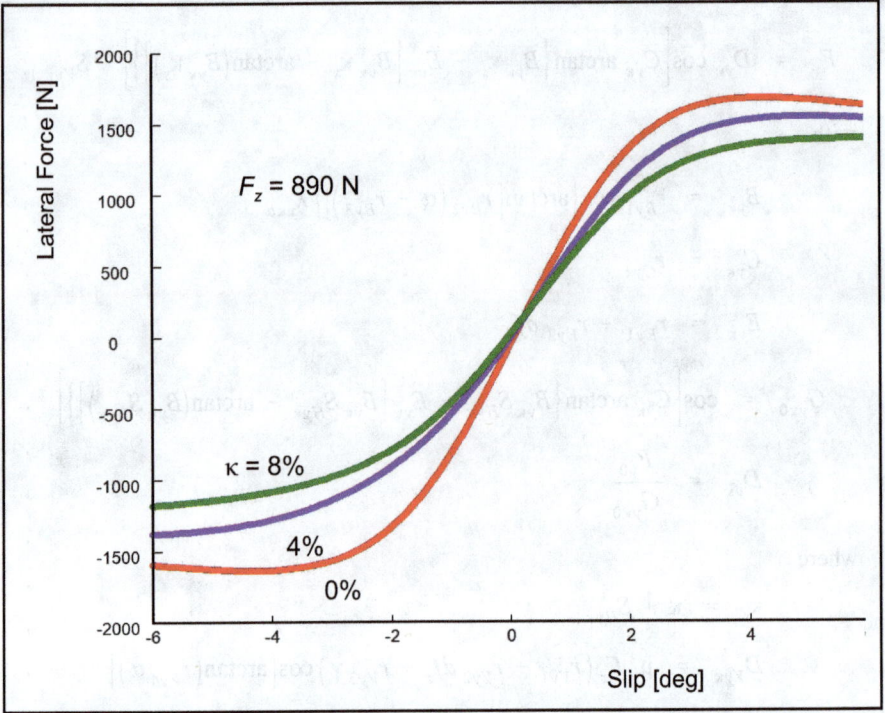

Figure (2.40): EXAMPLE LATERAL FORCE VARIATION WITH LONGITUDINAL SLIP BASED ON MF 5.2 TIRE MODEL

loads. Except for the region at small loads, which is generally considered to be unimportant, the plot is fairly linear. The slope of a least squares linear curve fit through the data points yields the static spring rate.

It can be argued that the static tire test is not an accurate representation of actual operating conditions experience by the tire. Experiments performed on a vibration tire tester, such as the rotating drum apparatus schematically shown in Figure (2.42), are a better indication of tire performance. Tire properties determined in this fashion are referred to as dynamic properties.

As mentioned before, the tire spring and damping rates are semi-empirical constants. This means that the dynamic tire properties depend not only on the experimental results, but also the mathematical model on which they are based. For instance, using the linear spring and damper model, the basic equation describing the force acting on the load cell depicted in Figure (2.42) is

$$F(t) = b\,\dot{x}_D + k\,x_D \tag{2.52}$$

where x_D is the rotating drum vertical displacement. One possible circumferential shape for the rotating drum evolves from superimposing a sinusoidal curve on top of a base circle. In this case the drum vertical motion is given by

$$x_D = X_D \sin(\omega t) \qquad (2.53)$$

and

$$\dot{x}_D = X_D \omega \cos(\omega t) \qquad (2.54)$$

Inserting Eqs (2.53) and (2.54) in Eq (2.52) results in

$$F(t) = X_D \sqrt{(\omega t)^2 + k^2} \sin(\omega t + \varphi) \qquad (2.55)$$

where

$$\varphi = \arctan\left(\frac{\omega b}{k}\right) \qquad (2.56)$$

The information provided by the tire vibration test apparatus allows the force amplitude and phase angle to be deduced from physical measurements. Since the drum displacement amplitude and rotational speed are controlled operating parameters, the spring rate and damping rate can then be simultaneously solved from the force amplitude and phase angle, i.e.,

$$k = \frac{\dfrac{F}{X_D}}{\sqrt{1 + \tan^2\varphi}} \qquad (2.57)$$

and

$$b = \frac{\dfrac{F}{\omega X_D}\tan\varphi}{\sqrt{1 + \tan^2\varphi}} \qquad (2.58)$$

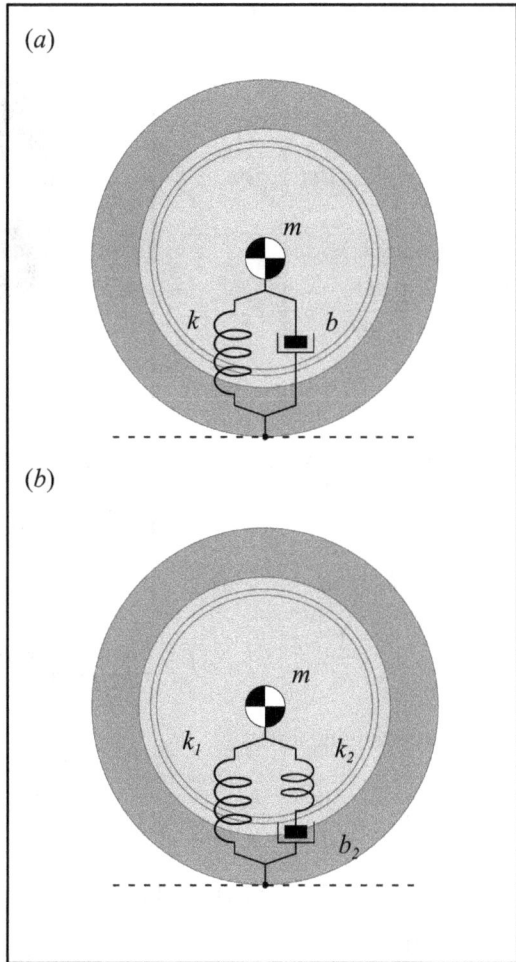

The analysis of the viscoelastic model is not quite as

(a)

(b)

Figure (2.41): TIRE RIDE MODELS (a) LINEAR SPRING MODEL AND (b) VISCOELASTIC MODEL

straight forward, but is still manageable. In this case the force sensed by the load cell is assumed to be of the form

$$F(t) = F_1 + F_2 \tag{2.59}$$

The force F_1 is due to the single spring, i.e., $F_1 = k_1 x_D$, and the force F_2 is created by the spring and damper series combination. An expression for the force F_2 in terms of the drum displacement is derived by noting that

$$x_D = x_s + x_d \tag{2.60}$$

where x_s is the displacement of the series spring and x_d is that of the damper. Differentiating Eq (2.60) with respect to time produces

$$\frac{dx_D}{dt} = \frac{dx_s}{dt} + \frac{dx_d}{dt}$$

or

$$\frac{dx_D}{dt} = \frac{1}{k_2}\frac{dF_2}{dt} + \frac{1}{b_2}F_2 \tag{2.61}$$

Using Eq (2.59), the previous equation can also be written as

$$\frac{dx_D}{dt} = \frac{1}{k_2}\left(\frac{dF}{dt} - k_1\frac{dx_D}{dt}\right)$$

$$+ \frac{1}{b_2}\left(F - k_1 x_D\right)$$

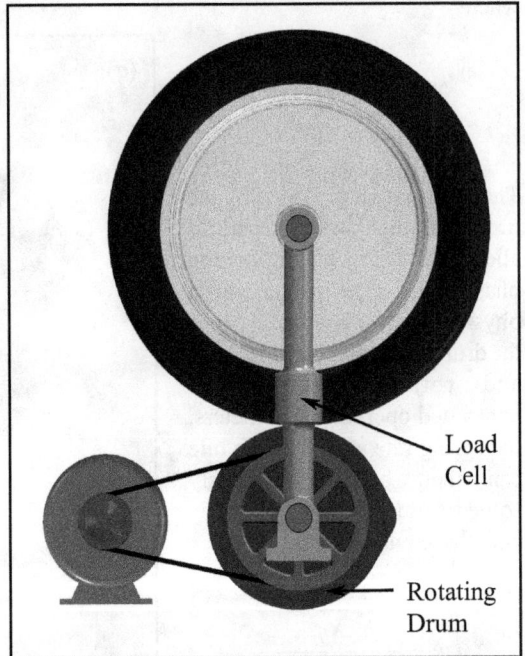

Load Cell

Rotating Drum

Figure (2.42): SCHEMATIC OF ROTATING DRUM TIRE TEST APPARATUS

or, upon rearranging terms,

$$\frac{dF}{dt} + \frac{k_2}{b_2}F = \left(k_1 + k_2\right)\frac{dx_D}{dt} + \frac{k_1 k_2}{b_2}x_D \tag{2.62}$$

Placing Eqs (2.53) and (2.54) into Eq (2.62) yields the differential equation

$$\frac{dF}{dt} + \frac{k_2}{b_2}F = X_D\left[\left(k_1 + k_2\right)\omega\cos(\omega t) + \frac{k_1 k_2}{b_2}\sin(\omega t)\right] \tag{2.63}$$

This is the basic relationship which describes the interaction between force and displacement in a viscoelastic tire. The steady-state solution of Eq (2.63) is

$$F(t) \; = \; X_D \frac{b_2 \omega}{\sqrt{1 + (\tau\omega)^2}} \sin(\omega t + \varphi) + X_D k_1 \sin(\omega t) \qquad (2.64)$$

where

$$\tau = \frac{b_2}{k_2} \qquad (2.65)$$

and

$$\varphi = \arctan(\tau\omega) \qquad (2.66)$$

A strategy used to evaluate tire properties for the viscoelastic tire model is to first determine the spring rate k_1 by static means. Then using Eqs (2.64), (2.65) and (2.66), the spring rate k_2 and damping rate b_2 can be deduced from amplitude and phase angle measurements obtained from the tire vibration test apparatus.

Chapter 3:
ACCELERATION AND BRAKING

When automotive enthusiasts refer to the performance of a vehicle they usually want to know how well the car accelerates, as opposed to how well the car handles in cornering. In the context of vehicle dynamics, the subject of "performance" encompasses both acceleration and braking behavior of a vehicle moving in a straight path. It is fairly easy to develop analytical models which provide reasonable estimates of vehicle performance. Using these models the design engineer can study the affect of altering the power train, e.g., changing the engine size or transmission ratios, on acceleration and top speed. The models can also be used as an aid in developing design specifications for the braking system. Although not strictly a performance characteristic, the body/chassis pitch motion of squat and dive observed during acceleration and braking (or more accurately the suspension design for anti-squat and anti-dive control) are usually included as a topic for discussion in acceleration and braking.

(3.1) DYNAMICS OF ACCELERATION

There are a variety of vehicle design parameters and characteristics which affect the acceleration performance of a car. These parameters range from tire performance characteristics and engine/drivetrain performance characteristics to the aerodynamic behavior of the vehicle body. All of these are of interest to the engineer in developing either a concept car design or re-designing an existing vehicle. The first task in studying the acceleration performance of a car is to develop basic equations of motion. The second task is to use these equations to develop an expression, or algorithm, to estimate the forward acceleration as a function of the vehicle's geometric parameters and operational variables.

The kinematic analysis of vehicle acceleration tends to be somewhat specific to the vehicle's drivetrain configuration. In particular, the direction of the tire tractive forces depend on whether we are considering a rear-wheel-drive vehicle, a front-wheel-drive vehicle, or an all-wheel-drive vehicle. The illustration in Figure (3.1) depicts a two-dimensional representation of a rear-wheel-drive sports car accelerating up an inclined plane. Also shown in the figure are the external forces acting on the sports car, which also includes the inertia forces appropriate for a translating xz-coordinate system attached to the vehicle's sprung mass center of gravity. These forces fall into five general categories:

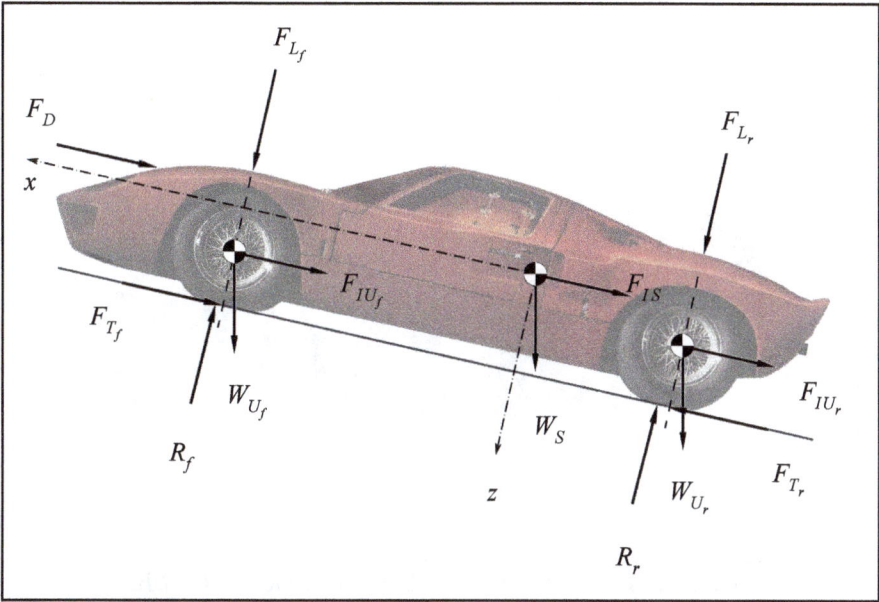

Figure (3.1): FORCES ACTING ON AN ACCELERATING VEHICLE

(1) front and rear tire tractive forces (applied at the road surface):

$$F_{T_f}, \quad F_{T_r}$$

(2) body/chassis (sprung weight) and front/rear (unsprung weight) gravitational forces:

$$W_S; \quad W_{U_f}, \quad W_{U_r}$$

(3) aerodynamic downforce, or anti-lift, (front and rear axle planes) and aerodynamic drag:

$$F_{L_f}, \quad F_{L_r}; \quad F_D$$

(4) road surface reactions (front and rear axle planes):

$$R_f, \quad R_r$$

and (5) sprung and unsprung (front and rear axle planes) inertia forces:

$$F_{IS}; \quad F_{IU_f}, \quad F_{IU_r}$$

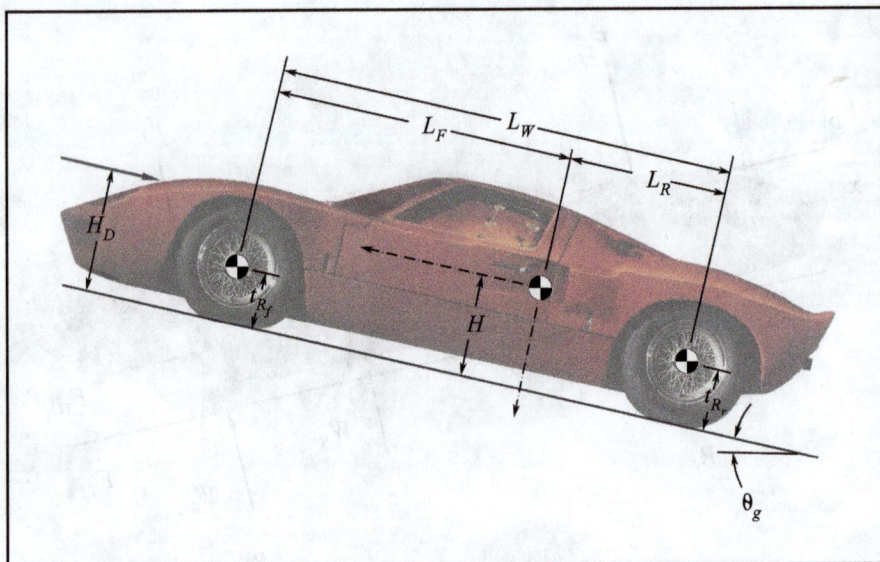

Figure (3.2): GEOMETRIC PARAMETERS OF ACCELERATING VEHICLE

Prior to developing the equations of motion, additional geometric parameters must be specified as defined in Figure (3.2). The definition of the various geometric parameters are as follows:

L_W ≡ vehicle wheelbase,

L_F, L_R ≡ longitudinal distance of the sprung mass center of gravity (CG) from the front and rear axle planes,

H ≡ height of the sprung mass CG above the ground plane,

H_D ≡ height of the aerodynamic drag pressure center above the ground plane,

t_{R_f}, t_{R_r} ≡ front (rear) tire effective rolling radius, and

θ_g ≡ grade angle (uphill positive)

If the vehicle is in existence, either in production or prototype, most of these parameters can be experimentally measured or determined from specifications of the vehicle. However, if the analysis is being performed on a concept vehicle which does not physically exist, then reasonable estimates must be made before an acceleration performance analysis can be performed. Appendix B provides an example of the methods used to estimate the geometric location of the sprung mass CG of a concept vehicle. Aerodynamic considerations pertaining to ground vehicles is discussed later in this chapter. The concept of dynamic equilibrium[1] can be used as an approach to develop the equations of motion, which is facilitated by dividing the sports car up into

---

[1] *A review of basic rigid body dynamics is provided in Appendix C.*

three, coupled, free-body-diagrams: (1) the vehicle body and chassis, (2) the front wheel assembly, and (3) the rear wheel assembly. We will start by analyzing the front wheel assembly shown in Fig (3.3). Upon closer examination of Fig (3.3), it should be noted that suspension reaction forces F_{S_f} and R_{S_f} have been added to compensate for the "push" influence of the body/chassis on the front wheel assembly, along with the dimension a_f to adjust for the forward displacement of the road reaction due to the tire rolling resistance.

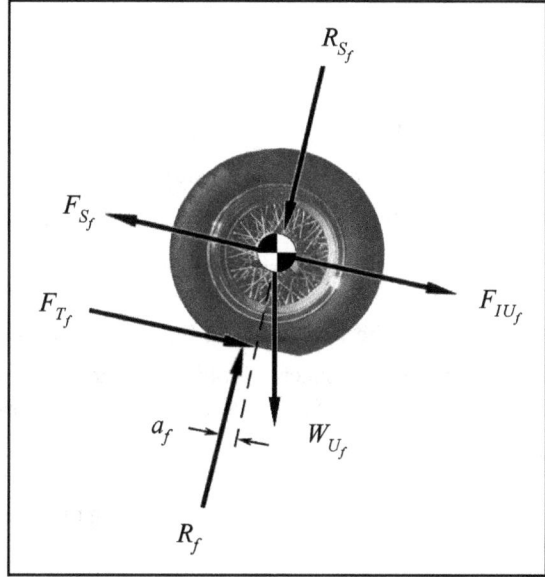

Figure (3.3): FRONT WHEEL ASSEMBLY FREE-BODY-DIAGRAM

The basic equations of motion applied to a translating *xyz*-coordinate system attached to the unsprung mass CG of the front wheel assembly are given by:

(1)

$$\Sigma F_x = 0$$
$$= F_{S_f} - F_{T_f} - W_{U_f} \sin\theta_g - F_{IU_f}$$

or

$$F_{S_f} = F_{T_f} + W_{U_f} \sin\theta_g + F_{IU_f} \qquad (3.1)$$

(2)

$$\Sigma F_z = 0$$
$$= R_{S_f} - R_f + W_{U_f} \cos\theta_g$$

or

$$R_{S_f} = R_f - W_{U_f} \cos\theta_g \qquad (3.2)$$

and (3)

$$\Sigma M_y = -\left(\Sigma I_f\right)\dot{\omega}_f$$
$$= -F_{T_f} t_{R_f} + R_f a_f$$

or

$$F_{T_f} t_{R_f} = R_f a_f + \left(\Sigma I_f \right) \dot{\omega}_f \tag{3.3}$$

where ΣI_f is the polar moment of inertia summation for all components rotating about the front axle spin axis and $\dot{\omega}_f$ is the angular acceleration of the front wheel assembly. For a non-sliding tire the angular acceleration is related to the forward vehicle acceleration by the geometric relationship $\dot{\omega}_f = \dfrac{A_x}{t_{R_f}}$, where A_x is the vehicle's forward acceleration. The term $R_f a_f$ in Eq (3.3) corresponds to the tire rolling resistance moment. Using the concept of rolling resistance force introduced in Section (2.2.4), the rolling resistance force can be alternately expressed as $R_f a_f = 2 F_{R_f} t_{R_f}$. {Note: since we are using a two-dimensional representation of the vehicle, a factor of two must be included in the previous equation to compensate for the fact we actually have two tires contributing to the rolling resistance.} Using these relationships, Eq (3.3) can then be rewritten as

$$F_{T_f} = 2 F_{R_f} + \frac{\Sigma I_f}{t_{R_f}^2} A_x \tag{3.4}$$

One final observation is to note that the inertia force of the front axle unsprung mass is also related to the forward acceleration by

$$F_{IU_f} = \frac{W_{U_f}}{g} A_x \tag{3.5}$$

Using Eqs (3.4) and (3.5), Eq (3.1) becomes

$$F_{S_f} = 2 F_{R_f} + W_{U_f} \sin \theta_g + \left(g \frac{\Sigma I_f}{t_{R_f}^2} + W_{U_f} \right) \frac{A_x}{g} \tag{3.6}$$

The rear wheel assembly, Fig (3.4), is analyzed in the same manner as the front wheel. Once again we start with the basic equations of motion:

(1)
$$\Sigma F_x = 0$$
$$= - F_{S_r} + F_{T_r} - W_{U_r} \sin \theta_g - F_{IU_r}$$

or, using

$$F_{IU_r} = \frac{W_{U_r}}{g} A_x$$

and rearranging

$$F_{S_r} = F_{T_r}$$

$$- W_{U_r} \sin\theta_g$$

$$- W_{U_r} \frac{A_x}{g}$$

$$(3.7)$$

(2)

$$\Sigma F_z = 0$$

$$= R_{S_r} - R_r$$

$$+ W_{U_r} \cos\theta_g$$

or

$$R_{S_r} = R_r$$

$$- W_{U_r} \cos\theta_g$$

$$(3.8)$$

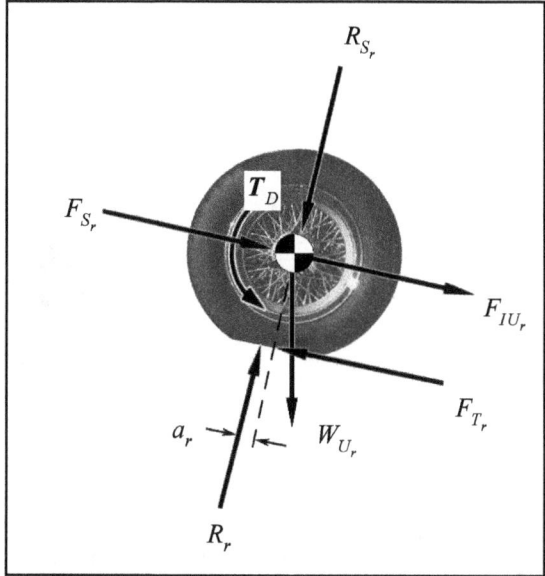

Figure (3.4): REAR WHEEL ASSEMBLY FREE-BODY-DIAGRAM

and (3)

$$\Sigma M_y = -\left(\Sigma I_r\right)\dot{\omega}_r$$

$$= F_{T_r} t_{R_r} + R_r a_r - T_D$$

The term T_D is the drive line torque applied to the driving wheels. As before, the angular acceleration is related to the forward vehicle acceleration by the geometric relationship $\dot{\omega}_r = \dfrac{A_x}{t_{R_r}}$ and the rolling resistance term $R_r a_r$ can be replaced with $2 F_{R_r} t_{R_r}$. Upon rearrangement, the previous equation becomes

$$F_{T_r} = \frac{T_D}{t_{R_r}} - 2 F_{R_r} - \frac{\Sigma I_r}{t_{R_r}^2} A_x \qquad (3.9)$$

Back substituting the tractive effort given by Eq (3.9) into Eq (3.7) finally yields the longitudinal chassis/suspension reaction force:

$$F_{S_r} = \frac{T_D}{t_{R_r}} - 2 F_{R_r} - W_{U_r} \sin\theta_g - \left(g \frac{\Sigma I_r}{t_{R_r}^2} + W_{U_r} \right) \frac{A_x}{g} \qquad (3.10)$$

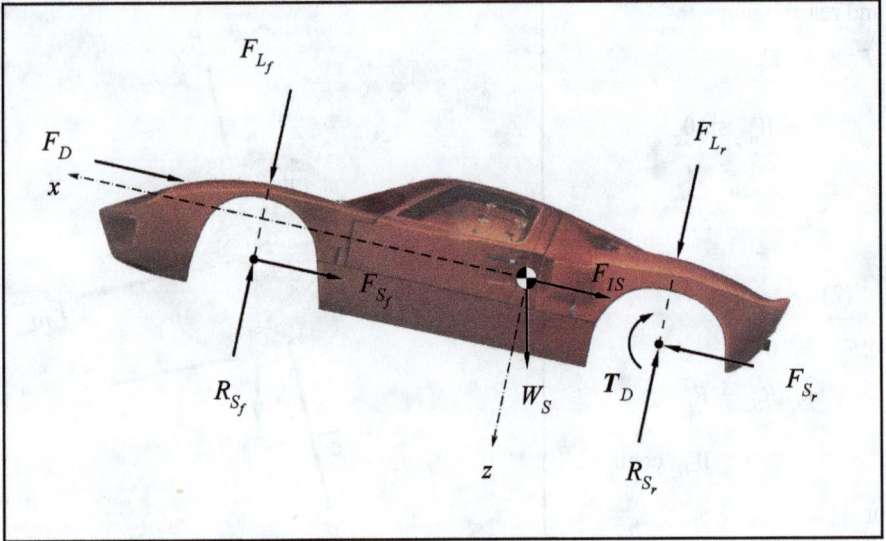

Figure (3.5): BODY/CHASSIS FREE-BODY-DIAGRAM

The last bit of analysis is applying basic equations of motion to the body and chassis (sprung mass). The corresponding free-body-diagram is shown in Fig (3.5). Let's start with:

$$\Sigma F_x = 0$$
$$= -F_D - F_{S_f} + F_{S_r} - W_S \sin\theta_g - F_{IS} \qquad (3.11)$$

where

$$F_{IS} = \frac{W_S}{g} A_x \qquad (3.12)$$

The suspension reaction forces F_{S_f} and F_{S_r} can be eliminated from Eq (3.11) by inserting Eqs (3.6) and (3.7). Doing this, along with Eq (3.12), results in:

$$\left[\left(W_{U_f} + W_{U_r} + W_S \right) + g \frac{\Sigma I_f}{t_{R_f}^2} \right] \frac{A_x}{g} =$$

$$F_{T_r} - F_D - 2 F_{R_f} - \left(W_{U_f} + W_{U_r} + W_S \right) \sin\theta_g \qquad (3.13)$$

The summation $W_{U_f} + W_{U_r} + W_S$ found in the previous equation is the total weight of the vehicle, which can be more compactly written as just W. The rear tractive force in Eq (3.13) can be alternately expressed in terms of the applied drive shaft torque using Eq (3.9). In this case Eq (3.13) finally becomes:

$$\frac{A_x}{g} = \frac{\dfrac{T_D}{t_{R_r}} - F_D - 2\left(F_{R_f} + F_{R_r}\right) - W \sin \theta_g}{W + g\dfrac{\Sigma I_f}{t_{R_f}^{\,2}} + g\dfrac{\Sigma I_r}{t_{R_r}^{\,2}}} \tag{3.14}$$

The previous equation provides quite a bit of useful information about vehicle acceleration, but it is not quite sufficient to calculate performance estimates. Although not quite evident, expressions for the road surface reactions must also be known. To accomplish this task the dynamic analysis of the vehicle continues. The next equation to examine is:

$$\Sigma F_z = 0$$
$$= F_{L_f} + F_{L_r} + W_S \cos\theta_g - R_{S_f} - R_{S_r}$$

or

$$R_{S_f} = F_{L_f} + F_{L_r} + W_S \cos\theta_g - R_{S_r} \tag{3.15}$$

The other equation we need is:

$$\Sigma M_y = 0$$
$$= F_D\left(H_D - H\right) + \left(R_{S_f} - F_{L_f}\right)L_f$$
$$-F_{S_f}\left(H - t_{R_f}\right) - \left(R_{S_r} - F_{L_r}\right)L_r + F_{S_r}\left(H - t_{R_r}\right) + T_D \tag{3.16}$$

Placing Eq (3.15) into (3.16) and rearranging results in:

$$R_{S_r} = F_D\left(\frac{H_D - H}{L_w}\right) + W_S\frac{L_f}{L_w}\cos\theta_g$$

$$- F_{S_f}\left(\frac{H - t_{R_f}}{L_w}\right) + F_{S_r}\left(\frac{H - t_{R_r}}{L_w}\right) + F_{L_r} + \frac{T_D}{L_w}$$

(3.17)

Back substitution of this equation into Eq (3.15) yields the front normal suspension reaction:

$$R_{S_f} = -F_D\left(\frac{H_D - H}{L_w}\right) + W_S\frac{L_r}{L_w}\cos\theta_g$$

$$F_{S_f}\left(\frac{H - t_{R_f}}{L_w}\right) - F_{S_r}\left(\frac{H - t_{R_r}}{L_w}\right) + F_{L_f} - \frac{T_D}{L_w}$$

(3.18)

After a considerable amount of algebraic manipulation, along with the assistance of Eqs (3.2), (3.6), (3.8) and (3.10), we can generate the following set of equations describing the road surface reaction forces:

$$R_r = F_{L_r} + F_D\frac{H_D}{L_w} + W_S\left(\frac{L_f}{L_w}\cos\theta_g + \frac{H}{L_w}\sin\theta_g\right) + W_{U_f}\frac{t_{R_f}}{L_w}\sin\theta_g$$

$$+ W_{U_r}\left(\cos\theta_g + \frac{t_{R_r}}{L_w}\sin\theta_g\right) + 2\left(F_{R_f}\frac{t_{R_f}}{L_w} + F_{R_r}\frac{t_{R_r}}{L_w}\right)$$

$$+ \left[W_S\frac{H}{L_w} + \left(g\frac{\Sigma I_f}{t_{R_f}^2} + W_{U_f}\right)\frac{t_{R_f}}{L_w} + \left(g\frac{\Sigma I_r}{t_{R_r}^2} + W_{U_r}\right)\frac{t_{R_r}}{L_w}\right]\frac{A_x}{g}$$

(3.19)

and

$$R_f = F_{L_f} - F_D\frac{H_D}{L_w} + W_S\left(\frac{L_r}{L_w}\cos\theta_g - \frac{H}{L_w}\sin\theta_g\right) - W_{U_r}\frac{t_{R_r}}{L_w}\sin\theta_g$$

$$+ W_{U_f}\left(\cos\theta_g - \frac{t_{R_f}}{L_w}\sin\theta_g\right) - 2\left(F_{R_f}\frac{t_{R_f}}{L_w} + F_{R_r}\frac{t_{R_r}}{L_w}\right)$$

$$- \left[W_S\frac{H}{L_w} + \left(g\frac{\Sigma I_f}{t_{R_f}^2} + W_{U_f}\right)\frac{t_{R_f}}{L_w} + \left(g\frac{\Sigma I_r}{t_{R_r}^2} + W_{U_r}\right)\frac{t_{R_r}}{L_w}\right]\frac{A_x}{g}$$

(3.20)

It is interesting to note that as the acceleration increases the rear road surface reaction increases while simultaneously the front reaction decreases. This is referred to as dynamic weight transfer, that is to say, there is a shift in weight from the front axle to the rear axle when the vehicle accelerates.

The coupled equations of motion relationships, i.e., Eqs (3.14), (3.19) and (3.20), we have developed thus far can be applied to any condition of acceleration. If we desire to study maximum acceleration as imposed by the adhesion limits of the driving tires, then we must introduce additional information.

(3.1.1) TRACTION LIMITED PERFORMANCE

Traction limited performance is the upper limit of vehicle acceleration when the force required to accelerate the vehicle exceeds the tire's grip on the road surface. This occurs at the point when maximum adhesion is achieved just before the tire skids. If we focus our attention on a rear wheel drive car, traction limited performance would be mathematically defined as:

$$F_{T_r} = \mu_p R_r \tag{3.21}$$

By virtue of Eq (3.9), the maximum applied torque at the wheel can be expressed as

$$\frac{T_{D,max}}{t_{R_r}} = \mu_p R_r + 2F_{R_r} + g\frac{\Sigma I_r}{T_{R_r}^{\,2}}\frac{A_{x,max}}{g} \tag{3.22}$$

An expression describing the traction limited maximum acceleration is derived by substituting Eqs (3.19) and (3.22) into Eq (3.14), which, after considerable manipulation, is:

$$\frac{A_{x,max}}{g} = \left\{ \left[\mu_p F_{L_r} - \left(1 - \mu_p \frac{H_D}{L_w}\right) F_D + \mu_p \left(W_S \frac{L_f}{L_w} + W_{U_r}\right) \right] \cos\theta_g \right.$$

$$- \left[W_S \left(1 - \mu_p \frac{H}{L_w}\right) + W_{U_f}\left(1 - \mu_p \frac{t_{R_f}}{L_w}\right) + W_{U_r}\left(1 - \mu_p \frac{t_{R_r}}{L_w}\right) \right] \sin\theta_g$$

$$\left. + 2\left[\left(\mu_p \frac{t_{R_f}}{L_w} - 1\right) F_{R_f} + \mu_p \frac{t_{R_r}}{L_w} F_{R_r} \right] \right\} \div$$

$$\left\{ \left(W_{U_f} + g \frac{\Sigma I_f}{t_{R_f}^2}\right)\left(1 - \mu_p \frac{t_{R_f}}{L_w}\right) - \mu_p \frac{g}{L_w}\frac{\Sigma I_r}{t_{R_r}} \right.$$

$$\left. + W_S\left(1 - \mu_p \frac{H}{L_w}\right) + W_{U_r}\left(1 - \mu_p \frac{t_{R_r}}{L_w}\right) \right\}$$

$$(3.23)$$

Although we have developed quite a few expressions describing acceleration performance, we are not quite ready to undertake a simulation. Lacking in our knowledge base are the influences of the power train (engine, transmission and differential) and aerodynamic downforce and drag. Consequently, we will temporarily depart from our discussion of acceleration to explore these topics.

(3.2) POWER TRAIN PERFORMANCE

The acceleration model as expressed by Eq (3.14) contains the term T_D which is the driving torque applied to the rear wheels of our sports car. Needless to say, the driving torque originates at the car's power plant, be it internal combustion engine or electric motor. In order to develop an expression for the driving torque a closer examination of the power train is necessary.

Figure (3.6) depicts the flow of power through the engine, transmission and differential/drive shaft system, which is customarily referred to as the power train. The power provided by the engine is given by $T_E \omega_E$ where T_E is the output torque at the flywheel and ω_E is the engine rotational speed. The engine output power becomes the input power for the transmission; however, not all of the input power is converted to useful transmission output power, $T_T \omega_T$. A portion of the input power is consumed by frictional and windage losses and is dissipated in the form of heat, $\dot{Q}_T$. Another portion of the input power becomes rotational kinetic energy resulting from the spinning

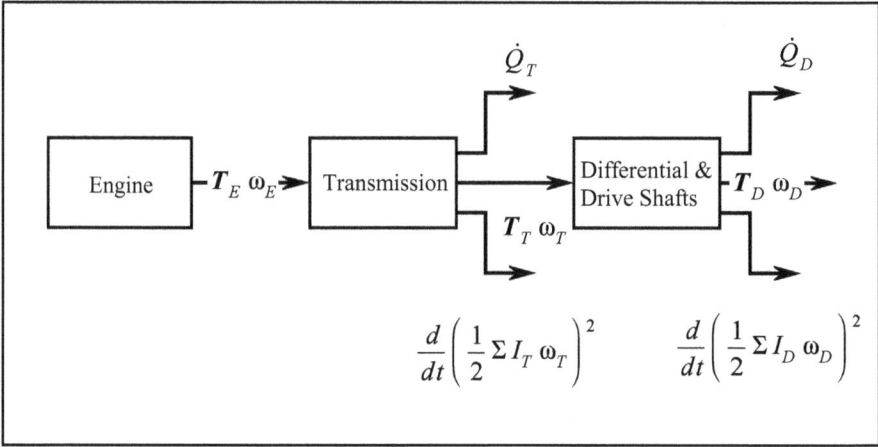

Figure (3.6): POWER FLOW DIAGRAM

of the internal gears and shafts, $\dfrac{d}{dt}\left(\dfrac{1}{2}\Sigma I_T\,\omega_T\right)^2$, where ΣI_T is the polar moment summation of the internal components. Balancing the input and output energy of the transmission leads to:

$$T_E\,\omega_E = \dot{Q}_T + T_T\,\omega_T + \frac{d}{dt}\left(\frac{1}{2}\Sigma I_T\,\omega_T\right)^2 \qquad (3.24)$$

The mechanical efficiency of the transmission, η_T , can be used in describing the frictional heat loss, that is, $\dot{Q}_T = \left(1 - \eta_T\right)T_E\,\omega_E$. Using this relationship in Eq (3.24) and rearranging results in

$$T_T\,\omega_T = \eta_T\,T_E\,\omega_E - \frac{d}{dt}\left(\frac{1}{2}\Sigma I_T\,\omega_T\right)^2 \qquad (3.25)$$

Applying the same logic to the differential and drive shaft assembly we can show

$$T_D\,\omega_D = \eta_D\,T_T\,\omega_T - \frac{d}{dt}\left(\frac{1}{2}\Sigma I_D\,\omega_D\right)^2 \qquad (3.26)$$

The rotational speeds ω_E, ω_T and ω_D are not independent of each other, but are constrained by the gear ratios used in the design of the transmission and differential, i.e., $\omega_T = G_D\,\omega_D$ and $\omega_E = G_T\,\omega_T$, or $\omega_E = G_T\,G_D\,\omega_D$. In addition, the drive shaft rotational speed is not independent of the vehicle forward velocity. The two are related by $\omega_D = \dfrac{V_x}{t_{R_r}}$. Using these relationships, Eqs (3.25) and (3.26) can be combined into:

$$T_D \frac{V_x}{t_{R_r}} = \eta_T \, \eta_D \, G_T \, G_D \, T_E \frac{V_x}{t_{R_r}}$$

$$- \eta_D \frac{d}{dt}\left[\frac{1}{2}\Sigma I_T \left(G_D \frac{V_x}{t_{R_r}}\right)^2\right] - \frac{d}{dt}\left[\frac{1}{2}\Sigma I_D \left(\frac{V_x}{t_{R_r}}\right)^2\right]$$

Completing the differentiating process and noting $\dfrac{dV_x}{dt} = A_x$ leads to the result:

$$\frac{A_x}{g} = \frac{\eta_T \, \eta_D \, G_T \, G_D \dfrac{T_E}{t_{R_r}} - F_D - 2\left(F_{R_f} + F_{R_r}\right) - W \sin\theta_g}{W + g\dfrac{\Sigma I_f}{t_{R_f}^{\,2}} + g\dfrac{\Sigma I_r}{t_{R_r}^{\,2}} + \eta_D \, G_D^{\,2} \, g\dfrac{\Sigma I_T}{t_{R_r}^{\,2}} + g\dfrac{\Sigma I_D}{t_{R_r}^{\,2}}} \tag{3.28}$$

The previous equation can be used in Eq (3.14) to eliminate the drive shaft torque in favor of the engine torque and gear ratio specifications:

$$\frac{T_D}{t_{R_r}} = \eta_T \, \eta_D \, G_T \, G_D \frac{T_E}{t_{R_r}} - \left(\eta_D \, G_D^{\,2} \, g\frac{\Sigma I_T}{t_{R_r}^{\,2}} + g\frac{\Sigma I_D}{t_{R_r}^{\,2}}\right)\frac{A_x}{g} \tag{3.27}$$

At this point it would appear that we have only complicated matters in developing an expression for vehicle acceleration. But, this is not the case. By introducing engine torque and gear ratio specifications we have directly incorporated vehicle design into it's acceleration performance. However, before proceeding further, we will need to gain an understanding of the power train design.

(3.2.1) ENGINE MODEL FOR VEHICLE PERFORMANCE STUDIES
In order to estimate vehicle acceleration performance and the subsequent specification of transmission and differential gear ratios, the engine torque and power performance curves must be known. As is often the case during the preliminary design phase of the vehicle drivetrain, the engine may not have yet been specified. The only avenue left to overcome this obstacle is to use a fictitious engine in our vehicle. This is not as difficult as it may seem. With all of the research and advanced developments in internal combustion engine design it is not surprising that most modern engines have about the same brake mean effective pressure (BMEP) performance. Figure (3.7) illustrates a BMEP curve representative of a modern medium performance, four-stroke,

spark ignited engine. The engine torque output is related to the BMEP by the relationship:

$$T_E = BMEP \frac{V_d}{4\pi} \tag{3.29}$$

where V_d is the engine displacement. The engine power is then determined from $P_E = T_E \omega_E$. Using this approach the design engineer can study the affect of various engines on vehicle performance by simply selecting reasonable engine displacements.

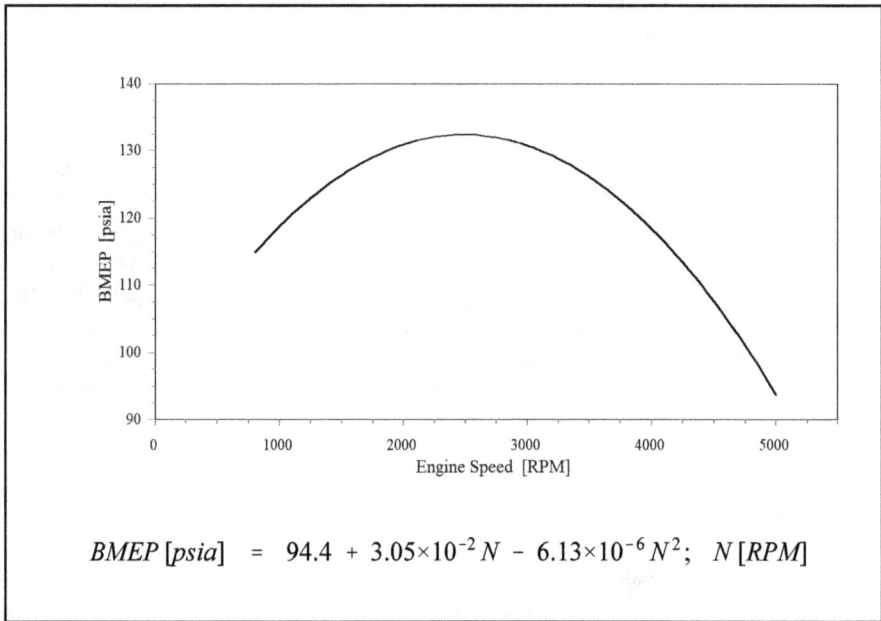

$$BMEP\,[psia] = 94.4 + 3.05\times10^{-2}\,N - 6.13\times10^{-6}\,N^2; \quad N\,[RPM]$$

Figure (3.7): TYPICAL BRAKE MEAN EFFECTIVE PRESSURE CURVE FOR MODERN MEDIUM PERFORMANCE ENGINES

(3.2.2) ESTIMATING TRANSMISSION AND DIFFERENTIAL GEAR RATIOS

Looking at the analytical analysis of acceleration presented thus far, we are now faced with the challenge of specifying transmission and differential gear ratios before we can proceed further. Once again, if the vehicle under investigation exists this is not a problem as the gear ratios are known. However, if we are in the vehicle design stage then we must devise a scheme that will us to select initial transmission and differential gear ratio values appropriate for the intended use of the vehicle.

The beginning point of our scheme starts with a basic equation relating engine speed to vehicle road speed. The general form of this equation is:

$$V_x = \frac{t_{R_r}}{G_T G_D} \omega_E \qquad (3.30)$$

or, using the USCS system of units:

$$G_T G_D [--] = \frac{\pi}{528} \frac{t_{R_r} [\text{in}] \, N_E [\text{RPM}]}{V_x [\text{MPH}]} \qquad (3.31)$$

The rear tire effective rolling radius, engine speed and ground speed are all design values or target values that can be specified for our vehicle. But, as the previous equation indicates, only the combination of transmission and differential gear ratios can be related to vehicle parameters. To overcome this problem we will consider vehicle operation when the transmission has been shifted into it's highest gear. Typically a gear ratio of $G_T = 1.0$ is chosen in this situation. This is a direct drive ratio which minimizes energy loss in the transmission. With this specification, we can calculate the differential gear ratio from Eq (3.31).

There are two general approaches that may be used to develop a strategy to calculate gear ratios: the economy method and the performance method. The economy method attempts to maintain the engine speed in a range of maximum torque, where the engine has it's best efficiency and consequently minimum fuel consumption. The performance method using an engine speed range where maximum power output from the engine is obtained. Focusing on our current task of calculating a differential gear ratio, it is up to the designer to select which method most closely matches the intended use of the vehicle.

Eq (3.31) is used in both methods, but the interpretation, or specification, of engine speed and road speed is different between the two models. For the economy method the engine speed chosen is near that obtained for maximum engine torque, and the road speed is the design "cruising" speed of the vehicle. On the other hand, using the performance method we would select an engine speed where maximum power is obtain to drive the car at the design "top" speed (which is not necessarily the maximum speed). Lets consider the two methods by means of an example. In both cases we will use the same base vehicle, which has the following characteristics:

Weight:	2683 lbs
Weight distribution:	44% front, 56% rear
Wheelbase :	95 inch
CG height:	14.8 inch
Tire diameter:	30.4 inch (effective rolling radius: 14.7 in)

For the economy method we will use a 3.2L V6 (196 cu in) engine, which, using the BMEP curve shown in Fig (3.7), has the torque characteristics illustrated in Fig (3.8). Upon closer examination of Fig (3.8), it is noticed that maximum torque occurs around 2500 RPM. This is a logical choice for engine operation at cruising speed as engine torque is closely related to engine efficiency. We will also select 70 MPH as our design cruising speed. (Both engine speed and road speed are discretionary

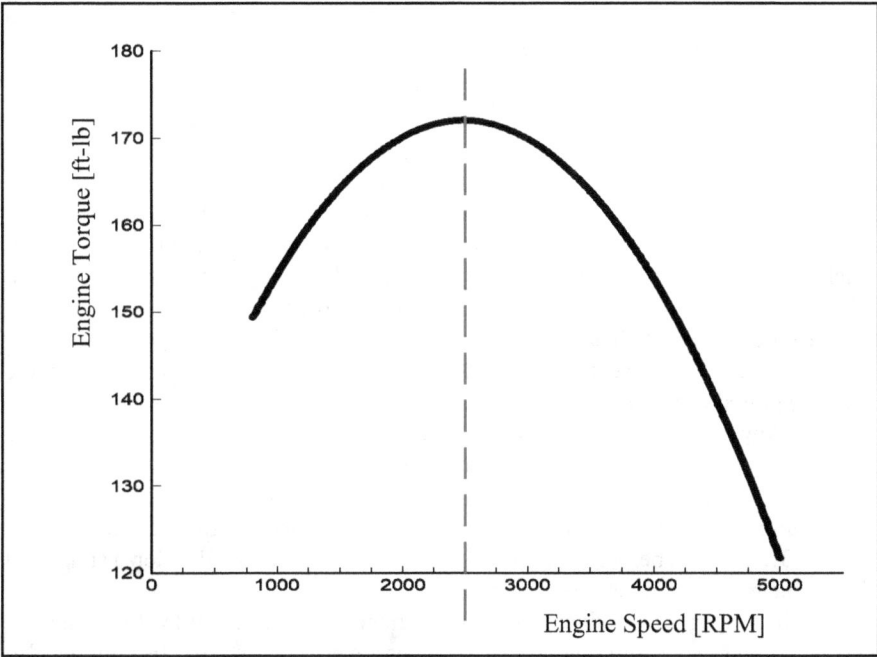

Figure (3.8): 3.2L V6 ENGINE TORQUE CURVE

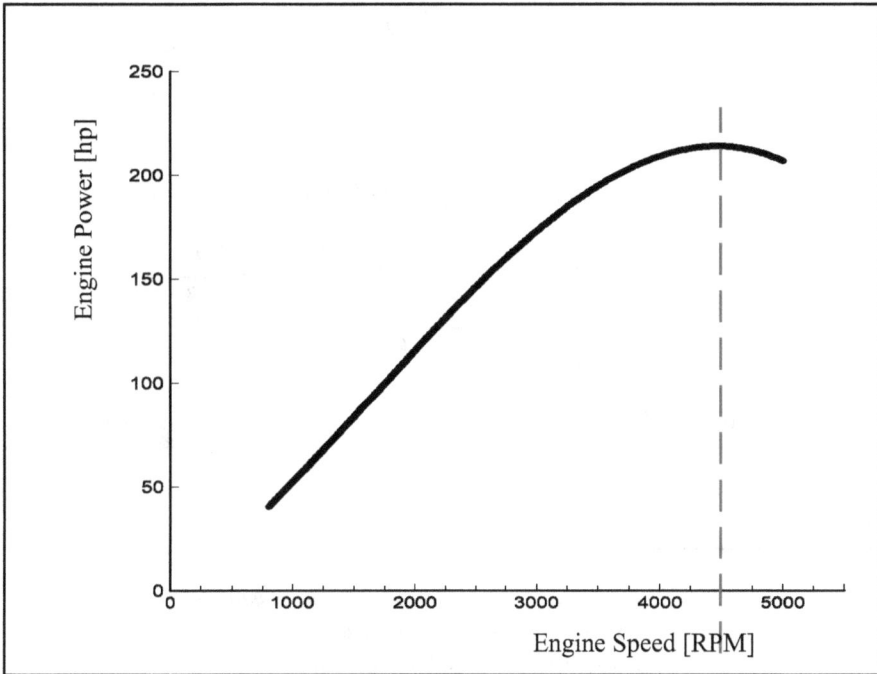

Figure (3.9): 5.7L V8 ENGINE POWER CURVE

specifications left to the judgement of the designer, and are not universal.) Using Eq (3.31), we calculate a differential gear ratio value of $G_D = 3.12$[2].

The other approach used in determining the differential gear ratio is the performance method. In this approach the designer is more interested in top speed rather than fuel economy. In this example we will select a 5.7L V8 (350 cu in) engine as our engine of choice. The power curve, based on Fig (3.7), is shown in the figure below. According to Fig (3.9), the maximum power output of the engine occurs at around 4500 RPM, and becomes our design engine speed. Our goal for top speed, which also becomes a design specification, is 140 MPH. Using these design values, the differential gear ratio from Eq (3.31) is calculated to be $G_D = 2.81$. A word of caution is warranted. Although a top speed has been specified, at this point there is no guarantee that the engine can produce enough power to achieve this speed. This topic will be explored further on in this chapter.

Now that the differential gear ratio has been determined we are ready to move on to the transmission. It is convenient to start with the transmission's lowest gear ratio (highest numerically), i.e., first gear. One approach of determining the first gear ratio is to examine the limitations that occur at maximum tractive effort as discussed in Section (3.1.1). But, before doing so we will find it useful to first develop a simplified form of the maximum tractive effort equations.

In this regard, consider a vehicle experiencing a low forward velocity moving on level ground such that the aerodynamic forces are essentially zero. In addition, also consider the sprung weight of the vehicle to be significantly greater than either the front axle or rear axle unsprung weight. With these considerations, Eq (3.23) reduces to:

$$\frac{A_{x,\max}}{g} \cong \frac{\mu_p \dfrac{L_f}{L_w}}{1 - \mu_p \dfrac{H}{L_w}} \tag{3.32}$$

Eq (3.28) can be simplified by introducing a mass factor, γ_m, defined as:

$$\gamma_m = 1 + \frac{g \dfrac{\Sigma I_f}{t_{R_f}^2} + g \dfrac{\Sigma I_r}{t_{R_r}^2} + \eta_D\, G_D{}^2\, g \dfrac{\Sigma I_T}{t_{R_r}^2} + g \dfrac{\Sigma I_D}{t_{R_r}^2}}{W} \tag{3.33}$$

Using this definition along with the previous considerations, allows us to express Eq (3.28) as:

---

[2] *It should be noted that in an economy application a transmission overdrive gear ratio could be used at cruising speed. Overdrive ratios are usually in the range of 0.6 to 0.8. If an overdrive ratio of, say, 0.78 was used, then the differential gear ratio changes from a value of 3.12 to 4.00.*

$$\frac{A_{x,\max}}{g} = \frac{\eta_T \, \eta_D \, G_{T1} \, G_D \, T_{E,\max}}{\gamma_m \, W \, t_{R_r}} \tag{3.34}$$

It should be noted that we are applying the previous equation to just the transmission's first gear ratio G_{T1}. Combining Eqs (3.32) and (3.34) yields:

$$G_{T1} = \frac{\gamma_m \, W \, t_{R_r} \left(\dfrac{\mu_p \dfrac{L_f}{L_w}}{1 - \mu_p \dfrac{H}{L_w}} \right)}{\eta_T \, \eta_D \, G_D \, T_{E,\max}} \tag{3.35}$$

It is interesting to note that, as a practical matter, the mass factor values usually are within the range of 1.08 to 1.20. Also, transmission and differential mechanical efficiencies are around 86% to 94%. Consequently, calculation of the transmission's first gear ratio, or at least an estimate, can be made by making a reasonable guesses of the mass factor and mechanical efficiencies.

Lets continue the two examples we started. First, for the V6 economy car the peak friction coefficient would be around 0.8 for a standard automotive tire. From the torque curve this engine produces about 172 ft-lb of torque at maximum output. Then from Eq (3.35) the transmission's first gear ratio evaluates as 3.28. In a similar manner, for our V8 sports car, having more of a racing tire rubber compound, we will use a value of 1.2 for the peak friction coefficient. Based on BMEP curve the V8 engine produces about 307 ft-lbs of maximum torque. Using these values the first gear ratio for our sports car is 3.30.

At this point we have estimates for the first and final gear ratio of the transmission, as well as the differential gear ratio. The next step is to determine how many gears our transmission should have and their respective values. To accomplish this task the geometric progression rule can used.

The geometric progression rule states that the engine operates within the same engine speed range for each combination of transmission and differential gear ratio. This is illustrated in Fig (3.10), where N_{EU} represents the upper design limit of engine operation and N_{EL} the lower. At the first shift point, V_{x1}, the relationship between engine speed and road speed can be expressed as:

$$V_{x1} = 2\pi \frac{t_{R_r}}{G_{T1} \, G_D} N_{EU} = 2\pi \frac{t_{R_r}}{G_{T2} \, G_D} N_{EL}$$

which simplifies to:

$$\frac{N_{EU}}{N_{EL}} = \frac{G_{T1}}{G_{T2}}$$

At the second shift point, V_{x2}, using the same approach, it can be shown that:

$$\frac{N_{EU}}{N_{EL}} = \frac{G_{T2}}{G_{T3}}$$

The approach can be continued for any number of gears, n, to arrive at the conclusion: The term on the left-hand-side of the previous equation is defined as the transmission

$$\frac{N_{EU}}{N_{EL}} = \frac{G_{T1}}{G_{T2}} = \frac{G_{T2}}{G_{T3}} = \ ... \ = \frac{G_{Tn-1}}{G_{Tn}} \qquad (3.36)$$

gear factor, i.e.,

$$K_{G} \equiv \frac{N_{EU}}{N_{EL}} \qquad (3.37)$$

such that Eq (3.36) becomes

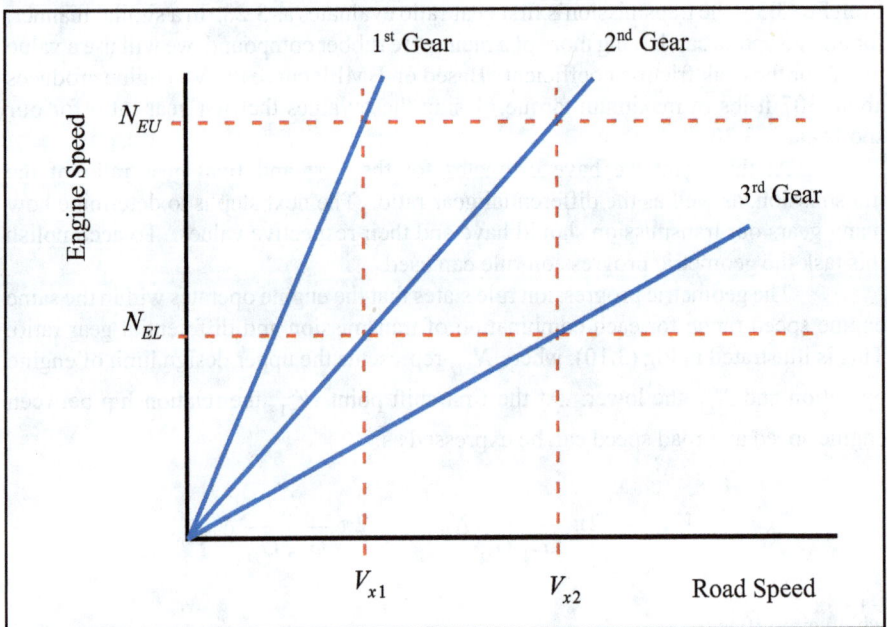

Figure (3.10): GEOMETRIC PROGRESSION RULE

$$K_G = \frac{G_{T1}}{G_{T2}} = \frac{G_{T2}}{G_{T3}} = \ldots = \frac{G_{Tn-1}}{G_{Tn}} \tag{3.38}$$

Eq (3.38) can be written in a more compact form by noting that

$$G_{T2} = \frac{G_{T1}}{K_G}$$

$$G_{T3} = \frac{G_{T2}{}^2}{G_{T1}} = \frac{G_{T1}{}^2}{K_G{}^2 G_{T1}} = \frac{G_{T1}}{K_G{}^2}$$

$$\vdots$$

$$G_{Tn} = \frac{G_{T1}}{K_G{}^{n-1}}$$

$$\tag{3.39}$$

Eq (3.39) is the mathematical basis of the geometric progress rule. To study how this equation is actually used, we will return to our current example.

    The first step is to determine a value for the transmission factor suitable for the intended use of our vehicle. Previously introduced was the concept of the power method and the economy method. This philosophy also extends to aid us in the evaluation of the remaining gear ratios. It is our intention that our V6 engine operate as efficiently as possible. In this regard we initially select a torque band, such that a range of engine speed, where the torque is at least 97.5% of maximum. From the torque curve illustration in Fig (3.8) this corresponds to a range of 1750 to 3220 RPM. From the transmission factor definition, Eq (3.37), we calculate $K_G = 1.84$.

    The next step is to calculate the required number of gears for the transmission. The last portion of Eq (3.39) can be rewritten as:

$$n = 1 + \frac{\ln\left(\dfrac{G_{T1}}{G_{Tn}}\right)}{\ln K_G} \tag{3.40}$$

Although we do not know the number of gears, we do know that the highest gear ratio is 1.0 from our previous efforts in calculating the differential gear ratio. Using this information, $n = 2.95$. Obviously a fractional value for the number of gears is not possible; therefore, we round up the number of gears to 3.

    The last step is make a few adjustments and then calculate the remaining gear ratios. The first adjustment is to re-evaluate the transmission gear factor, by rearranging the last equation in the series found in Eq (3.39):

$$K_G = \left(\frac{G_{T1}}{G_{Tn}} \right)^{\frac{1}{n-1}} \qquad (3.41)$$

With n = 3, K_G becomes 1.81, which also results in a new engine speed range of 1780 to 3220 RPM if centered about the design speed of 2500 RPM for maximum engine torque. Finally from Eq (3.39) the second gear ratio is calculated to be $G_{T2} = 1.81$.

The second example we looked at was the sports car with the V8 engine. In this example, the performance method is being used. Instead of selecting an engine speed range where maximum torque occurs, we are now interested in selecting a range close to maximum power. However, centering a power band around the engine speed for maximum power is not practical. As shown in Fig (3.9) this speed appears to be close to engine redline, that is, the maximum allowable engine speed. To alleviate this problem the upper engine speed is chosen to be 4500 RPM where maximum power is generated. The lower engine speed is initially selected at 3500 RPM where the engine produces 92% of maximum power.

Using an engine speed range of 3500 to 4500 RPM produces a transmission factor of 1.29. Following the same procedure as before, we calculate the number of gears to be 5.74, or rounded down to be 5. The transmission factor would then adjust to 1.35 with a corresponding engine speed range of 3330 to 4500 RPM. The transmission gear ratios then calculated from Eq (3.39) are: $G_{T2} = 2.45$, $G_{T3} = 1.82$ and $G_{T4} = 1.35$.

Summarizing the results we have for the two different scenarios:

V6 Engine
Differential Ratio: 3.12
Transmission
    1st Gear Ratio: 3.28
    2nd Gear Ratio: 1.81
    3rd Gear Ratio: 1.00

V8 Engine
Differential Ratio: 2.81
Transmission
    1st Gear Ratio: 3.30
    2nd Gear Ratio: 2.45
    3rd Gear Ratio: 1.82
    4th Gear Ratio: 1.35
    5th Gear Ratio: 1.00

It must be remembered that a considerable amount of engineering judgement was used in obtaining the foregoing results. These results are by no means the only possible outcome as different vehicle designers may use different approaches to the selection and determination of gear ratios. In addition, the gear ratio values are initial estimates only. Gear ratio values will most certainly change when modeling vehicle acceleration to obtain optimum performance. It must also be realized that the mechanical design of the transmission and differential has not been considered, which may also necessitate a change in gear ratios to accommodate fabrication and construction. Nor, has the subjective judgement of the driver been included as to how well the vehicle performs. There is a lot work design work still left to do, but, at least, we have a starting point.

(3.3) AERODYNAMIC FORCES AND MOMENTS

The aerodynamic forces created as a result of a vehicle moving through air have a significant influence on acceleration and cornering performance, as well as fuel economy, at moderate to high ground speeds. If one is to accurately predict vehicle performance, a method must be devised to account for these forces in vehicle dynamic studies. The two general classes of aerodynamic forces are lift and drag. Although most production cars produce a certain amount of lift at moderate to high speeds, the affect is undesirable. Lift at either axle reduces the normal force acting on the tire contact path, which in turn decreases the tires performance ultimately leading to instability. As vehicle designers we are more interested in minimizing lift and focusing on anti-lift or downforce improvements. Consequently, our discussion is focused mainly on the aerodynamic characteristics of downforce and drag.

To gain a better understanding of aerodynamic behavior it is convenient to think of the vehicle as being a stationary object placed in a moving air stream instead of the vehicle moving through stagnant air; although it is necessary to make an adjustment to compensate for the road surface. In general the nature of air flow around a vehicle is quite complex. At the design stage the most accurate method to determine aerodynamic forces is by experimental means through the use of a wind tunnel. In this approach direct measurements of downforce and drag are taken on a scale model placed in an airstream within a wind tunnel. The classic method of representing drag and downforce data is to use dimensionless coefficients. That is, a drag coefficient defined as:

$$C_D = \frac{F_D}{\frac{1}{2}\rho V^2 A_p}$$

and front/rear axle lift[3] coefficients:

$$C_{L_f} = \frac{F_{L_f}}{\frac{1}{2}\rho V^2 A_p} \qquad\qquad C_{L_r} = \frac{F_{L_r}}{\frac{1}{2}\rho V^2 A_p}$$

These coefficients are correlated with the Reynolds number, a dimensionless parameter describing the flow field, i.e.,

$$Re_L = \frac{\rho V L}{\mu}$$

where V is the vehicle velocity, A_p is the projected frontal area of the vehicle, L is the overall vehicle length, and ρ is the air density. Once obtained, the lift and drag coefficients can be applied to a variety of vehicle speeds encountered during performance, acceleration or cornering, studies.

---

[3] *Within the current text downforce and lift coefficients share the same definition; however, downforce coefficients are represented as being positive while lift coefficients are negative.*

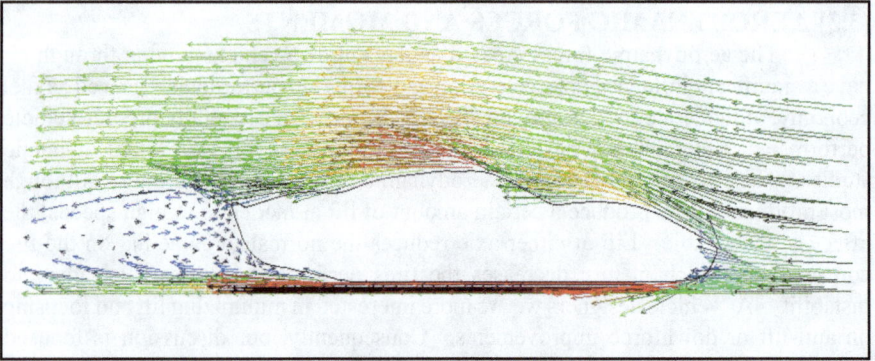

Figure (3.11): EXAMPLE CFD AIR VELOCITY VECTOR PLOT IN TWO DIMENSIONS

The experimental approach may not be practical in the early design stage as making changes to the scale model and re-testing is a time consuming process. An alternative approach is to use an analytical method where only a geometric description, or model, of the vehicle is needed. In this regard, a computational fluid dynamics (CFD) computer simulation may be utilized to study aerodynamic behavior. Figure (3.11) illustrates the results obtain from a CFD simulation of air flowing around a sports car body in two-dimensions. We should be point out that a the two-dimensional model is being used only for purposes of discussion. In actuality, the two-dimensional model is not particularly accurate as it neglects induced vortex motion off the sides of the vehicle body, as well as the influence of rotating tires on the disruption of the moving air field. These affects need to be included in the development of a practical simulation.

The post analysis of the CFD results requires a bit of background and explanation. Drag forces are a combination of different effects due to:
- form drag, caused by wake turbulence created by the shape of the vehicle body,
- skin drag, caused by the shear stresses acting on the exterior vehicle surface,
- air movement through cooling and ventilation passages, and
- virtual drag created by the addition of auxiliary aerodynamic devices such as wings.

In most practical situations vehicles operate in high Reynolds number ranges with viscous effects confined to boundary layers and wake regions. In instances as such, the form drag constitutes the majority of drag while the skin drag contributes only about 10% of the total. Cooling and ventilation passages are usually only a few percent of total and are therefore of lesser importance during the preliminary design of the vehicle. Aerodynamic devices are of such a specific application that general treatise is impractical. With these considerations, our current attention is focused first on aspects of pressure drag and skin drag.

In general, aerodynamic effects can be described in terms of forces at the solid body-ambient air interface. The forces acting on the vehicle are the result of "wall" shear stress due to fluid viscosity and velocity gradients, and normal stress due to

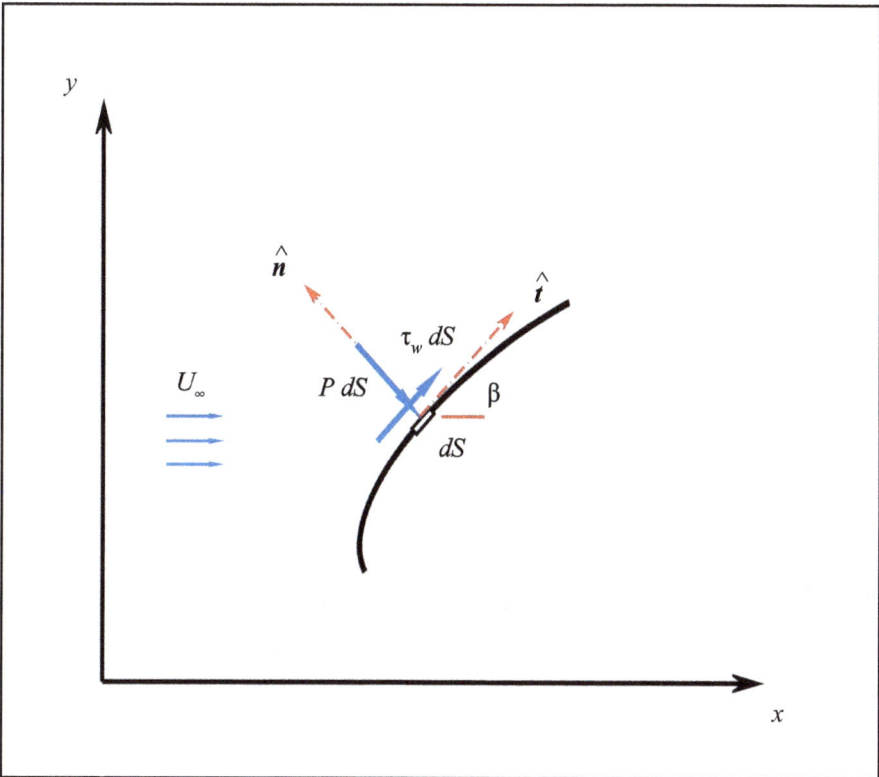

Figure (3.12): DIFFERENTIAL NORMAL AND TANGENTIAL AERODYNAMIC FORCES

staticpressure variations. Figure (3.12) depicts normal and tangential aerodynamic forces acting on a differential surface area element of a two-dimensional vehicle body. Considering our interest in modeling aerodynamic forces applied to vehicle performance studies, we will neglect the effect of wall shear stress. In this case the total force exerted on the vehicle can be approximated by the equation:

$$\vec{F} = \oint_{A_s} P \, d\vec{A} \tag{3.42}$$

where As is the surface area of the vehicle. The area vector can be described in terms arc length and width, and outward surface normal, by $d\vec{A} = -w \, dS \, \hat{n}$. The surface normal, $\hat{n}$, is related to the two-dimensional coordinate system by $\hat{n} = -\sin\beta \, \hat{i}_x + \cos\beta \, \hat{i}_y$. Using the two-dimensional coordinate system, the area vector can be expressed as:

$$d\vec{A} = w \sin\beta \, dS \, \hat{i}_x - w \cos\beta \, dS \, \hat{i}_y \tag{3.43}$$

However, it should be noted that $dy = \sin\beta \, dS$ and $dx = \cos\beta \, dS$. An alternate expression for the area vector is therefore

$$d\vec{A} = w \, dy \, \hat{i}_x - w \, dx \, \hat{i}_y \tag{3.44}$$

Substituting Eq (3.44) into Eq (3.42) gives us

$$F_x \hat{i}_x + F_y \hat{i}_y = w \oint_C P\left(dy \, \hat{i}_x - w \, dx \, \hat{i}_y\right) \tag{3.45}$$

The resultant force acting on the vehicle in the direction of the upstream velocity is the drag force, and the resultant force normal to the flow direction is the downforce (which we might think of as "anti-lift"). The mathematical description of lift and drag follows from Eq (3.45):

$$F_D = w \oint_C P \, dy \tag{3.46}$$

and

$$F_L = w \oint_C P \, dx \tag{3.47}$$

where C is the body contour.

To simplify the analysis it is convenient to describe the pressure distribution in terms of a pressure coefficient defined as

$$C_P \equiv \frac{P - P_\infty}{\frac{1}{2} \rho_\infty V^2} \tag{3.48}$$

where the ∞ subscript refers to the free stream, or far field, conditions. Rearranging the pressure coefficient definition yields

$$P = P_\infty + \frac{1}{2} C_P \rho_\infty V^2 \tag{3.49}$$

Substituting Eq (3.49) into Eq (3.46):

$$F_D = w P_\infty \oint_C dy + \frac{1}{2} \rho_\infty V^2 w \oint_C C_P \, dy$$

But $\oint_C dy = 0$, which allows to reduce the previous equation to

$$F_D = \frac{1}{2} \rho_\infty V^2 w \oint_C C_P \, dy \tag{3.50}$$

Recalling the definition of drag coefficient allows us to simplify the previous equation to

$$C_D = \frac{w}{A_P} \oint_C C_P \, dy \tag{3.51}$$

The last remaining step in modeling drag is to develop an expression for the drag pressure center, i.e., the distance above the ground plane where the drag force acts (the parameter HD in our acceleration model). We start by defining an xy-coordinate system which has an origin where the vertical front axle plane intersects with the horizontal road surface plane. The differential aerodynamic moment about this origin is

$$d\vec{M}_0 = \vec{r} \times d\vec{F} \tag{3.52}$$

The in terms of x and y components, the position vector can be expressed as

$$\vec{r} = x \, \hat{i}_x + y \, \hat{i}_y \tag{3.53}$$

From our previous work the differential aerodynamic force is a function of the pressure and differential surface area, and can be written as

$$d\vec{F} = w P \, dy \, \hat{i}_x - w P \, dx \, \hat{i}_y \tag{3.54}$$

Placing Eqs (3.53) and (3.54) into Eq (3.52) results in

$$d\vec{M}_0 = (-x w P \, dx - y w P \, dy) \, \hat{i}_z \tag{3.55}$$

As indicated in the previous equation, the aerodynamic moment consists of two terms. The first term on the right-hand-side represents the contribution of the lift force to the total moment, and, as you may surmise, the second term represents the drag force contribution. If we isolate just the drag component then we may write the drag moment in integral form as

$$M_D = F_D H_D = w \oint_C P y \, dy \tag{3.56}$$

From Eq (3.56), along with a bit of mathematical manipulation, we can show that the drag pressure center is also related to the pressure coefficient by

$$H_D = \frac{\displaystyle\oint_C C_P y \, dy}{\displaystyle\oint_C C_P \, dy} \tag{3.57}$$

By following the foregoing steps in a similar fashion, expressions for the lift force and lift pressure center can be developed, which are:

$$F_L = \frac{1}{2} \rho_\infty V^2 w \oint_C C_p \, dx \qquad (3.58)$$

and

$$H_L = \frac{\oint_C C_p \, x \, dx}{\oint_C C_p \, dx} \qquad (3.59)$$

However, this is not exactly the information we need in our acceleration performance model. We need to have expressions for the equivalent downforce (anti-lift) acting at the front and rear axle planes. This goal is accommodated by noting $F_{L_r} = F_L \dfrac{H_L}{L_w}$

and $F_{L_f} = F_L \left(1 - \dfrac{H_L}{L_w} \right)$. It will take a bit of effort, but we can eventually prove that:

$$C_{L_r} = \frac{w}{L_w A_p} \oint_C C_p \, x \, dx \qquad (3.60)$$

and

$$C_{L_f} = \frac{w}{A_p} \oint_C C_p \left(1 - \frac{x}{L_w} \right) dx \qquad (3.61)$$

It has taken a while, but we have developed expressions for lift and drag coefficients, as well as the drag pressure center, in terms of just the pressure coefficient. Returning to our CFD example, Figure (3.13) depicts the pressure coefficient results for our sports car. It must be pointed out that a continuous function describing the pressure coefficient around the contour of the sports car body has not been developed, but rather numerical values at nodes defined by discrete locations. Since a closed form solution of the integrals is not possible, a numerical approximation must be sought.

Lets consider an example where force is to be approximated from known pressure values acting along a two-dimensional surface which is represented by nodes as illustrated in Figure (3.14). The basic equation describing the x-component of force acting at node I is

$$F_{x,i} = w \int P_i \, dy \cong w P_i \Delta y = \frac{1}{2} w P_i \left(y_{i+1} - y_{i-1} \right) \qquad (3.62)$$

The total force acting on the surface is the summation of all the individual node forces along the contour, that is

$$F_x \cong \frac{1}{2} w \sum_C P_i \left(y_{i+1} - y_{i-1} \right)$$

(3.63)

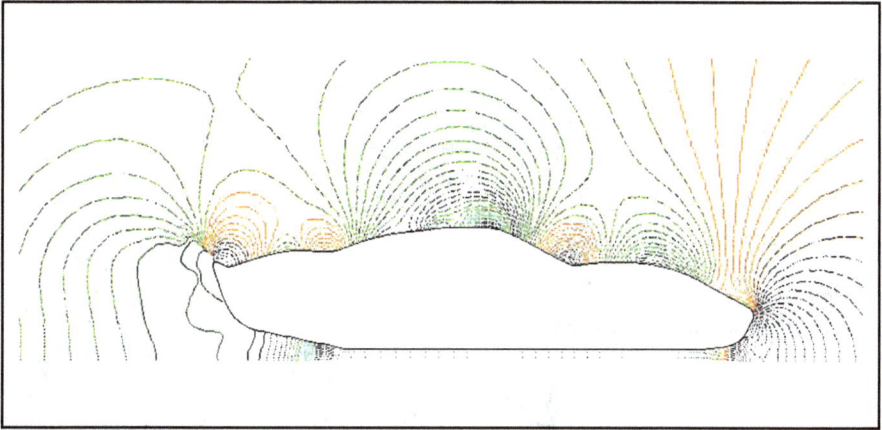

Figure (3.13): EXAMPLE CFD PRESSURE COEFFICIENT PLOT IN TWO DIMENSIONS

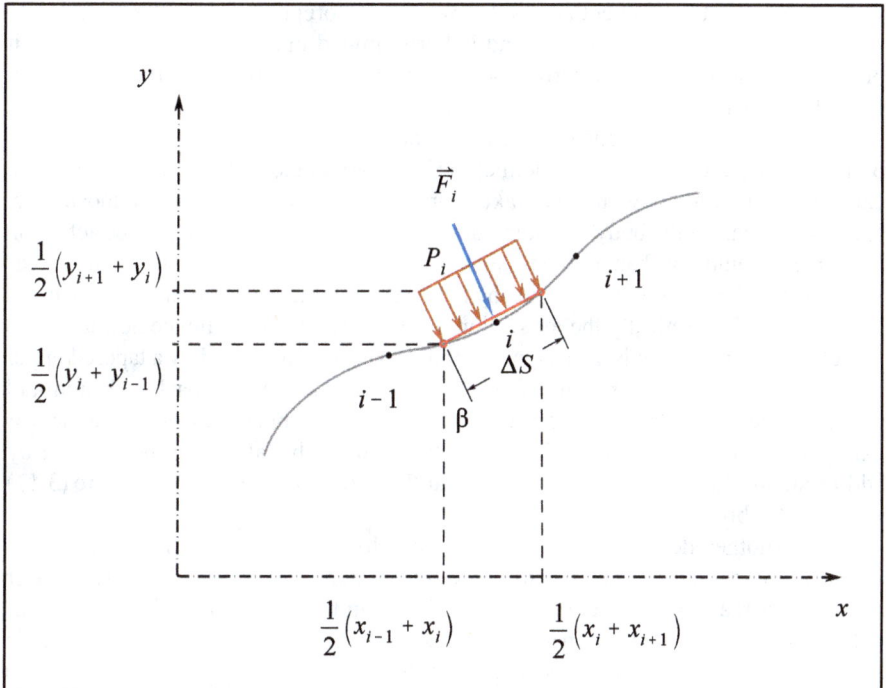

Figure (3.14): STATIC PRESSURE AT DISCRETE LOCATIONS ALONG BODY CONTOUR

Using a similar strategy the drag and lift (downforce) coefficients can be approximated by:

$$C_D = \frac{w}{2 A_p} \sum_C C_{P,i} \left(y_{i+1} - y_{i-1} \right) \tag{3.64}$$

$$C_{L_r} = \frac{w}{2 L_w A_p} \sum_C C_{P,i} \, x_i \left(x_{i+1} - x_{i-1} \right) \tag{3.65}$$

$$C_{L_f} = \frac{w}{2 A_p} \sum_C C_{P,i} \left(1 - \frac{x_i}{L_w} \right) \left(x_{i+1} - x_{i-1} \right) \tag{3.66}$$

and the drag pressure center by:

$$H_D = \frac{\displaystyle\sum_C C_{P,i} \, y_i \left(y_{i+1} - y_{i-1} \right)}{\displaystyle\sum_C C_{P,i} \left(y_{i+1} - y_{i-1} \right)} \tag{3.67}$$

We must remember that development of the foregoing analysis only applies to vehicle bodies in two dimensions, and is fairly limited in accuracy. Nevertheless, it provides the designer with a starting point, especially in examining airflow separation from the body in the aft section of the vehicle.

The main importance of airflow separation in the rear portion of the vehicle body is the separation line which defines the boundary of the turbulent wake behind the car. The local pressure within the wake is practically uniform and below atmospheric. If we cut off part of the body extending into the wake, the resultant drag force acting on the vehicle would not change. Consequently, terminating the body not far from the rear axle plane replaces natural separation by forced separation due to the sudden kink in the body shape. Theoretically the aft section of the body should terminate normal to the vehicle's longitudinal axis. This type of body shape, characterized by a tapered upper body surface and an abrupt cut-off tail, is referred to as a Kammback or Kamm-tail profile, named after Dr Wunibald I.E. Kamm [1]. In addition, lift, especially at the rear axle plane, can be produced at high vehicle speeds. This affect can be reduced by adding an up-tilted flap, or spoiler, at the rear the body. Our sports car in Figure (3.11) incorporates both of the these features.

Another design feature not to be overlooked in ground clearance. Most vehicles have a relatively flat undertray. At large ground clearances this creates a faster moving airstream over the upper body surface than the lower. The difference in air velocities causes a difference in static air pressure between the upper and lower vehicle surface creating lift. However, as the ground clearance decreases the air space between the undertray and road surface becomes a venturi [2], causing the air to move much faster underneath the body compared to that above. Subsequently, the magnitudes of

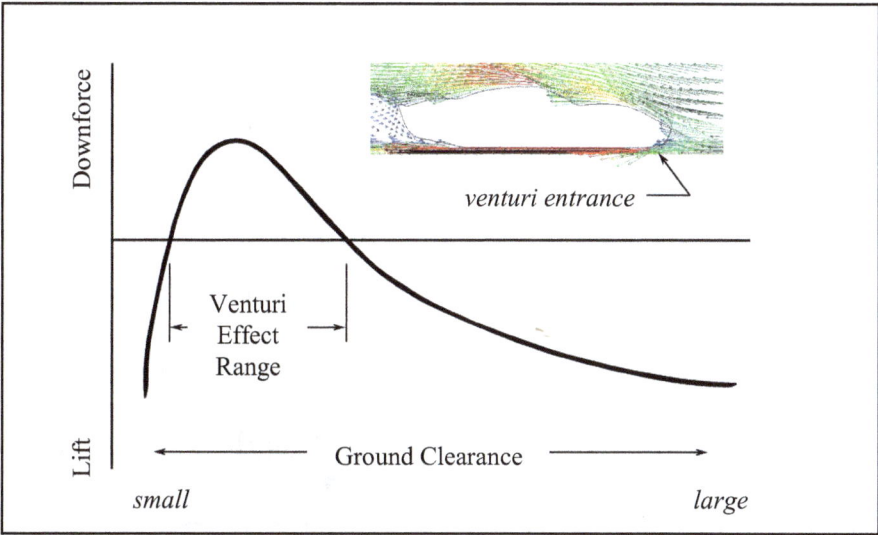

Figure (3.15): GROUND CLEARANCE AFFECT ON DOWNFORCE

static pressure reverse and downforce is created. A continued reduction in ground clearance leads to a situation where the air flow is restricted to such an extent that the static pressures reverse once again, creating lift. The affect of ground clearance on downforce (lift) is subjectively shown in Figure (3.15).

(3.4) ESTIMATING VEHICLE ACCELERATION

It has taken us a while to get here, but we have the necessary background to enable us to estimate vehicle acceleration performance. The general strategy used in a computer program or spreadsheet is illustrated by the flowchart shown in Figure (3.16). It is generally convenient to plot the acceleration, in terms of g's, against the vehicle forward velocity. This graph creates an acceleration map for the various transmission gear ratios. Also included in this map is the traction limited acceleration performance. A comparison of the estimated acceleration performance with the traction limit gives the designer objective feedback with regards to the size (power) of the engine being used in the design. If the estimated acceleration exceeds the traction limit, then the engine has too much power for the tires and drivetrain being used. That is, the tires break loose from the road surface resulting in a loss of control. The other graph of interest is created by plotting the forward velocity as a function of time.

Returning to our earlier V6 and V8 sports car example, a simulation was performed using the data in Tables (3.1) and (3.2). Starting with the V8 example, the acceleration and velocity performance graphs are displayed in Figures (3.17) and 3.18). In a similar fashion the V6 results are depicted in Figures (3.19) and (3.20).

Now that we finally have simulation results to examine, it should be remembered that we have only completed the first design iteration. A few performance issues stand out. In both of the acceleration performance graphs the maximum acceleration in first gear is about 10% lower than the traction limit. This is due to the

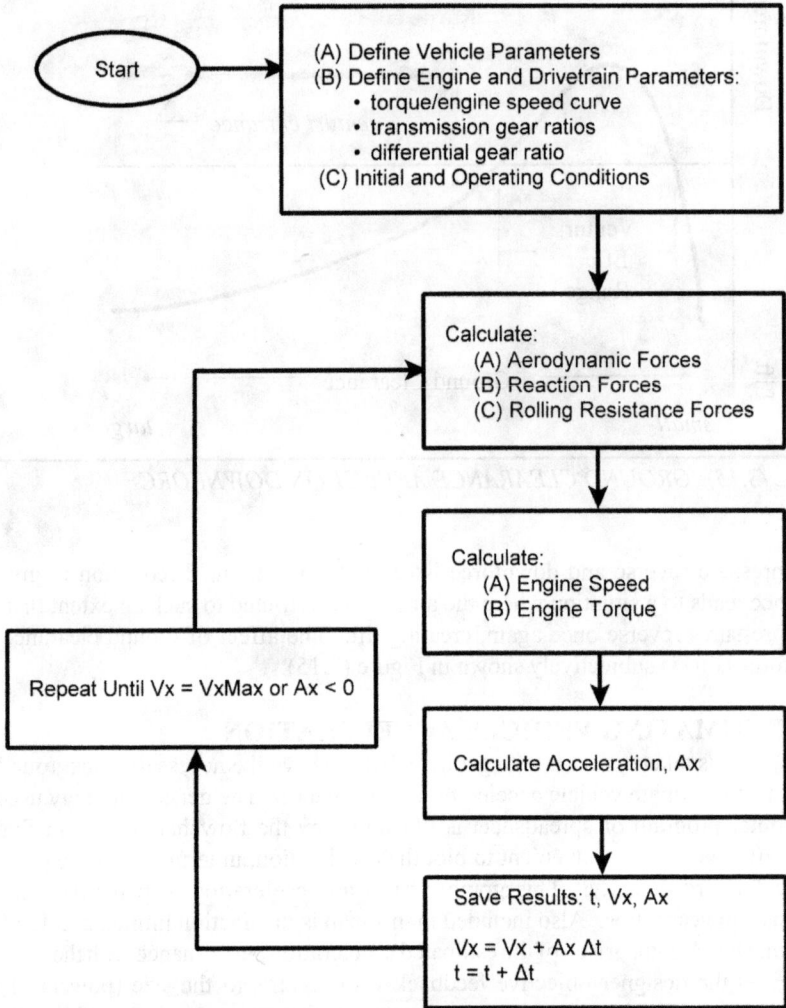

Figure (3.16): ACCELERATION CALCULATION FLOWCHART

fact that we used an approximation of the traction limit when calculating the first gear ratio. Now that we have a better value for the traction limit, a revised calculation can be made. Referring to the performance graphs of our V8 sports car, the five speed transmission provides relatively smooth acceleration and forward velocity, but the acceleration is quite a bit below the traction limit. In this case the engine is underpowered for the tires being used on the sports car; however, the tire compound (as represented by the peak traction coefficient value of 1.2) is too soft for extended use on highways. The next iteration for the V8 sports car, left to the discretion of the designer,

Table (3.1): EXAMPLE "VEHICLE PARAMETERS" FOR ACCELERATION PERFORMANCE SIMULATION

Vehicle Parameters		
	V8 Sports Car	V6 Sports Car
Wheelbase, Lw [in]	95.0	95.0
Sprung Weight, WS [lb]	1845	1845
Sprung Weight C.G.:		
Lf [in]	41.8	41.8
Lr [in]	53.2	53.2
H [in]	14.8	14.8
Unsprung Weight		
Front Axle, WUf [lb]	293	293
Rear Axle, WUr [lb]	545	545
Unsprung Polar Moment		
Front Axle, IUf [lb in2]	17740	17740
Rear Axle, IUr [lb in2]	32950	32950
Drag Coefficient, Cd [–]	0.39	0.39
Drag Pressure Center, Hd [in]	15.8	15.8
Downforce Coefficient:		
Front Axle, Clf [–]	0.075	0.075
Rear Axle, Clr [–]	0.112	0.112
Front Profile Area, Ap [ft2]	9.22	9.22
Tire Effective Radius		
Front, tRf [in]	14.7	14.7
Rear, tRr [in]	14.7	14.7
Peak Traction Coefficient, μP [–]	1.2	0.8
Tire Pressure [psi]	28	28

Table (3.2): EXAMPLE "ENGINE AND DRIVETRAIN PARAMETERS" AND "OPERATING CONDITIONS" FOR ACCELERATION PERFORMANCE SIMULATION

Engine and Drivetrain Parameters		
	V8 Sports Car	V6 Sports Car
Engine Displacement, Vd [in3]	350	196
Transmission Gear Ratios		
1st	3.30	3.28
2nd	2.45	1.84
3rd	1.82	1.00
4th	1.35	
5th	1.00	
Differential Gear Ratio	2.81	3.12
Mechanical Efficiency:		
Transmission, ηT [–]	0.9	0.9
Differential, ηD [–]	0.9	0.9
Polar Moment of Inertia:		
Transmission, IUT [lb in2]	300	300
Differential, IUD [lb in2]	590	590
Operating Conditions		
Atmospheric Pressure, P [psia]	14.696	14.696
Atmospheric Temperature [°F]	60	60
Road Grade [deg]	0	0

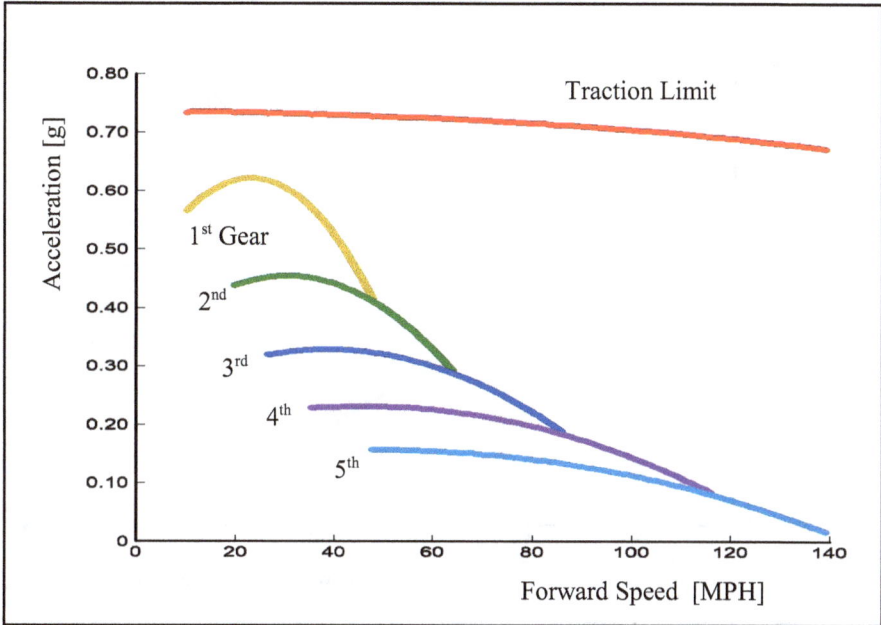

Figure (3.17): ACCELERATION PERFORMANCE OF EXAMPLE V8 SPORTS CAR

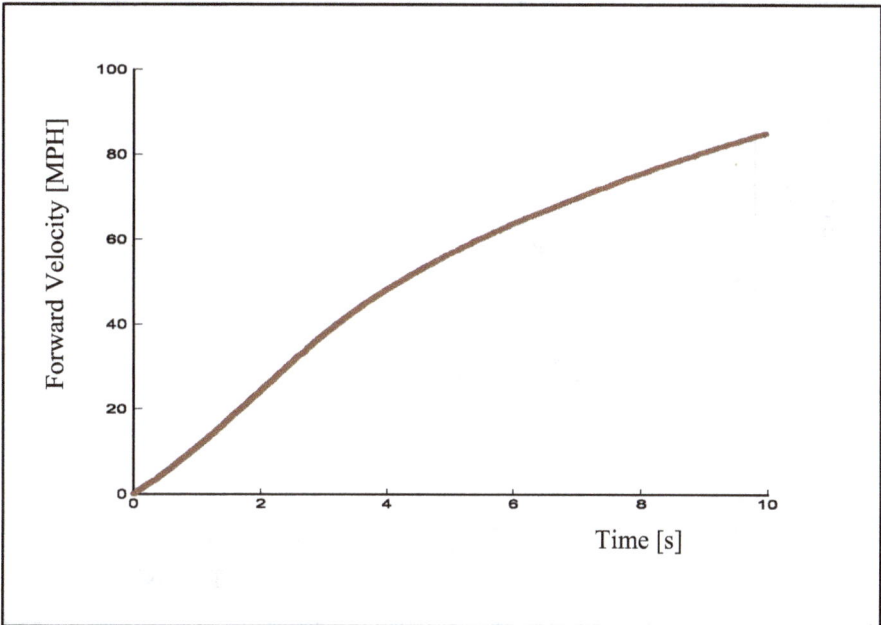

Figure (3.18): FORWARD VELOCITY RESULTS FOR EXAMPLE V8 SPORTS CAR

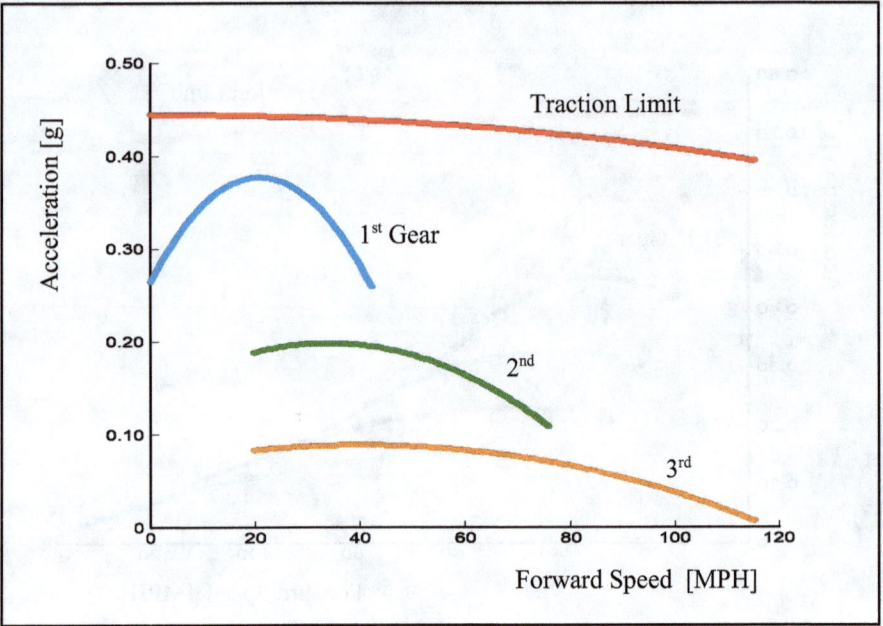

Figure (3.19): ACCELERATION PERFORMANCE EXAMPLE OF V6 SPORTS CAR

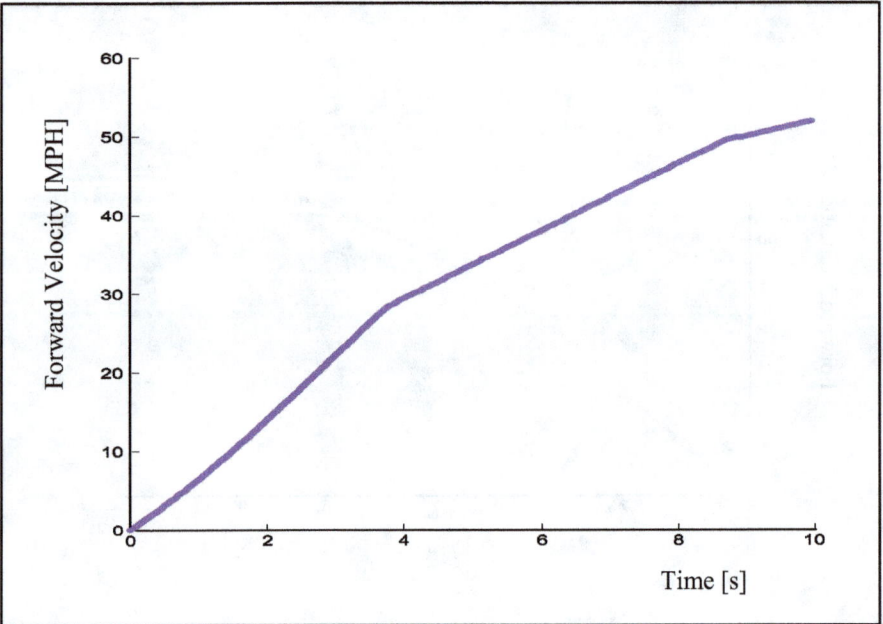

Figure (3.20): EXAMPLE VELOCITY RESULTS FOR V6 SPORTS CAR

may include a trial and error process of varying the engine size and tire compound to obtain better performance results.

The V6 sports car results are interesting; there are a few problems. One will notice, as previously discussed, that the acceleration curves are quite a bit less than the traction limit; however, since this car was designed with economy in mind, this is to be expected. The main problem with the design is evidenced by the distance between the acceleration curves, associated with the transmission gear ratios, resulting in "jerk" when shifting the transmission. This is also depicted in the discontinuity of the forward speed curve. The next iteration for this design, once again left to the discretion of the designer, may include a smaller engine speed band and adding a fourth transmission gear.

One final thought regarding our current study; the performance results previously presented are based on both vehicles having the same weight. This is not the case. Generally speaking more powerful engines lead to an increase in weight of the engine and drivetrain components, which should be accounted for in the acceleration performance simulation.

It may interest the reader to know that the example sports we have worked with is loosely based on the classic GT40 race car. Several models (MkI through MkIV) of the car were developed with various combinations of engine and transmission/differential gear ratios. Two interesting combinations are: 289 cu. in./350 bhp engine coupled with a five speed transmission (ratios: 2.23, 1.53, 1.21, 1.00, 0.81) and differential gear ratio of 3.20 {top speed: 165 MPH, 0 to 60 MPH in 5.3 s}; and, 427 cu. in./485 bhp engine coupled with a four speed transmission (ratios: 2.22, 1.43, 1.19, 1.00) and differential gear ratio of 2.78 {top speed: 186 MPH, 0 to 60 in 3.1 s}.

(3.5) BRAKE SYSTEMS AND PERFORMANCE

The foregoing sections of this chapter have been devoted to understanding the dynamics of vehicle acceleration. Equally important, perhaps more so, is understanding how to design a vehicle braking system which has the capacity to execute a controlled stop in a timely manner. In this section we will first study vehicle performance under conditions of deceleration due to braking. Our goal here is to establish the performance criteria which the design of the brake system must meet. Then we will examine a hydraulic brake system typical of modern practice, and finally undertake a general analysis of disc brake components.

(3.5.1) DYNAMICS OF DECELERATION

The objective of the present study is to determine the influence of major design parameters on vehicle deceleration performance. A summary of forces acting on a decelerating vehicle on a level road surface are shown in Figure (3.21). A conservative approach in applying appropriate forces is to neglect any force which assists in braking, which are the rolling resistance forces and the aerodynamic forces. In addition, the forward shift in of the vehicle's center of gravity due to pitching is neglected. Under these restrictions, the equations of motion are:

$$\sum F_x = 0$$

$$= F_I - F_{bf} - F_{br} \tag{3.68}$$

$$\sum F_z = 0$$

$$= W - R_f - R_r \tag{3.69}$$

and

$$\sum M_{CG} = 0$$

$$= \left(F_{bf} + F_{br} \right) H + R_r L_r - R_f L_f \tag{3.70}$$

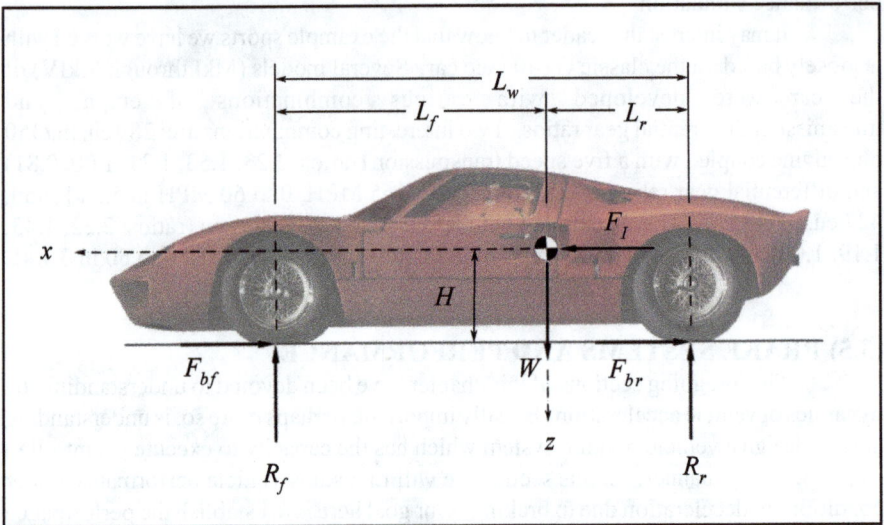

Figure (3.21): GEOMETRY AND FORCE DIAGRAM OF DECELERATING VEHICLE

The inertia force, F_I, acting at the sprung mass center of gravity can be expressed in terms of an equivalent vehicle weight using the mass factor {Eq (3.33)} and total vehicle weight, i.e., $\gamma_m W$, and the deceleration of the vehicle coordinate system relative to an earth fixed coordinate system, D_x:

$$F_I = \gamma_m W \frac{D_x}{g} \tag{3.71}$$

Using Eq (3.71), the following relationships can be derived from the equations of motion:

$$F_{bf} + F_{br} = \gamma_m W \frac{D_x}{g} \tag{3.72}$$

$$R_f = W \left(\frac{L_r}{L_w} + \gamma_m \frac{D_x}{g} \frac{H}{L_w} \right) \tag{3.73}$$

and

$$R_r = W \left(\frac{L_f}{L_w} - \gamma_m \frac{D_x}{g} \frac{H}{L_w} \right) \tag{3.74}$$

The first term of the right-hand-side of Eqs (3.73) and (3.74) is the weight on their respective axles under static conditions. The second term represents the transfer of weight (increase on the front axle, decrease on the rear) caused by braking.

The interaction between the tires and the road surface must also be accounted for in the braking analysis. This can be accomplished by expressing the brake force at each axle in terms of a normal force, which has the same magnitude as the reaction force, and a coefficient of adhesion:

$$F_{bf} = \mu_f R_f$$
$$= \mu_f \left(\frac{L_r}{L_w} + \gamma_m \frac{D_x}{g} \frac{H}{L_w} \right) W \tag{3.75}$$

and

$$F_{br} = \mu_r R_r$$
$$= \mu_r \left(\frac{L_f}{L_w} - \gamma_m \frac{D_x}{g} \frac{H}{L_w} \right) W \tag{3.76}$$

A graphical interpretation of Eqs (3.75) and (3.76) is presented in Figure (3.22). At first glance these plots suggest that the braking forces are a linear function of the coefficient of adhesion, various vehicle geometric parameters, and deceleration. Actually this is not the case. When the brakes are actuated a resistive effort is applied to the wheels or drivetrain through the disc or drum components of the brake system. The braking force created at the tire contact patch is a reaction to the resistive effort. The braking forces at both the front and rear axles are therefore controlled by the brake system. Consequently, there is no mechanical guarantee that the front wheels will not lock-up before the rear wheels, and vice versa. We can, however, use Eqs (3.75) and (3.76) along with Eq (3.72) to examine limiting conditions where one pair of wheels is very near to lock-up, i.e., maximum braking.

If we focus our attention on the front axle for a moment, the front maximum brake force evaluated from Eq (3.75) is

$$F_{bf,\,\text{max}} = \mu_p \left(\frac{L_r}{L_w} + \gamma_m \frac{D_x}{g} \frac{H}{L_w} \right) W \qquad (3.77)$$

where μ_p is the peak coefficient of adhesion. From Eq (3.72) we also have

$$F_{bf,\,\text{max}} + F_{br} = \gamma_m W \frac{D_x}{g}$$

or

$$\frac{D_x}{g} = \frac{F_{bf,\,\text{max}} + F_{br}}{\gamma_m W} \qquad (3.78)$$

Substituting Eq (3.78) into Eq (3.77) results in

$$F_{bf,\,\text{max}} = \frac{\mu_p \dfrac{L_r}{L_w}}{1 - \mu_p \dfrac{H}{L_w}} W + \frac{\mu_p \dfrac{H}{L_w}}{1 - \mu_p \dfrac{H}{L_w}} F_{br} \qquad (3.79)$$

Similarly, the rear maximum brake for is

$$F_{br,\,\text{max}} = \frac{\mu_p \dfrac{L_f}{L_w}}{1 + \mu_p \dfrac{H}{L_w}} W - \frac{\mu_p \dfrac{H}{L_w}}{1 + \mu_p \dfrac{H}{L_w}} F_{bf} \qquad (3.80)$$

Eqs (3.72), (3.79) and (3.80) mathematically describe the basic braking performance of any vehicle having two axles. These equations can be arranged on a brake performance map, which is simply a graph of the rear axle brake force versus the front axle brake force, as shown in Figure (3.23). The maximum brake forces described by Eqs (3.79) and (3.80) establish the braking performance boundaries, or upper limits. These boundaries are dependent only on vehicle weight, center of gravity location, and peak coefficient of adhesion. The boundaries are used to divide the braking performance map into four regions, also shown in Figure (3.23).

Region I is the desired region for efficient braking and vehicle stability. In Region II the front wheels are in a locked condition. Although there is some loss in directional control when the front wheels lock up, yaw stability is maintained. This is not the case inside Region III where the rear wheels are locked and the front wheels are not. Any perturbation in yaw, which is certain to occur from irregularities in the road surface, will force the vehicle into an unstable condition causing the vehicle to initiate a spin. The vehicle will continue to spin until the front axle "swaps ends" with the rear

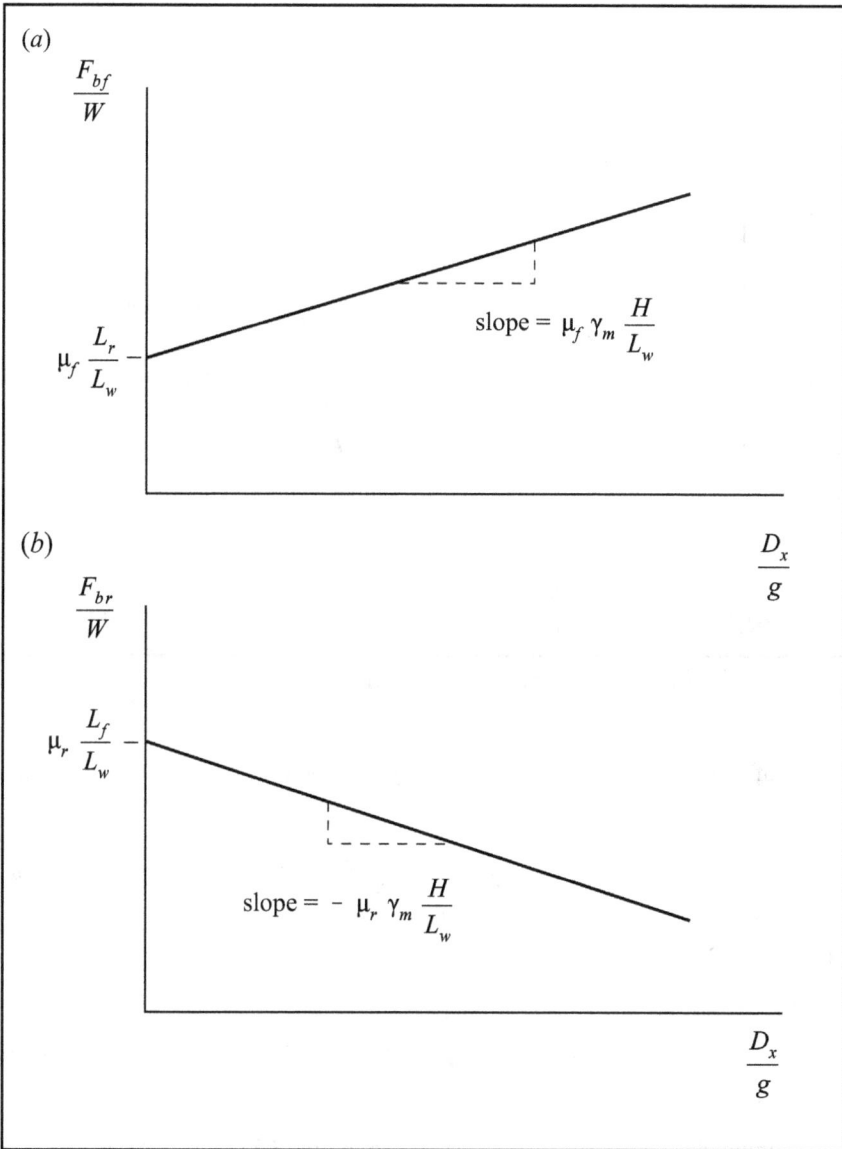

Figure (3.22): GRAPHICAL INTERPRETATION OF (a) FRONT BRAKE FORCE AND (b) REAR BRAKE FORCE AS A FUNCTION OF DECELERATION

axle. Essentially the latter attitude puts us back in Region II. The wheels of both axles, front and rear, are locked in Region IV. Braking in this region is very inefficient, (remember, the coefficient of adhesion rapidly decreases from it's peak value to it's sliding value), but stability is still maintained.

The intersection of the maximum brake force boundaries defines a unique point of maximum braking, which in turn defines the maximum deceleration a vehicle is

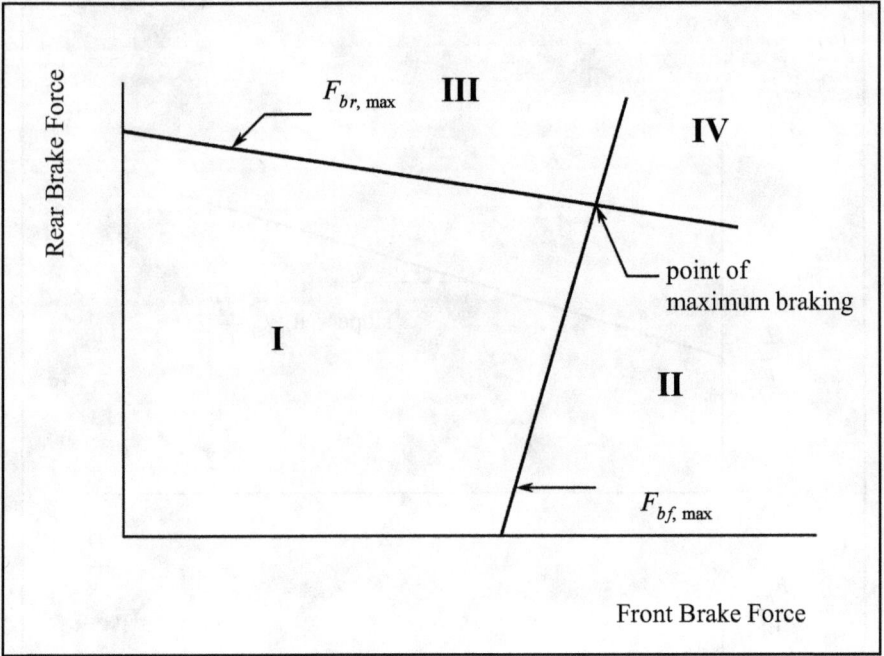

Figure (3.23): BRAKE PERFORMANCE MAP

capable of achieving. The maximum braking point on the brake performance map is given by

$$F_{bf, mbp} = \mu_p \left(\frac{L_r}{L_w} + \mu_p \frac{H}{L_w} \right) W \qquad (3.81)$$

and

$$F_{br, mbp} = \mu_p \left(\frac{L_f}{L_w} - \mu_p \frac{H}{L_w} \right) W \qquad (3.82)$$

The maximum deceleration is found by substituting Eqs (3.81) and (3.82) in Eq (3.72), which leads to the conclusion that $\dfrac{D_{x, max}}{g} = \dfrac{\mu_p}{\gamma_m}$.

Thus far in our analysis we have treated μ_p as an operational constant depending solely on tire characteristics and road surface conditions. This implies that there is just one pair of maximum brake force boundaries, which is true only at the time the brakes are actuated. Obviously over a period of time tire performance and road surface conditions will change. Therefore, when analyzing brake performance of a vehicle it is more appropriate to treat μ_p as a variable. With this in mind we can generate a series of maximum brake force boundaries, each with its own unique maximum deceleration point. The curve which passes through the locus of these points is referred to as the optimum brake performance curve.

For purposes of brake system design, it is convenient to mathematically describe the optimum brake performance curve in terms of an optimum rear brake force/optimum front brake force function. One method of deriving this function starts by evaluating the maximum front brake force at the point of maximum deceleration, i.e., from Eq (3.81)

$$F_{bf,\,opt} = \mu \left(\frac{L_r}{L_w} + \mu \frac{H}{L_w} \right) W \qquad (3.83)$$

which can be rewritten as

$$\frac{H}{L_w} \mu^2 + \frac{L_r}{L_w} \mu - \frac{F_{bf,\,opt}}{W} = 0 \qquad (3.84)$$

The solution to Eq (3.84) for the coefficient of adhesion, is

$$\mu = \frac{-\dfrac{L_r}{L_w} + \sqrt{\left(\dfrac{L_r}{L_w}\right)^2 + 4\dfrac{H}{L_w}\dfrac{F_{bf,\,opt}}{W}}}{2\dfrac{H}{L_w}} \qquad (3.85)$$

Using Eq (3.72) the maximum rear brake force at the point of maximum deceleration is

$$F_{br,\,opt} = \mu W - F_{bf,\,opt} \qquad (3.86)$$

Finally, the optimum brake performance curve is found by inserting Eq (3.85) into Eq (3.86):

$$F_{br,\,opt} = \left\{ \frac{-\dfrac{L_r}{L_w} + \sqrt{\left(\dfrac{L_r}{L_w}\right)^2 + 4\dfrac{H}{L_w}\dfrac{F_{bf,\,opt}}{W}}}{2\dfrac{H}{L_w}} \right\} W - F_{bf,\,opt} \qquad (3.87)$$

The optimum brake performance curve, as its name implies, represents the best braking performance and balance between the front and rear brakes the vehicle is capable of achieving under various road conditions. This curve can also be plotted on a brake performance map along with the maximum brake force boundaries. For example, consider a sports car with the following specifications:

Weight: 2683 lb	**Lw: 95.0 in**
ym: 1.09	**Lf: 53.2 in**
μp: 1.00	**Lr: 41.8 in**
	H: 14.8 in

Using this data we are able to construct the optimum brake performance curve and map as illustrated in Figure (3.24). This graph establishes the optimum performance relationships between the front axle and rear axle brake forces which result in maximum deceleration of our sports car. The actual vehicle deceleration, however, is not necessarily the maximum obtainable. Actual braking performance depends on the design and operating characteristics of the brake system incorporated into the vehicle. We have reached the point now where furthering our understanding of brake performance and brake system design must be preceded by learning about brake components and typical system layout.

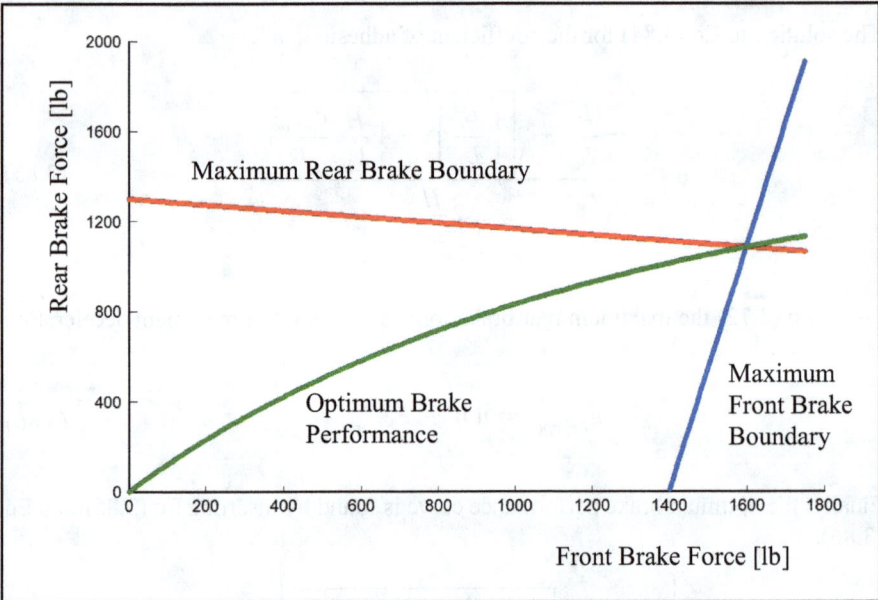

Figure (3.24): EXAMPLE SPORTS CAR BRAKE PERFORMANCE MAP

(3.5.2) BRAKE COMPONENTS AND LAYOUT

Brake systems are generally classified according to their method of actuation, i.e., mechanical, hydraulic, or air (pneumatic) brakes; and then further designated asbeing either manual or power assist. Hydraulic brakes are universally used for automobile and light truck applications. Hydraulic brakes and air brakes are used on trucks, and mechanical systems are used mainly on agricultural equipment. We will not examine all of the different types of brake systems and applications, but rather focus on typical hydraulic brake systems, components and operation.

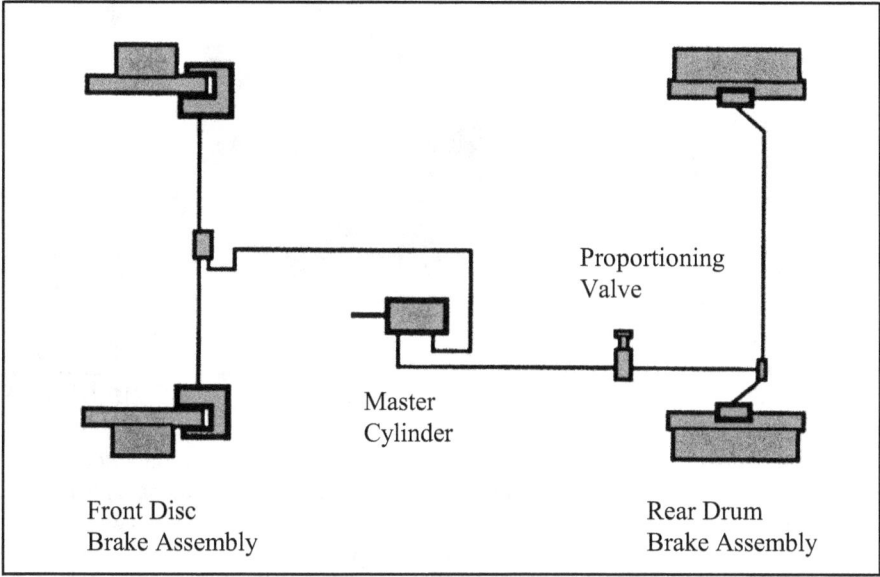

Figure (3.25): TYPICAL FOUR-WHEEL FRONT DISC/REAR DRUM BRAKE LAYOUT

A hydraulic brake system layout common to four-wheel vehicles is shown in Figure (3.25). The major components of the hydraulic brake system are the master cylinder and the remote slave cylinders found in each of the drum or disc brake assemblies. The operation of the hydraulic brake system follows Pascal's Law (1658), which states that the pressure of a confined liquid acts equally in all directions and with equal force on equal areas. Thus, the pressure created in the master cylinder, by the application of a force through a foot actuated lever, can be transmitted equally through steel tubing and high pressure hoses to the slave cylinders. The slave cylinders in turn utilize the pressure to force brake shoes or caliper pads against a brake drum or disc rotor, respectively.

In years past, drum brakes have been used mainly on solid rear axle suspension systems. The drum brake assembly shown in Figure (3.26) is a "duo-servo" design which is mounted directly to the axle housing. Two brake shoes are used in this design: a primary shoe toward the front of the vehicle and a secondary shoe toward the rear. The top of the shoes fit into a wheel (slave) cylinder and rests against an anchor pin attached to the brake flange plate. The bottom of the shoe fit into grooves of the adjustment (shoe-to-drum clearance) mechanism, and are secured by means of a connecting spring.

The duo-servo brake design is referred to as being self-energizing, which is an operational feature of the design. When the brakes are applied the resulting hydraulic pressure in the brake forces the two pistons at opposite ends of the wheel cylinder in an outward direction. In turn, the movement of the pistons pushes the brake shoes against the drum. The frictional drag tends to rotate the brake shoes when they contact the drum.

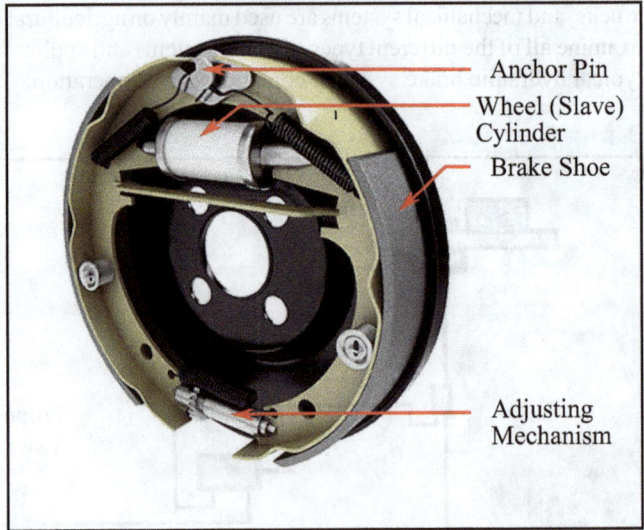

Figure (3.26): DUO-SERVO DRUM BRAKE COMPONENTS

The primary shoe moves away from the anchor pin while exerting a force on the secondary shoe through the adjustment mechanism. Simultaneously, the secondary shoe rotates until it comes into contact with the anchor pin. The total force acting of the secondary shoe is the combination of the piston applied force and the friction reaction force of the primary shoe. It is the combination of forces acting on the secondary shoe that accounts for the self-energizing feature of duo-servo drum brakes. The result is a high brake torque for a relatively low amount of physical effort at the brake pedal. The disadvantage of the duo-servo design is a high variation in brake torque caused by a small change in the lining/drum coefficient of friction.

There are other performance problems associated with drum brakes that are especially noticeable on sports and racing cars. All drum brakes are susceptible to "brake fade." Brake fade is a loss of brake torque caused by

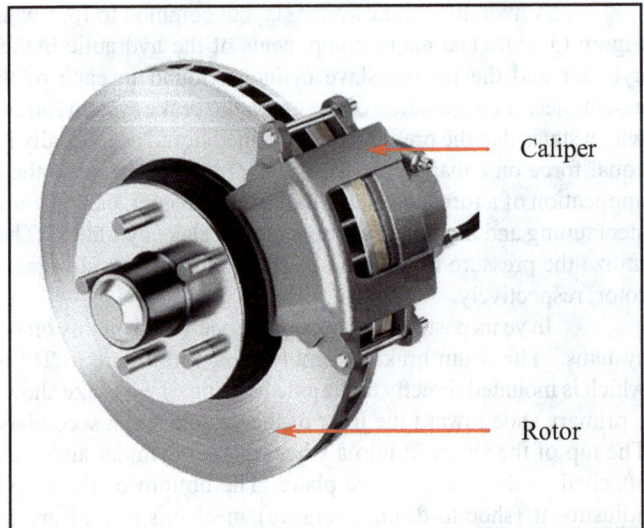

Figure (3.27): FLOATING CALIPER DISC BRAKE COMPONENTS

excessive heat dissipation of continued braking. The heat dissipation results in an increase operating temperature which, in turn, causes a drop in the coefficient of friction between the lining and drum. Thermal expansion of the drum, also caused by excessive hear dissipation, is another cause of brake fade. In this case the drum expands away from the shoe linings. The expansion creates a distortion or difference in curvature between the drum and brake shoe, which results in a loss of lining and drum contact surface area. The sensitivity of drum brakes to temperature limits the maximum operating temperature to about 750 °F (400 °C). Disc brakes, on the other hand, can operate at temperatures of 1500 to 1600 °F without fading.

Disc brakes, Figure (3.27), have been used on the front suspension of American production cars since the late 1960's; however, difficulties in incorporating a manual emergency brake in the caliper has precluded their use in rear suspension designs until recent years. For front suspension applications the brake rotor is attached to the wheel hub and fits inside the wheel rim. This is referred to as an "outboard" design. On independent rear suspensions it is possible, and advantageous when attempting to minimize the unsprung weight, to mount the brake rotor to an axle flange on the differential unit. This is an "inboard" design.

In either case, a stationary caliper exerts a clamping action on the rotor when the brakes are applied. There are two basic caliper designs: fixed and floating. The design name is derived from the method in which the caliper is secured to the caliper mounting bracket. Fixed calipers are rigidly attached to the mounting bracket using conventional fasteners such as hex bolts. Floating calipers are fitted to the mounting bracket through the use of pins which are threaded to the mounting bracket. The pins allow the caliper to slide in an axial direction relative to the axle centerline, but are restrained from moving in either the radial or tangential directions. Fixed caliper designs normally have two or four pistons evenly split between the left and right half of the caliper. This piston arrangement produces a fairly low amount of drag on the rotor, and also has an even inner and outer pad wear characteristics. Floating calipers have one or two pistons located inboard relative to the rotor. These calipers have the potential for increased rotor drag and produce uneven and tapered pad wear; but, they have the advantage of being easier to package inside the wheel rim.

As shown in Figure (3.25), the master brake cylinder is one of the major brake system components. The diagram also indicates that the master cylinder provides pressure to two fluid circuits (a front brake circuit and a rear brake circuit) and is referred to as a dual master cylinder design. The strategy of two circuits is common in modern brake systems. In essence this design allows the front and rear brake systems to operate independently of each other, but still provides common actuation. In the event of a malfunction or failure in either the front or rear brake circuit limited braking ability is provided by the opposite set of brakes.

Under normal operating conditions the fluid pressure developed in the front and rear portions of the master cylinder are the same. However, in order to achieve brake balance, as noted in our earlier optimum brake performance map, it is generally required that the front and rear brake circuits have different pressures. To provide this difference in pressure, a proportioning valve is normally incorporated into the rear brake circuit between the master cylinder and wheel cylinders. A proportioning valve functions similar to a pressure regulator in the sense that once the input pressure reaches a set point, the output pressure becomes metered in proportion to the input pressure. The proportion percentage is fixed based on the internal construction of the valve, but ranges from about 43% to 57% of the input pressure. Proportioning valves are either of the fixed or adjustable design. The fixed design applies to the brake system requirements of a specific vehicle. An example of operating characteristics of an

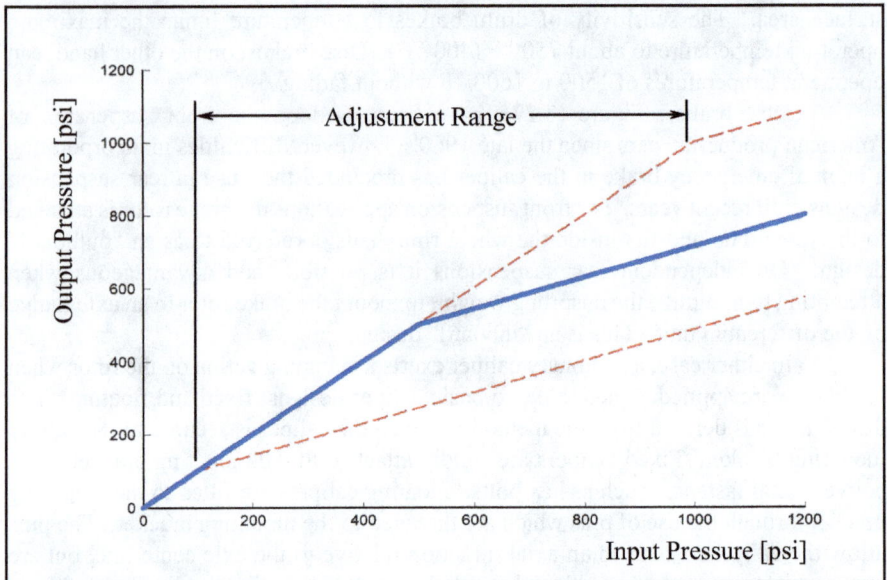

Figure (3.28): ADJUSTABLE PROPORTIONING VALVE PERFORMANCE

adjustable proportioning valve is shown in Figure (3.28). The graph indicates that the set point pressure can be adjusted from 100 to 1000 psi with a metered output pressure of 43%.

(3.5.2A) Disc Brake Analysis

Considering the performance advantages of disc brake systems, we will focus our attention on the specification of disc brake components. In particular the suspension designer us involved in the selection of the rotor configuration, i.e., solid or vented, diameter and location, and also in the caliper specifications. These specifications are not arbitrary as the performance of the caliper and rotor must match the braking requirements of a particular vehicle. In order for us to achieve a certain degree of proficiency in selecting and specifying these components it is beneficial for us to have a general understanding of caliper and rotor performance characteristics.

The braking performance of disc brake systems is directly attributed to the frictional force interaction between the rotor and caliper pad. Figure (3.29) depicts a cross-sectional view of a floating caliper and rotor. The force exerted on the caliper pads, F_P, is a function of the line pressure, P_L, created by the master cylinder, and the caliper piston area, A_C: $F_P = P_L A_C$. This force when uniformly distributed over the surface area of the pad, A_p, represents the pressure which is exerted on the rotor by the pad, i.e.,

$$P_R = \frac{F_P}{A_P}$$

$$= P_L \frac{A_C}{A_P} \qquad (3.88)$$

A differential amount of sliding friction force created by the pad pressure is given by

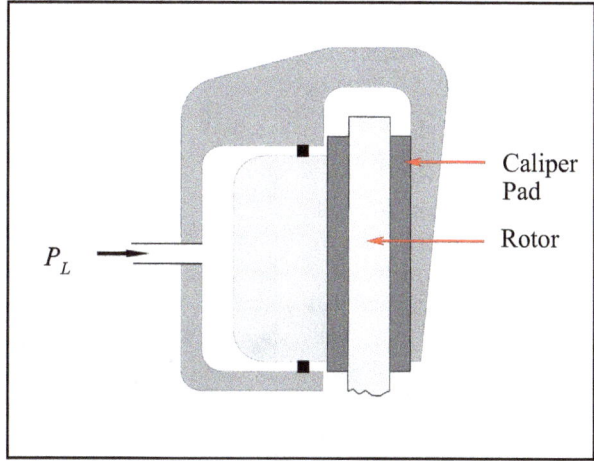

Figure (3.29): CALIPER AND ROTOR CROSS-SECTION

$$dF_f = \mu P_R dA = \mu P_L \frac{A_C}{A_P} dA \qquad (3.89)$$

where μ is the coefficient of sliding friction between the pad and rotor, and dA is a differential amount of rotor surface area. Since the friction force acts on the rotor at a distance, r, from the rotor spin axis, a braking torque is created. In differential form the braking torque is expressed as

$$dT = r \, dF_f = \mu P_L \frac{A_C}{A_P} r \, dA \qquad (3.90)$$

The braking torque generated by one face of the rotor is found by integrating the previous equation, i.e.,

$$T = \mu P_L \frac{A_C}{A_P} \int_0^{A_P} r \, dA \qquad (3.91)$$

Eq (3.91) can be slightly simplified by using an effective rotor radius defined as

$$r_e = \frac{1}{A_P} \int_0^{A_P} r \, dA \qquad (3.92)$$

Using this definition, Eq (3.91) becomes

$$T = \mu P_L A_C r_e \qquad (3.93)$$

It is noted that the effective rotor radius is solely geometry dependent; therefore, a braking torque relationship more explicit than Eq (3.93) can only be found if specific caliper pad and rotor geometries are considered. An industry standard for a caliper pad shape does not exist. This means that caliper designs must be evaluated on an individual basis. As an example, consider the caliper pad and rotor geometry shown in Figure (3.30). By aligning the radial displacement along the y-axis a differential amount of rotor surface area, as defined by the caliper footprint, can be expressed as

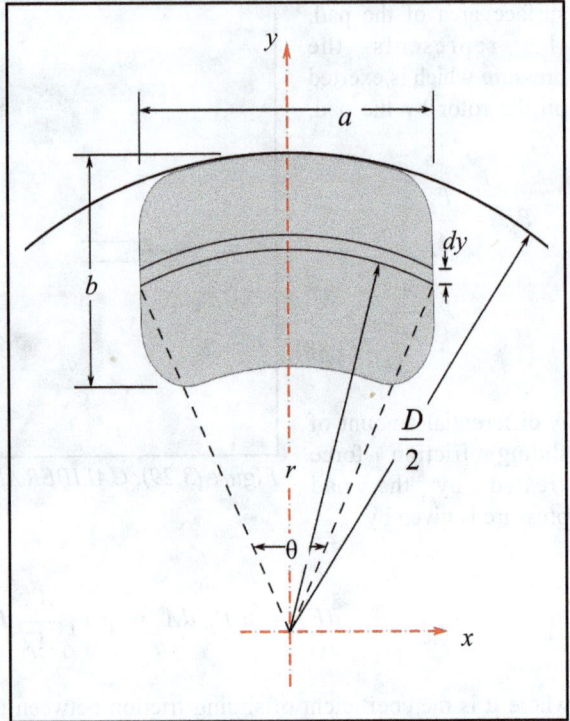

Figure (3.30): CALIPER PAD FOOTPRINT ON ROTOR

$$dA = \theta\, y\, dy \qquad (3.94)$$

where

$$\theta = 2 \arctan\left(\frac{a}{2y}\right) \qquad (3.95)$$

Using Eqs (3.94) and (3.95), along with a bit of extensive integration, the caliper pad footprint area evaluates to:

$$A_P = \int_{\frac{D}{2}-b}^{\frac{D}{2}} 2 \arctan\left(\frac{a}{2y}\right) y\, dy$$

$$= -\frac{\left(D^2 - 4bD + 4b^2\right)}{4} \arctan\left(\frac{a}{D-2b}\right) + \frac{a^2}{4} \arctan\left(\frac{D-2b}{a}\right)$$

$$- \frac{a^2}{4} \arctan\left(\frac{D}{a}\right) + \frac{D^2}{4} \arctan\left(\frac{a}{D}\right) + \frac{ab}{2}$$

$$(3.96)$$

In a similar fashion, Eq (3.93) becomes:

$$
T = \mu P_L \frac{A_C}{A_P} \int_{\frac{D}{2}-b}^{\frac{D}{2}} 2 \arctan\left(\frac{a}{2y}\right) y^2 \, dy
$$

$$
= \mu P_L \frac{A_C}{A_P} \left[\frac{(D-2b)^3 \arctan\left(\dfrac{a}{2b-D}\right) + D^3 \arctan\left(\dfrac{a}{D}\right)}{12} \right.
$$

$$
\left. + \frac{a^3}{24} \ln\left(\frac{a^2 + 4b^2 - 4bD + D^2}{a^2 + b^2}\right) + \frac{ab}{6}(D-b) \right] \tag{3.97}
$$

At first glance the braking torque for one rotor face appears to be a very complex function of caliper pad dimensions and rotor diameter. But, when we consider the relatively small range of rotor diameter that the suspension designer has to work with when packaging the caliper and rotor assembly inside the wheel rim, the braking torque has a nearly linear variation with rotor diameter. We can demonstrate this point by examining the brake torque results when numeric values are inserted into Eq (3.93).

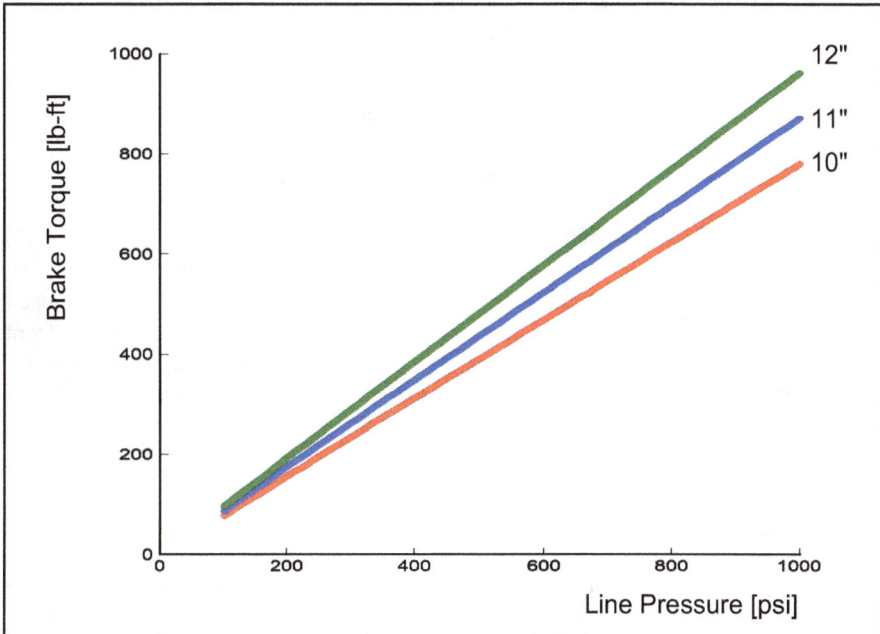

Figure (3.31): EXAMPLE SINGLE-FACE ROTOR BRAKE FORCE PERFORMANCE

Chapter 3: ACCELERATION AND BRAKING

Specifically, consider a caliper which has a piston area of 4.9 in², and caliper pad dimensions of a = 4.75 in and b = 2.05 in. Modern caliper pad materials provide a fairly constant sliding coefficient of about 0.4 up to temperature of approximately 950 °F over a wide range of line pressure. The brake torque performance graph shown in Figure (3.31) was generated using rotor diameters of 10, 11 and 12 inches. If we select a constant value of line pressure we notice that a linear increase in rotor diameter results in an almost linear increase in brake torque, which confirms our earlier statement. It should be noted that the results depicted in Figure (3.31) are for just one face of the rotor and would need to be multiplied by two to estimate the performance of the caliper/rotor assembly.

We will conclude our discussion of disc brake systems by returning to our earlier sports car example that involved developing a brake performance map. To continue this example we will now include the addition of a disc brake system on both the front and rear axle. In order to simplify the design we will use the same caliper on all wheels. As we discovered earlier, due to weight transfer to the front axle, the front axle braking ability needs to be greater than the rear. The difference in braking capability suggest that it is possible to use a smaller rotor diameter on the rear axle than the front, thus reducing unsprung weight. With these considerations in mind, we will consider a potential disc brake system for our sports car that has the following properties:

disc pad dimensions:
   a = 4.75 in   **b = 2.05 in**
disc pad coefficient of sliding friction: 0.4
caliper piston area: 4.9 in2
front rotor diameter: 11 in
rear rotor diameter: 10.5 in
proportioning valve:
   set-point pressure: 250 psi
   output pressure regulation: 43%
effective tire rolling radius (front and rear): 14.7 in

The estimated performance of the proposed disc brake system is shown in Figure (3.32) superimposed on top of the earlier brake performance map of Figure (3.24). Considering this is our first iteration of selecting brake components, our proposed design matches the optimum brake curve fairly well[4]. It is also noticed that as the front brake force increases the design eventually crosses the maximum brake force boundary into Region II where t he front brakes lock up before the rear, which also a desirable characteristic.

---

[4] *It should be noted that vehicles intended for commercial sale must comply with Federal Motor Safety Standards (FMVSS). For example, Standard No. 135 LIGHT VEHICLE BRAKE SYSTEMS, applies to multi-purpose passenger vehicles, trucks and buses with GVWR of 3500 kg or less manufactured after September 1, 2000.*

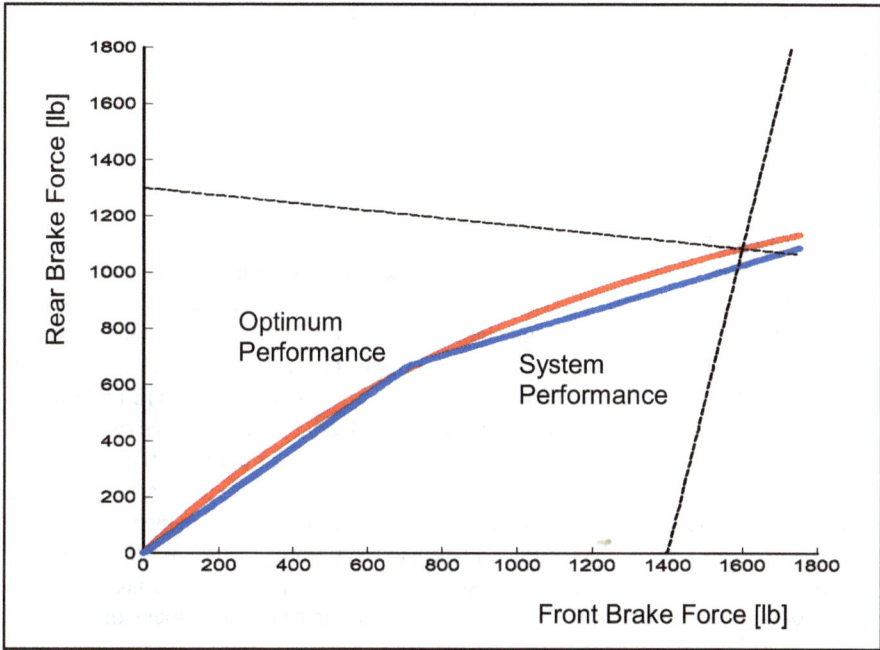

Figure (3.32): EXAMPLE SPORTS CAR DISC BRAKE SYSTEM PERFORMANCE

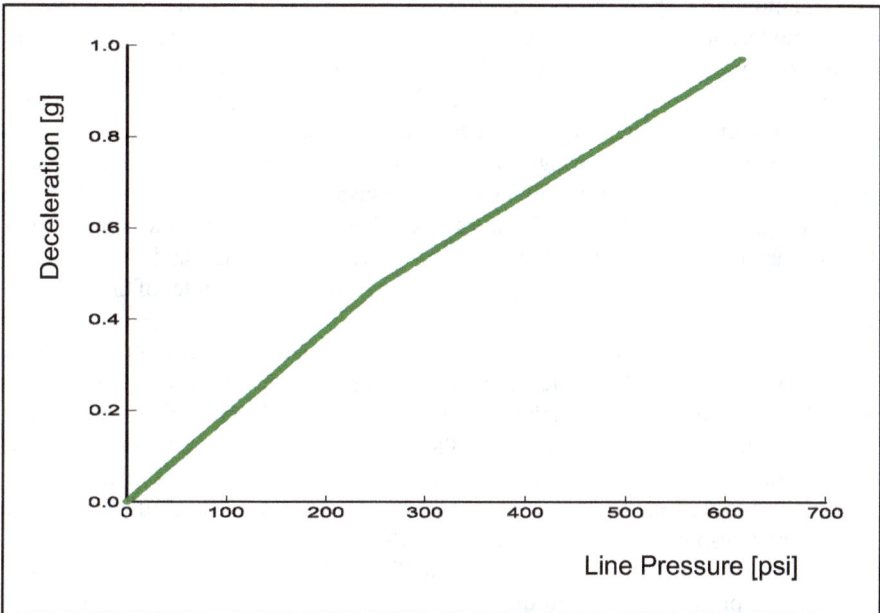

Figure (3.33): EXAMPLE SPORTS CAR DECELERATION RESULTS

An estimate of vehicle deceleration can also be presented in graphical form. The deceleration can be correlated against various vehicle and brake parameters, e.g., CG location, front brake force, etc., but one parameter, line pressure, is particularly useful in the selection of a master cylinder and ergonomic design of the brake actuation system. Figure (3.33) depicts the estimated deceleration of our sports car as a function of the required hydraulic line pressure.

(3.6) ANTI-SQUAT AND ANTI-DIVE PERFORMANCE

You may have noticed that when a car accelerates the front end has a tendency to rise while the rear moves slightly closer to the ground. When viewed from the side this appears as an aft rotation about the lateral axis of the vehicle body and chassis. This behavior is called "squat". A forward rotation is observed during deceleration or braking, which is referred to as "dive" behavior. In this section we will undertake a study of squat behavior as well as a study of dive. Both of these studies require us to specify the type of suspension we wish to analyze, but the analysis techniques are applicable to any suspension system. The purpose of these studies is to aide us in developing suspension design criteria. By careful design of the suspension linkage geometry it is possible to construct a suspension system to counteract the effects of acceleration and braking. The degree of success in controlling body and chassis rotation can be loosely called anti-squat and anti-dive performance of the suspension system.

(3.6.1) SQUAT ANALYSIS OF INDEPENDENT SUSPENSIONS WITH REAR DRIVE

As we discussed in Chapter 1, there are a variety of independent suspensions used in automotive applications. In spite of this diversity we can devise a universal vehicle model appropriate for squat analysis. First, we recognize that many independent suspension systems have the same functional equivalent as a single bar linkage connecting the wheel to the chassis. This is analogous to a trailing arm suspension except that in this case the pivot point located on the chassis is the instant center of the suspension linkage. The springs of the suspension systems can be mechanically represented by torsional springs acting about the suspension instant centers. Next, the left and right sides of the vehicle are squeezed together to form the simple two-dimensional model illustrated in Figure (3.34). The points labeled **B** and **C** are the front and rear suspension instant centers; points **A** and **D** represent the center of the front and rear tire contact patches.

As mentioned earlier, vehicle squat only occurs while accelerating. If we restrict our analysis to a vehicle accelerating from rest we need only consider two external forces acting on the vehicle: a tractive force, F_t, produced by the interaction of the rear wheels with the road surface, and an inertia force, F_I, opposing the forward motion. Forces due to aerodynamic drag and rolling resistance are insignificant during this initial stage, and can be neglected. Figure (3.35) shows a free body diagram of the front and rear suspensions, and sprung mass with forces and torques appropriate for an accelerating vehicle. We notice by inspecting Figure (3.35) that there are three torques acting on the sprung mass. The torques T_{sf} and T_{sr} are due to the torsional springs of the front and rear suspension, but what is the torque T_d ? This can be thought of as a reaction to the tractive effort at the rear wheels. It is referred to as the driveline torque,

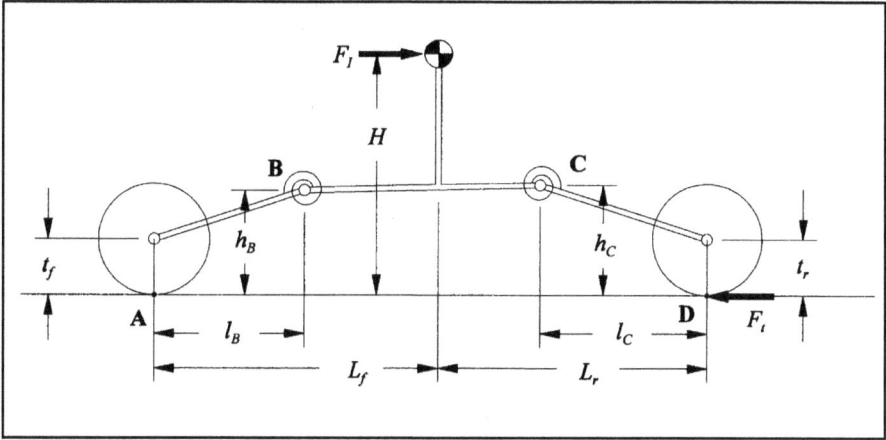

Figure (3.34): VEHICLE MODEL FOR SQUAT ANALYSIS

and, neglecting the rotational inertia of the driveline components, is approximated as $T_d = F_t t_R$, where t_R is the rear tire effective radius. For an independent rear suspension, the driveline torque is transmitted from the wheel hub through the drive axle (half-shafts) to the differential housing, and then passed on to the chassis through the housing mounting brackets. Although the differential housing is fixed to the chassis at a particular location, the effect of driveline torque on the sprung mass is not position dependent. We are, therefore, free to place the driveline torque anywhere on the sprung mass free body diagram.

The goal of performing a squat analysis is to understand the circumstances which lead to a change in vehicle pitch. To accomplish this goal it is necessary for us to study the change in equilibrium constraints from a stationary (static) vehicle to one which is accelerating (dynamic). The equations which describe static equilibrium are fairly simple and can be summarized as:

$$\left(T_{sf}\right)\big|_{static} = \left(W_f\right)\big|_{static} l_B \tag{3.98}$$

$$\left(T_{sr}\right)\big|_{static} = \left(W_r\right)\big|_{static} l_C \tag{3.99}$$

$$\left(W_f\right)\big|_{static} L_f = \left(W_r\right)\big|_{static} L_r \tag{3.100}$$

A study of the vehicle under dynamic conditions is a little more involved. We start by assuming that the change in vehicle attitude due to acceleration will be small. This

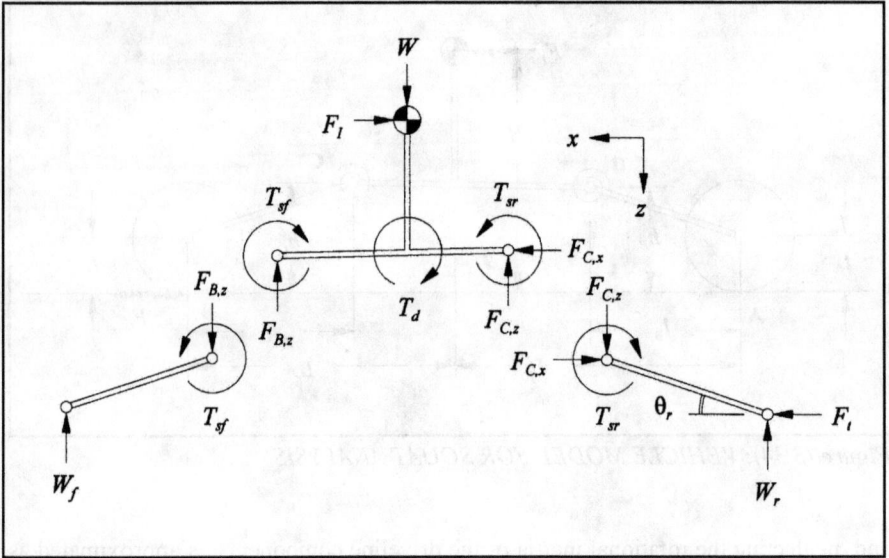

Figure (3.35): FREE-BODY-DIAGRAM FOR VEHICLE SQUAT MODEL

allows us to make the statement that vehicle geometric parameters, such as L_f and L_r, will remain approximately the same whether we consider static or dynamic conditions. For chassis and suspension systems of practical interest this is a fairly good assumption. Using this approximation, the basic dynamic equilibrium equations for each of the free body diagrams shown in Figure (3.35) are:

(*i*) for the front suspension

$$\sum F_z = 0 = F_{B,z} - W_f \tag{3.101}$$

$$\sum M_B = 0 = W_f l_B - T_{sf} \tag{3.102}$$

(*ii*) for the rear suspension

$$\sum F_x = 0 = F_t - F_{C,x} \tag{3.103}$$

$$\sum F_z = 0 = F_{C,z} - W_r \tag{3.104}$$

$$\sum M_C = 0 = -W_r l_C + F_t \left(h_C - t_r \right) + T_{sr} \tag{3.105}$$

and (*iii*) for the sprung mass

Chapter 3: ACCELERATION AND BRAKING 137

$$\sum M_{cg} = 0$$

$$= F_{B,z}\left(L_f - l_B\right) + T_{sf}$$

$$- F_{C,z}\left(L_r - l_C\right) + F_{C,x}\left(H - h_C\right) - T_{sr} + T_d$$

(3.106)

Since we are looking for a change in equilibrium constraints it is convenient for us to consider the axle reactions and torsional spring torques as the sum of a static component and a transitional component, i.e.,

$$W_f = \left(W_f\right)\Big|_{static} - \Delta W \tag{3.3.107a}$$

$$W_r = \left(W_r\right)\Big|_{static} + \Delta W \tag{3.3.107b}$$

$$T_{sf} = \left(T_{sf}\right)\Big|_{static} + \Delta T_{sf} \tag{3.3.107c}$$

$$T_{sr} = \left(T_{sr}\right)\Big|_{static} + \Delta T_{sr} \tag{3.3.107d}$$

The term ΔW represents the weight transferred from the front axle to the rear axle due to acceleration, and the ΔT terms are the changes in torsional spring torques required to maintain equilibrium. Using this notation along with the static relationships given by Eqs. (3.98) through (3.100), the foregoing dynamic equilibrium equations substantially reduce to just:

$$\Delta T_{sf} = - F_t \frac{H}{L_w} l_B \tag{3.108}$$

and

$$\Delta T_{sr} = F_t l_C\left[\frac{H}{L_w} - \frac{\left(h_C - t_r\right)}{l_C}\right]$$

$$= F_t l_C\left[\frac{H}{L_w} - \tan\theta_r\right] \tag{3.109}$$

where $L_w = L_f + L_r$.

The two previous equations define the changes which occur in the front and rear torsional spring torques of our vehicle model when accelerating from rest or constant speed. A negative change in torque means that the torsional spring relaxes. The geometrical change which must accompany this relaxation necessitates a decrease in the included angle between the suspension linkage and the chassis (sprung mass). An increase in the included angle is observed for positive changes in spring torque. The deviation of the front and rear included angles from their static position causes squat.

Eq (3.108) indicates that the change in front spring torque will always be negative. This means that the front end of the vehicle will always exhibit some amount of rise when accelerating. On the other hand, there are three different possibilities for the change in rear spring torque: positive, negative, or zero. Each of these possibilities has a different effect, which we will discuss shortly, on vehicle squat. From Eq (3.109) we conclude that it is the combination of the parameters H, L_w and θ_r that actually determines the outcome for the change in rear spring torque. For any particular vehicle there is not a great deal of latitude in making substantial changes in the center of gravity height or the wheelbase. Consequently the design of the rear suspension has the greatest influence on squat behavior. For instance, if it is desired to completely eliminate the change in rear spring torque, then during the design of the rear suspension system we would specify that the angle θ_r should correspond with:

$$\tan \theta_r = \frac{H}{L_w} \qquad (3.110)$$

Using this design, the chassis and rear suspension would appear to pivot as a single unit about the rear axle when the vehicle accelerates. Thus, in this case the rear end of the vehicle actually lifts a slight amount. Although somewhat of a misnomer, the specification given by Eq (3.110) is referred to as a 100 percent anti-squat design.

It is not usually desirable or practical to design the rear suspension for 100 percent anti-squat characteristics. As we will learn in the next section when we study dive, compromises must be made to achieve a balance between anti-squat and anti-dive performance. The amount of desired anti-squat can be introduced into the design of the rear suspension geometry by modifying Eq (3.110) as follows:

$$\tan \theta_r = \frac{H}{L_w} \left(\frac{\% \text{ anti-squat}}{100} \right) \qquad (3.110)$$

If we specify more than 100 percent anti-squat, then the change in rear torque will be negative. This causes the rear of the vehicle to rise. It is interesting to note that if the appropriate anti-squat design specification is made, the change in front and the rear spring torques will exactly balance causing both the front and rear of the chassis to rise. In this case the pitch of the vehicle will not change, but it is at the expense of increasing the center of gravity height. Specifying less than 100 percent anti-squat results in a positive rear spring torque change. The effect of this change causes the rear of the chassis to dip.

(3.6.2) DIVE ANALYSIS OF INDEPENDENT SUSPENSIONS WITH OUTBOARD BRAKES

In order to develop suspension design criteria to counteract the effects of dive, we are primarily interested in studying dive behavior when it is caused by the application of brakes. The methods we will use to study dive are essentially the same as those used to investigate squat. We begin with the simple two-dimensional vehicle model developed in the previous section. The geometrical parameters are the same, but the forces must be altered to represent braking conditions. The force F_{bf} shown in

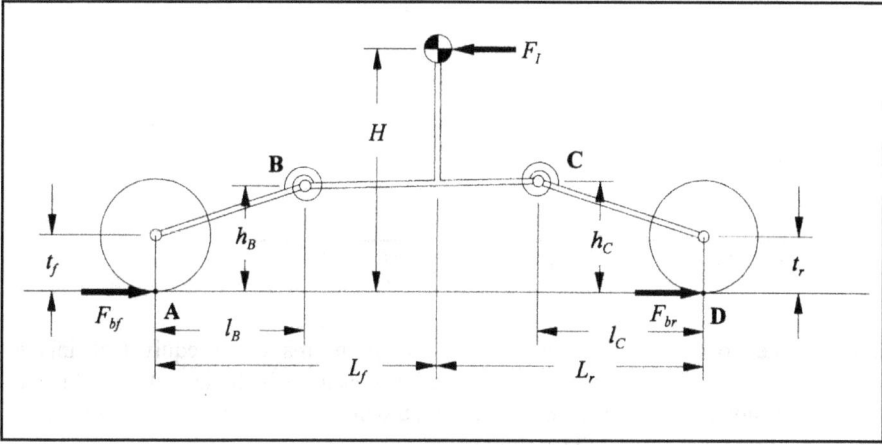

Figure (3.36): VEHICLE MODEL FOR DIVE ANALYSIS

Figure (3.36) is the front braking effort due to the combined effect of the left and right wheels, and F_{br} is the combined rear wheel braking effort. The next step is to create free body diagrams of the chassis, front suspension and rear suspension.

The placement of forces on the free body diagrams is straight forward, as is the torsional spring torques; but, we must also account for reactive torques due to the braking effort at Points **A** and **D**. The correct application of these reactive torques, however, depends on the location of the brakes on the vehicle, i.e., inboard or outboard. For example, let's consider a vehicle which has outboard brakes on both front and rear axles. When the brakes are applied, e.g., the caliper grabbing the rotor, a reactive torque is transmitted to the suspension linkage through the mounting brackets associated with the brake assembly. A free body diagram of the front suspension linkage derived from Figure (3.36) under these conditions appears in Figure (3.37).

Neglecting the rotational inertia of the wheel/hub assembly, the front reactive brake torque can be approximated as $T_{bf} = F_{bf} t_f$, where t_f is the front tire effective rolling radius. With these considerations a simpler, but equivalent, linkage system can be obtained by modifying the geometry of the front end of the suspension linkage to extend down from the wheel hub to Point **A**. The modified suspension with applicable forces is shown in Figure (3.38). The line defined by $\overline{\textbf{A B}}$ is the line of

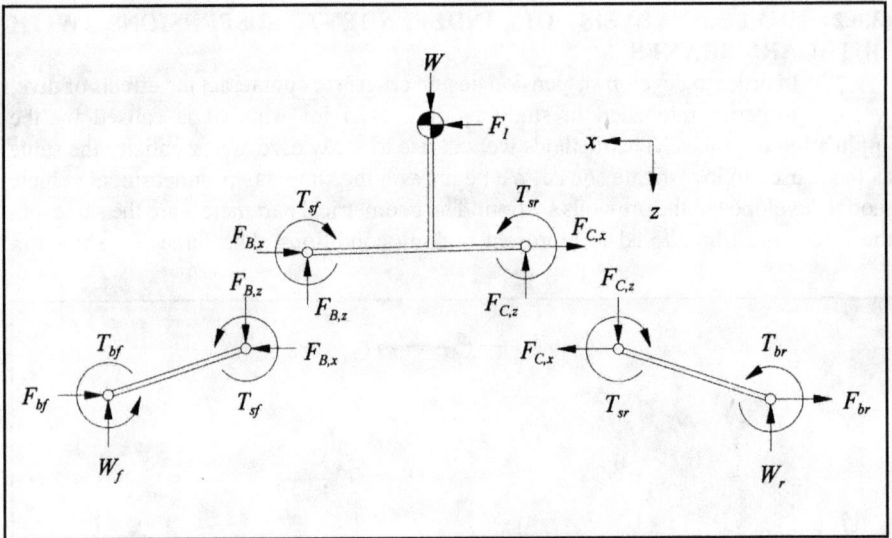

Figure (3.37): FREE-BODY-DIAGRAM FOR DIVE VEHICLE MODEL

action of the front wheel braking effort on the sprung mass. An equivalent linkage modification can also be performed on a rear suspension which has outboard brakes. This is also shown in Figure (3.38). Similarly, the line $\overline{CD}$ is the line of action of the rear wheel braking effort on the sprung mass.

The static equilibrium equations for the dive vehicle model are identical to those previously developed during the squat analysis, i.e., Eqs (3.98), (3.99), and (3.100). Just as in the squat analysis, we start the dynamic analysis by making an analogous assumption that the change in vehicle attitude due to deceleration will be small. As a consequence, the change in vehicle geometrical parameters from static to dynamic conditions can be neglected. Then, referring to Figure (3.38), the basic dynamic equilibrium equations for each of the free body diagrams are:

(*i*) for the front suspension

$$\sum F_x = 0 = -F_{bf} + F_{B,x} \tag{3.112}$$

$$\sum F_z = 0 = F_{B,z} - W_f \tag{3.113}$$

$$\sum M_B = 0 = W_f l_B - T_{sf} - F_{bf} h_B \tag{3.114}$$

(*ii*) for the rear suspension

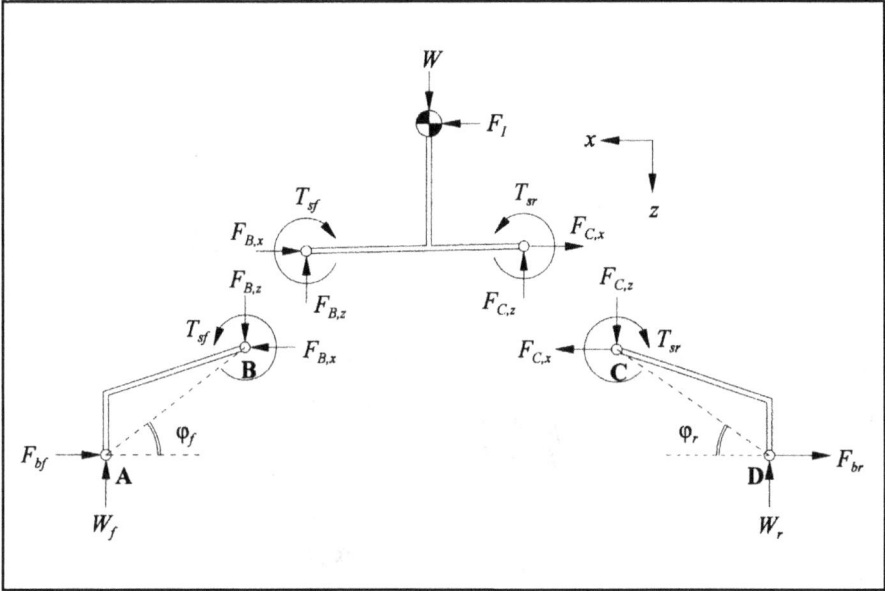

Figure (3.38): FREE-BODY-DIAGRAM FOR MODIFIED DIVE VEHICLE MODEL

$$\sum F_x = 0 = -F_{br} + F_{C,x} \tag{3.115}$$

$$\sum F_z = 0 = F_{C,z} - W_r \tag{3.116}$$

$$\sum M_C = 0 = -W_r l_C + T_{sr} - F_{br} h_C \tag{3.117}$$

and (*iii*) for the sprung mass

$$\sum M_{cg} = 0 = F_{B,z}\left(L_f - l_B\right) - F_{B,x}\left(H - h_B\right) + T_{sf}$$

$$- F_{C,z}\left(L_r - l_C\right) - F_{C,x}\left(H - h_C\right) - T_{sr} \tag{3.118}$$

Using Eqs (3.98) through (3.100) together with Eqs (3.107), the previous dynamic equilibrium equations simplify to just

$$\Delta T_{sf} = F_I l_B \left[\frac{H}{L_w} - \frac{F_{bf}}{F_I} \frac{h_B}{l_B} \right]$$

$$= F_I l_B \left[\frac{H}{L_w} - \frac{F_{bf}}{F_I} \tan \varphi_f \right] \tag{3.119}$$

and

$$\Delta T_{sr} = F_I l_C \left[\frac{H}{L_w} - \frac{F_{br}}{F_I} \frac{h_C}{l_C} \right]$$

$$= F_I l_C \left[\frac{H}{L_w} - \frac{F_{br}}{F_I} \tan \varphi_r \right] \tag{3.120}$$

We can further simplify Eqs (3.119) and (3.120) by defining a braking proportion parameter, p, as the ratio of the front wheel braking effort to the total braking effort. Then,

$$\Delta T_{sf} = F_I l_B \left[\frac{H}{L_w} - p \tan \varphi_f \right] \tag{3.121}$$

and

$$\Delta T_{sr} = F_I l_C \left[\frac{H}{L_w} - (1 - p) \tan \varphi_r \right] \tag{3.122}$$

The results as expressed by Eqs (3.121) and (3.122) are similar to those we obtained in the squat analysis. Except in this case these equations define changes in the front and rear torsional spring torques caused by deceleration or braking. The change in torsional torques causes the included angles between the suspension linkages and chassis to deviate from their static position. This is turn causes a change in vehicle attitude which results in dive. If we choose to completely eliminate dive, which is called

100% anti-dive performance, then both ΔT_{sf} and ΔT_{sr} must equal zero. From Eqs (3.121) and (3.122) we would specify the angles φ_f and φ_r to correspond to

$$\tan \varphi_f = \frac{H}{p L_w} \tag{3.123}$$

and

$$\tan \varphi_r = \frac{H}{(1 - p) L_w} \tag{3.124}$$

It is interesting to note from Eqs (3.123) and (3.124) that the angles φ_f and φ_r necessary to achieve 100% anti-dive are not functions of the point locations **B** and **C**, but rather the center of gravity location, wheelbase and proportion of front wheel braking effort. From a suspension design viewpoint, this means that with Points **A** and **D** fixed, we are free to place Points **B** and **C** anywhere along lines of slope $\tan \varphi_f$ and $\tan \varphi_r$, respectively, and still maintain 100% anti-dive. However, generally speaking, it is not advisable to design a suspension system for 100% anti-dive. The problem stems from the front suspension geometry. When the front wheel moves up due to an irregularity in the road surface it swings an arc about the instant center **B**. This motion throws the wheel forward into the road disturbance, which has a tendency to jar the vehicle. The effect of this creates a harsh ride. In addition, drivers depend on a certain amount of dive as a feedback mechanism to maintain control of their vehicle while braking. Generally, the amount of anti-dive should not exceed 50% for sports cars and, due to their considerably stiffer suspension, about 20% for race cars. Although it is possible to construct a suspension system which has instant centers above the center of gravity, this is also undesirable. Instant centers above the center of gravity produce excessive amount of roll steer when the vehicle executes a cornering maneuver.

Eqs. (3.123) and (3.124) can be easily modified to allow for variable amounts of desired anti-dive, i.e.,

$$\tan \varphi_f = \frac{H}{p L_w} \left(\frac{\% \text{ anti-dive}}{100} \right) \tag{3.125}$$

and

$$\tan \varphi_r = \frac{H}{(1 - p) L_w} \left(\frac{\% \text{ anti-dive}}{100} \right) \tag{3.126}$$

These are the final design equations for the angles φ_f and φ_r. It should be noted that these design equations were derived for suspension systems with outboard brakes. As you may suspect, inboard brake suspension systems will have a different set of equations.

Chapter 4:
RIDE CHARACTERISTICS

The primary purpose of any suspension system, regardless of its mechanical configuration, is to isolate the chassis and vehicle body from the road surface. Complete isolation is not possible because a road-suspension-chassis interaction always exists while the vehicle is moving. The reaction of the chassis and vehicle body to traveling over bumps and dips in the road is referred to as *vehicle ride*. Attributes of vehicle ride are defined in terms of *bounce* and *pitch*. Bounce is a linear displacement which describes the up and down (normal axis direction) movement of the chassis relative to the road surface. Pitch is a fore and aft rocking motion about the lateral axis expressed as an angular displacement. One of the suspension goals is to control the bounce and pitch motion of the vehicle. In this chapter we will study several vehicle and suspension mathematical models for predicting bounce and pitch behavior. We will also examine analytical means of evaluating vehicle ride. These are excellent tools for the design engineer is specifying spring and damping rates of prototype suspension systems; however, how well the suspension system meets the design goals is eventually answered, subjectively, by the driver's perception.

(4.1) HUMAN PHYSIOLOGICAL CONSIDERATIONS OF RIDE

The vehicle motion transferred to the driver (and passengers) can be characterized as a vibrational input of random amplitude and frequency. This vibration can be divided into four frequency ranges [1]:

(1) **Ride Motion** - Frequencies up to 80 cycles per minute (CPM) or 1.33 Hz. {*This is the basic oscillatory motion of the sprung mass on the suspension springs.*}

(2) **Shake** - Frequencies between 80 and 200 CPM (1.33 to 3.33 Hz).

(3) **Harshness** - Frequencies primarily in the range of 200 to 1000 CPM (3.33 to 16.67 Hz), but as high as 5000 CPM (83.33 Hz). {*Uncontrolled unsprung mass oscillation, i.e., wheel hop, is one form of harness in the range of 360 to 600 CPM.*}

(4) **Road Noise** - Frequencies of 5000 CPM (83.33 Hz) and higher. {*These frequencies are high enough such that they are not usually felt, but can be heard since the vibration is at the lower end of the audible range.*}

The frequency range of approximately 0 to 80 Hz is of special importance to suspension engineers as it most influences the human perception of ride quality. However, this perception is due primarily to human physical characteristics.

A considerable amount of research has been done in determining human sensitivities to vibration frequency and amplitude. Studies [2] of human beings subjected to vertical vibration while standing or seated on a platform have determined that: (1) the array of organs located inside the chest cavity have a mechanical resonance between 4 and 6 Hz for a seated individual and between 5 and 12 Hz while standing, (2) between 20 and 30 Hz the head exhibits a mechanical resonance, and (3) between 60 and 90 Hz the eyeball has a natural response. In addition, at low frequencies in the range of about 0.5 and 0.75 Hz the vestibular mechanism of the inner ear resonates producing dizziness and the sensation of "seasickness."

Attempting to quantitatively correlate human physical characteristics to suspension design criteria has proven to be quite a task. Early researchers include Lanchester (1907), Rowell (1923), Reiher and Meister (1931), and Jacklin and Liddell (1933) [3, 4]. It was not until 1947, however, the Dieckmann [5] developed a correlation between frequency, maximum vertical displacement (which is referred to as amplitude) and fatigue time. Dieckmann defined the three subjective comfort zones shown in Figure (4.1) corresponding to the definitions and equations below. The level of comfort is based on a K factor which is a simple function of frequency and amplitude. A value of K = 100 was arbitrarily selected as being intolerable.

(Zone 1) Imperceptible to slightly uncomfortable: $0.1 \leq K \leq 1$

(Zone 2) Slightly disagreeable with a fatigue time of one hour: $1 \leq K \leq 10$

(Zone 3) Very disagreeable to intolerable with a fatigue time of one minute: $10 \leq K \leq 100$

The Dieckmann factor K is calculated from the relationships:

$$K = af^2 \qquad ; 0 \leq f \leq 5\,\text{Hz} \qquad (4.1a)$$

$$K = 5\,af \qquad ; 5 \leq f \leq 40\,\text{Hz} \qquad (4.1b)$$

$$K = 200\,a \qquad ; 40 \leq f \leq 200\,\text{Hz} \qquad (4.1c)$$

where $a \equiv$ amplitude [mm] and $f \equiv$ frequency [Hz].

At approximately the same time, Janeway (1948) had independently developed a ride relationship from available published data [3,6]. Janeway's concept was to place a limit on the maximum amplitude acceptable for a satisfactory ride, without regard to fatigue time. He divided the spectrum of frequencies into three ranges which produced the following relationships:

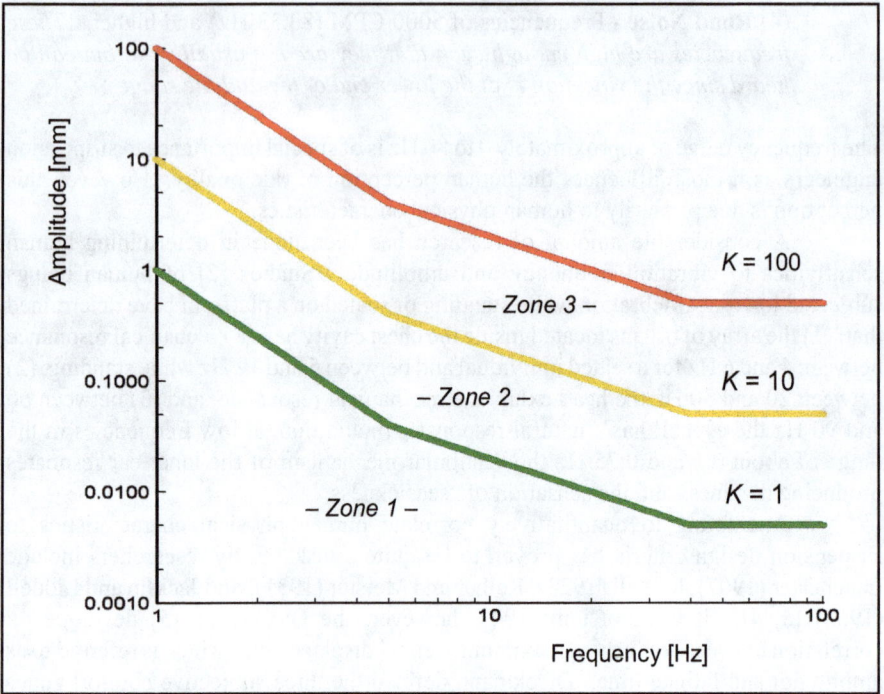

Figure (4.1): DIECKMANN COMFORT ZONES

$$J = \frac{1}{6} a f^3 \qquad\qquad ; \; 1 \le f \le 6 \, \text{Hz} \qquad\qquad (4.2\text{a})$$

$$J = a f^2 \qquad\qquad ; \; 6 \le f \le 20 \, \text{Hz} \qquad\qquad (4.2\text{b})$$

$$J = 20 \, af \qquad\qquad ; \; 20 \le f \le 60 \, \text{Hz} \qquad\qquad (4.2\text{c})$$

where $a \equiv$ amplitude [in] and $f \equiv$ frequency [Hz]. The constant J is defined as the Janeway "limit" or comfort criteria and is given as $J = \frac{1}{3}$. The Janeway relationships are graphically summarized in Figure (4.2).

As we will learn, the root-mean-squared (RMS) acceleration is a more useful parameter in suspension design than the vibration amplitude. Early test results of both Dieckmann and Janeway were derived from shake table experiments. Their work concentrated primarily on vertical sinusoidal vibration. Since the vertical displacement is in the form of a sinusoidal wave input, the RMS acceleration can be calculated from the previous amplitude-frequency relationships using the equation:

$$\text{RMS acceleration} = 2\sqrt{2} \, \pi^2 f^2 a \qquad\qquad (4.3)$$

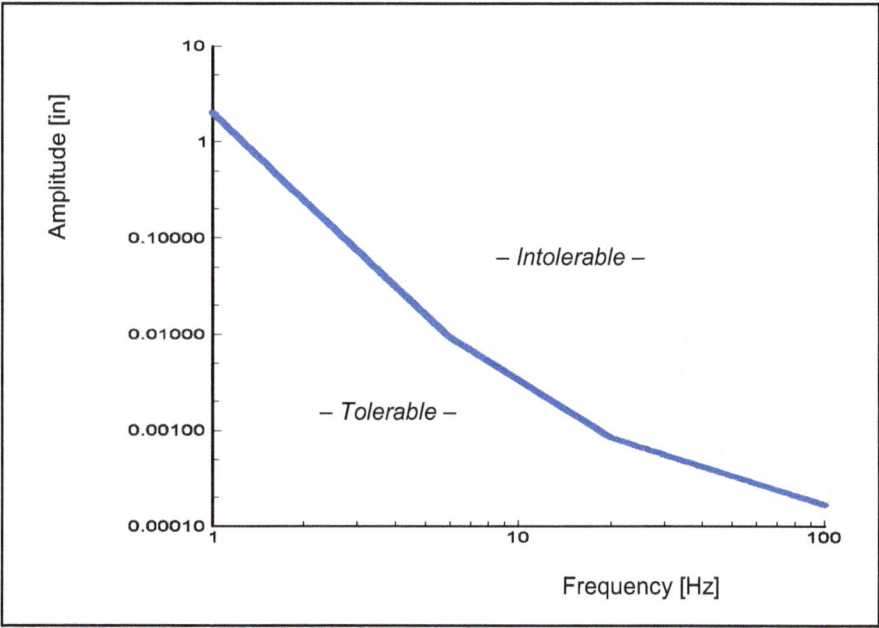

Figure (4.2): JANEWAY COMFORT LIMIT

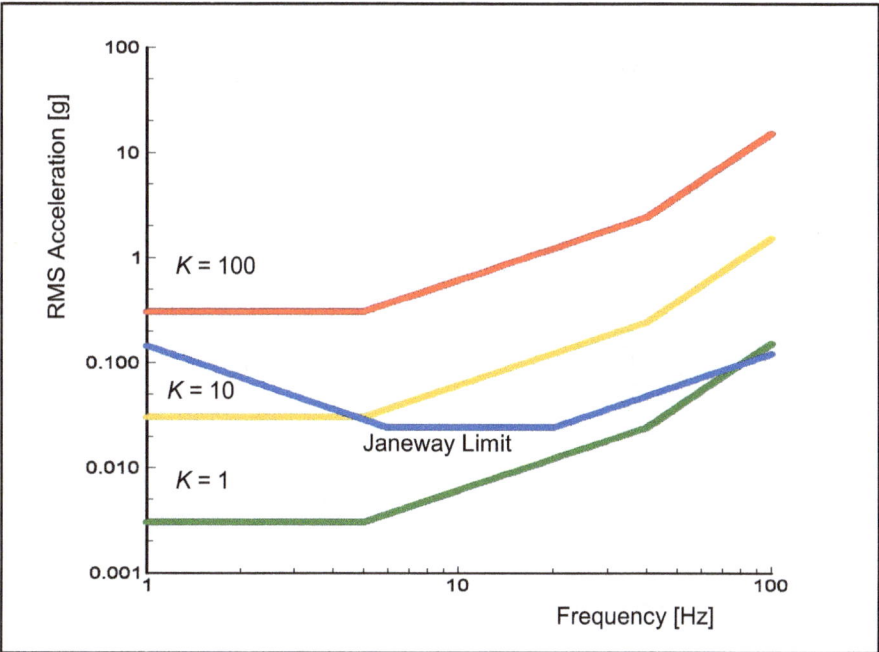

Figure (4.3): DIECKMANN and JANEWAY RMS ACCELERATION LIMITS

A graphical comparison between the Dieckmann and Janeway approaches using the RMS acceleration limits is illustrated in Figure (4.3). More recently, the international community has established recommended guidelines for fatigue time evaluation of whole body vibration in the range of 1 to 80 Hz. The guidelines applicable to vehicle vibration are contained in International Standards ISO 2631/1-1985 (E) and ISO 2631/3-1985 (E).

As previously stated, early investigations were based on experimental evidence from pure sinusoidal vibration of a single frequency. These results, however, have limited application to the broad range of vibration frequencies normally associated with road surface irregularities. The ISO standards have been devised in such a way that they apply to narrow band vibration consisting of frequencies within a one-third octave range about a center frequency, as well as single frequencies.

Several RMS acceleration-frequency relationships describing the "fatigue-decreased proficiency boundary" can be obtained by a simple curve fit of numerical values tabulated in ISO Standard 2631/1-1985 (E). Equations for 16 hour, 8 hour, and 4 hour limits[1] are given below for both normal and longitudinal (or lateral) RMS acceleration.

Normal RMS Acceleration
16 hour boundary:

$$A = \frac{0.424}{\sqrt{f}} \qquad\qquad ; 1 \le f \le 4\ \text{Hz} \qquad\qquad (4.4a)$$

$$A = 0.212 \qquad\qquad ; 4 \le f \le 8\ \text{Hz} \qquad\qquad (4.4b)$$

$$A = 0.212\left(\frac{f}{8}\right) \qquad\qquad ; 8 \le f \le 80\ \text{Hz} \qquad\qquad (4.4c)$$

8 hour boundary:

$$A = \frac{0.630}{\sqrt{f}} \qquad\qquad ; 1 \le f \le 4\ \text{Hz} \qquad\qquad (4.5a)$$

$$A = 0.315 \qquad\qquad ; 4 \le f \le 8\ \text{Hz} \qquad\qquad (4.5b)$$

---

[1] *The ISO standard gives numeric information for time limits of 24 hr, 16 hr, 8 hr, 4 hr, 2.5 hr, 1 hr, 25 min, 16 min, and 1 min. Consequently, other relationships could be developed.*

$$A = 0.315 \left(\frac{f}{8} \right) \qquad ; \ 8 \le f \le 80 \text{ Hz} \qquad (4.5c)$$

4 hour boundary:

$$A = \frac{1.060}{\sqrt{f}} \qquad ; \ 1 \le f \le 4 \text{ Hz} \qquad (4.6a)$$

$$A = 0.530 \qquad ; \ 4 \le f \le 8 \text{ Hz} \qquad (4.6b)$$

$$A = 0.530 \left(\frac{f}{8} \right) \qquad ; \ 8 \le f \le 80 \text{ Hz} \qquad (4.6c)$$

Longitudinal (Lateral) RMS Acceleration
16 hour boundary:

$$A = 0.150 \qquad ; \ 1 \le f \le 2 \text{ Hz} \qquad (4.7a)$$

$$A = 0.150 \left(\frac{f}{2} \right) \qquad ; \ 2 \le f \le 80 \text{ Hz} \qquad (4.7b)$$

8 hour boundary:

$$A = 0.224 \qquad ; \ 1 \le f \le 2 \text{ Hz} \qquad (4.8a)$$

$$A = 0.224 \left(\frac{f}{2} \right) \qquad ; \ 2 \le f \le 80 \text{ Hz} \qquad (4.8b)$$

4 hour boundary:

$$A = 0.335 \qquad ; \ 1 \le f \le 2 \text{ Hz} \qquad (4.9a)$$

$$A = 0.335 \left(\frac{f}{2} \right) \qquad ; \ 2 \le f \le 80 \text{ Hz} \qquad (4.9b)$$

In the previous equations, $A \equiv$ RMS acceleration [m/s^2] and $f \equiv$ single frequency or center frequency of one-third octave band [Hz]. The "fatigue-decreased proficiency boundary" relationships for normal acceleration are graphed in Figure (4.4), and the relationships for longitudinal acceleration are shown in Figure (4.5).

All of the foregoing ISO based relationships can be adjusted to represent different limits. In particular, the "exposure limit" is obtained by multiplying the RMS acceleration by 2, and the "reduced comfort boundary" is found by dividing the RMS acceleration by 3.15.

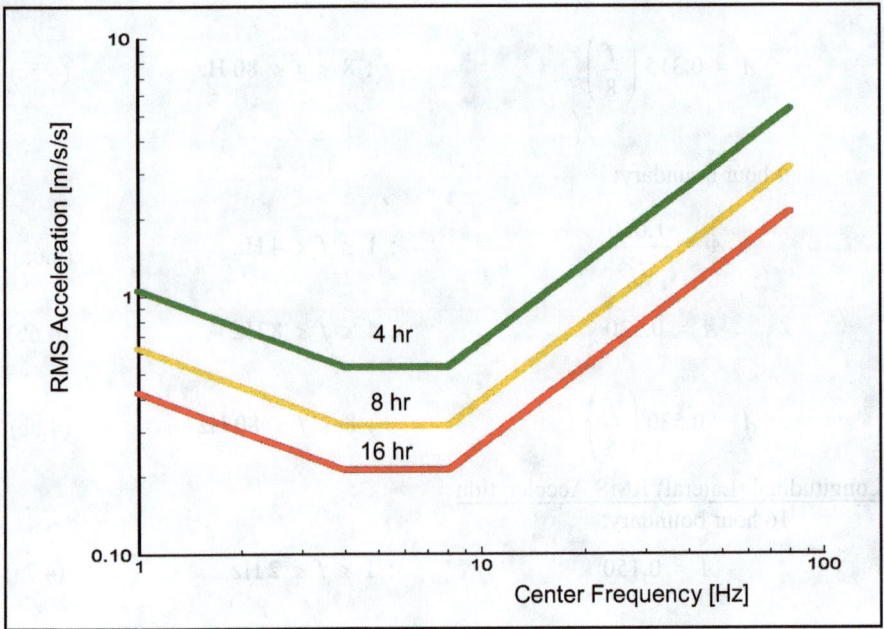

Figure (4.4): ISO NORMAL "FATIGUED-DECREASED PROFICIENCY" LIMITS

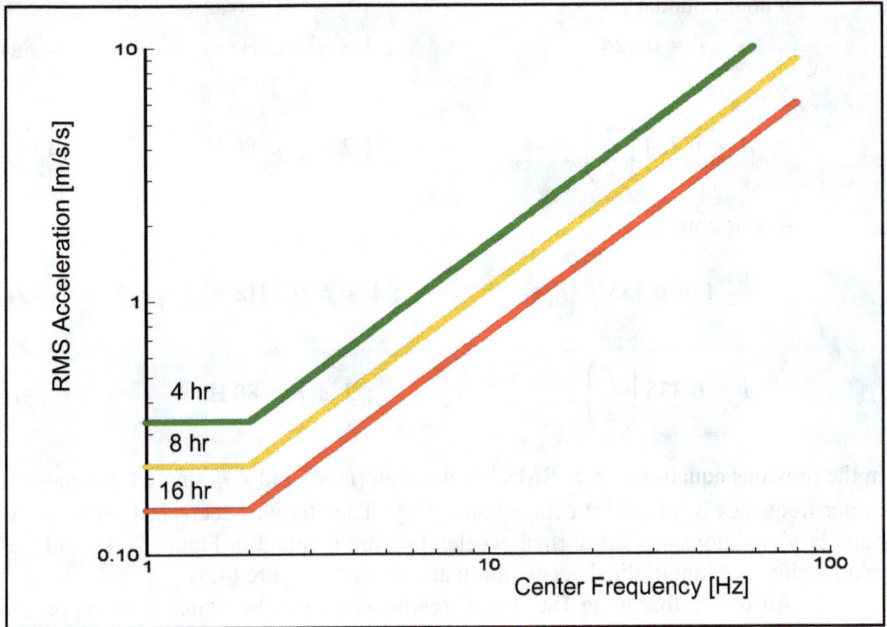

Figure (4.5): ISO LONGITUDINAL (LATERAL) "FATIGUED-DECREASED PROFICIENCY" LIMITS

One of the main uses of the ISO standard, as far as the suspension designer is concerned, is an evaluation of how well a new suspension will be perceived by the driver. An analytical model or experimental prototype of the vehicle suspension is first analyzed to ascertain the magnitude of accelerations which might be transmitted to the driver at various frequencies. Next, the designer must select the appropriate ISO boundary, i.e., "exposure limit", "fatigue-decreased proficiency boundary", or "reduce comfort boundary", based on their judgement regarding the intended use of the vehicle. A comparison is then made by overlaying a graph of the anticipated acceleration on top of the ISO standards graph. The ISO time limit which is completely above the transmitted acceleration curve is the time that the driver can reasonably be expected to operate the vehicle without experiencing discomfort.

(4.2) VEHICLE RIDE MODELING

At some point in the prototype design of the vehicle suspension system the effective spring and damping rates must be specified. In consideration of the purpose for which the vehicle is intended, one of the design criteria imposed by the suspension designer may be human perception of ride quality as governed by an appropriate standard, such as the ISO standard previously discussed. To use this standard requires basic knowledge of the magnitude of amplitude and acceleration likely to be experience by the driver. Without experimental data to guide the engineer, analytical means are sought to generate ride characteristics of the vehicle. Specifically, bounce and pitch response to variations in the road surface.

(4.2.1) ELEMENTARY MOTION OF VEHICLE SPRUNG MASS

The adjective "elementary" employed in the section title refers to the rough approximations we will make in modeling the vehicle and suspension system. These approximations will become obvious to the reader when comparing the mechanical system schematic, on which the mathematical models are based, to an actual vehicle. Better vehicle models will be developed, but for now our objective is to develop an elementary vehicle ride model which will enable us to gain insight into bounce behavior.

(4.2.1A) Undamped System Behavior

A simple mechanical schematic representing a vehicle traveling over a particular road surface is shown in Figure (4.6). The vehicle is modeled as a sprung and unsprung mass separated by a spring, without any shock absorbers (damping). The spring in this model has a stiffness, assumed linear, equivalent to the combined effect of all four suspension springs. The unsprung mass at each wheel is lumped together into a singe unsprung mass which, for the present analysis, remains in contact with the road surface. The model is classified as having a single degree of freedom: it has only one mass and is free to move in only one coordinate direction. Although this simple model has limitations in studying ride characteristics, especially since it does not contain damping, it is a good starting point. This model also allows us the opportunity to introduce a few definitions pertaining to harmonic motion and vibration.

Contrary to the appearance of the sprung mass coordinate in Figure (4.6). The datum line for z_s is not arbitrary. Under static conditions the spring deflects from it's

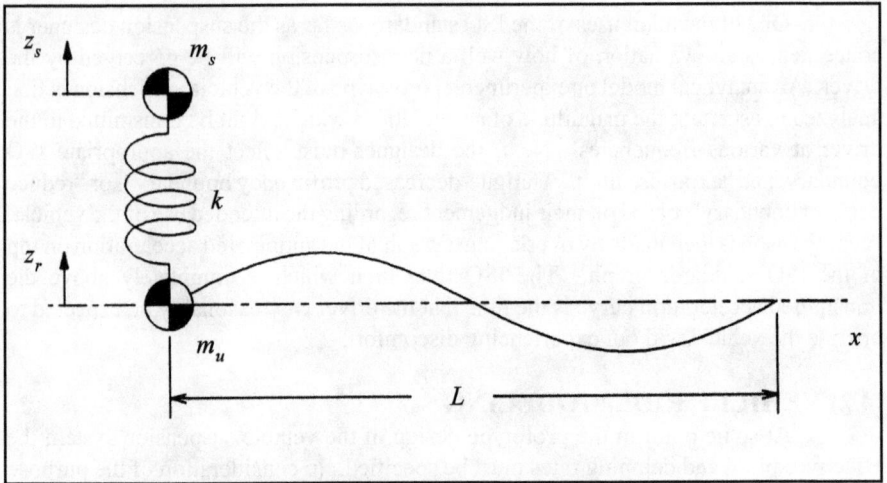

Figure (4.6): SIMPLE UNDAMPED VEHICLE MODEL

unstressed length to counteract gravity acting on the sprung mass. The implication of this definition is that the only force which acts on the sprung mass under dynamic conditions is that due to the spring, i.e., the gravitational body force has no affect on the acceleration of the sprung mass. The equation of motion for the sprung mass is then

$$m_s \frac{d^2 z_s}{dt^2} + k z_s = k z_r \qquad (4.10)$$

The road surface model selected for the present analysis is sinusoidal; therefore, the road surface displacement is given by

$$z_r(x) = Z_r \sin\left(\frac{2\pi}{L} x\right) \qquad (4.11)$$

where $Z_r \equiv$ road surface amplitude and $L \equiv$ road surface wavelength.

When we look closely at Eqs (4.10) and (4.10) we notice that there are two independent variables, t and x. In order to solve the differential equation we need to resolve these variables into a single independent variable. According to the chain rule of derivatives the sprung mass acceleration can be expressed as

$$\frac{d^2 z_s}{dt^2} = \frac{d}{dx}\left[\frac{dz_s}{dx}\frac{dx}{dt}\right]\frac{dx}{dt} \qquad (4.12)$$

The forward velocity of the vehicle, which we will assume to be constant, is defined as

$$V \equiv \frac{dx}{dt} \qquad (4.13)$$

Using this definition, Eq (4.12) becomes

$$\frac{d^2 z_s}{dt^2} = V^2 \frac{d^2 z_s}{dx^2} \qquad (4.14)$$

Inserting Eqs (4.11) and (4.14) into Eq (4.10), along with rearranging the terms, results in

$$\frac{d^2 z_s}{dx^2} + \frac{k}{m_s V^2} z_s = \frac{k Z_r}{m_s V^2} \sin\left(\frac{2\pi}{L} x\right) \qquad (4.15)$$

You may recall from applied mathematics that the natural circular frequency for a spring-mass mechanical system is determined from

$$\omega_{ns} = \sqrt{\frac{k}{m_s}} \qquad (4.16)$$

Also, a road surface frequency can be defined as

$$\omega \equiv \frac{V}{\tau} \; ; \qquad \tau \equiv \frac{L}{2\pi} \qquad (4.17)$$

We can make a direct substitution of the sprung mass circular natural frequency into Eq (4.15); however, to make use of the road frequency we must perform a bit of algebraic manipulation by scaling each of the terms in Eq (4.15) as shown below:

$$\frac{\left(\frac{1}{\tau}\right)}{\left(\frac{1}{\tau}\right)^2} \frac{d^2 z_s}{dx^2} + \frac{\left(\frac{1}{\tau}\right)}{\left(\frac{1}{\tau}\right)^2} \frac{\omega_{ns}^2}{V^2} z_s = \frac{\left(\frac{1}{\tau}\right)}{\left(\frac{1}{\tau}\right)^2} \frac{\omega_{ns}^2 Z_r}{V^2} \sin\left(\frac{x}{\tau}\right) \qquad (4.18)$$

Using the definition of road frequency, the previous equation becomes

$$\frac{\left(\frac{1}{\tau}\right)}{\left(\frac{1}{\tau}\right)^2} \frac{d^2 z_s}{dx^2} + \left(\frac{1}{\tau}\right)\left(\frac{\omega_{ns}}{\omega}\right)^2 z_s = \left(\frac{1}{\tau}\right)\left(\frac{\omega_{ns}}{\omega}\right)^2 Z_r \sin\left(\frac{x}{\tau}\right) \qquad (4.19a)$$

or

$$\frac{d^2\left(\frac{z_s}{\tau}\right)}{d\left(\frac{x}{\tau}\right)^2} + \left(\frac{\omega_{ns}}{\omega}\right)^2 \left(\frac{z_s}{\tau}\right) = \left(\frac{\omega_{ns}}{\omega}\right)^2 \left(\frac{Z_r}{\tau}\right) \sin\left(\frac{x}{\tau}\right) \qquad (4.19b)$$

Eq (4.19b) can be simplified by introducing the following dimensionless parameters:

$$z_s^* \equiv \frac{z_s}{\tau} \qquad x^* \equiv \frac{x}{\tau} \qquad Z_r^* \equiv \frac{Z_r}{\tau} \qquad \omega_{rs} \equiv \frac{\omega}{\omega_{ns}}$$

Then Eq (4.19b) reduces to

$$\frac{d^2 z_s^*}{dx^{*2}} + \frac{1}{\omega_{rs}^2} z_s^* = \frac{Z_r^*}{\omega_{rs}^2} \sin(x^*) \tag{4.20}$$

The previous equation is a linear, second order differential equation which can be solved using classic differential calculus techniques. It has the general solution

$$z_s^* = C_0 \sin\left(\frac{x^*}{\omega_{rs}}\right) + C_1 \cos\left(\frac{x^*}{\omega_{rs}}\right) + \frac{Z_r^*}{1 - \omega_{rs}^2} \sin(x^*) \tag{4.21}$$

The coefficients C_0 and C_1 are constants of integration and are determined from the initial conditions of the spring-mass system. For our analysis we will assume that the sprung mass in initially at rest, i.e.,

$$z_s^*(0) = 0 \qquad \text{and} \qquad \frac{dz_s^*}{dx^*}(0) = 0$$

From these initial conditions we find

$$C_0 = -\frac{\omega_{rs} Z_r^*}{1 - \omega_{rs}^2} \qquad \text{and} \qquad C_1 = 0$$

Eq (4.21) then simplifies to

$$z_s^* = \frac{Z_r^*}{1 - \omega_{rs}^2} \left[\sin(x^*) - \omega_{rs} \sin\left(\frac{x^*}{\omega_{rs}}\right) \right] \tag{4.22a}$$

or in dimensional form, albeit still using the sprung mass frequency ratio,

$$\frac{z_s}{Z_r} = \frac{1}{1 - \omega_{rs}^2} \left[\sin\left(2\pi \frac{x}{L}\right) - \omega_{rs} \sin\left(\frac{2\pi}{\omega_{rs}} \frac{x}{L}\right) \right] \tag{4.22b}$$

Now that we have developed an analytical model for the sprung mass displacement, what can we learn? From the standpoint of suspension specifications the only parameter we are free to vary is the spring stiffness. Reviewing the development of this analysis we find that the spring stiffness governs the sprung mass natural frequency, which in turn affects the frequency ratio. Sprung mass displacement per road surface amplitude, i.e., a normalized displacement, is plotted in Figure (4.7) for three values of frequency ratio. The normalized displacement is plotted against traveled distance normalized with respect to road surface wavelength.

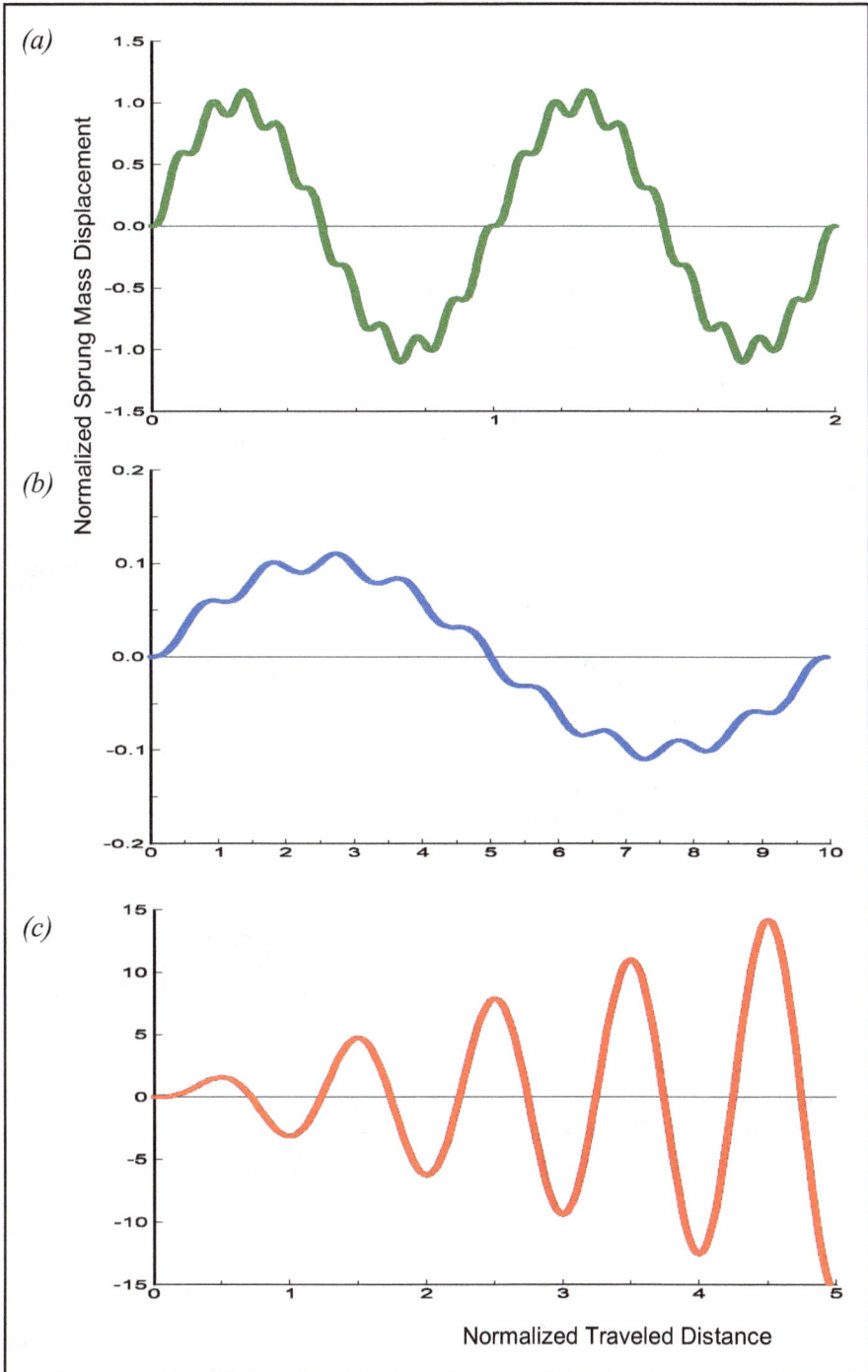

Figure (4.7): SIMPLE RIDE MODEL SIMULATION RESULTS

The graph for frequency ratio of 0.1 is representative of the results obtained for very large sprung mass frequencies. (Remember, by definition ω_{rs} is inversely proportional to ω_{ns}.) Except for a slight ripple in the curve, the sprung mass exactly follows the road contour. Since the suspension design goal is to isolate the vehicle body and chassis from the road, selecting a spring stiffness that yields a large natural frequency is not a good choice.

A very small sprung mass natural frequency gives quite different results. Examining the displacement plot in Figure (4.7b) we notice that it has a small peak amplitude. Therefore, we control bounce harmonic motion by appropriately specifying spring stiffness rates to yield a low natural frequency.

The graph of Eq (4.22b) shown in Figure (4.7c) illustrates that when the road frequency is close to the natural frequency the magnitude of the peak displacement becomes very large. In fact, when the frequency ratio is equal to one the peak displacement becomes infinity large. The means the suspension system becomes unstable and we lose control of the vehicle. It should be recognized that in all cases once the sprung mass is set into motion, the motion continues unabated. The only way to overcome this problem is by modifying the suspension to include a damping device (shock absorber). Obviously, if we add a new component to the current suspension system we need to develop new governing equations to simulate vehicle bounce.

(4.2.1B) Damped System Behavior

Figure (4.8) illustrates the simple vehicle schematic as modified by the addition of a shock absorber. For our elementary analysis we will assume that the shock absorber has linear characteristics. In this case the differential equation governing the motion of the sprung mass is

$$m_s \frac{d^2 z_s}{dt^2} + b \frac{dz_s}{dt} + k z_s = b \frac{dz_r}{dt} + k z_r$$

(4.23)

Following the same development steps detailed in Section (4.2.1A), Eq (4.23) can also be expressed in nondimensional form as

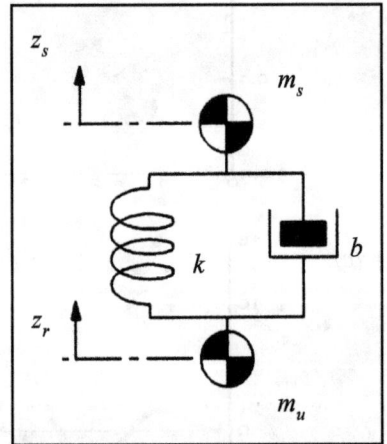

Figure (4.8): SIMPLE DAMPED VEHICLE MODEL

$$\frac{d^2 z_s^*}{dx^{*2}} + \frac{b}{m_s} \frac{1}{\omega} \frac{dz_s^*}{dx^*} + \frac{\omega_{ns}^2}{\omega^2} z_s^* = \frac{b}{m_s} \frac{Z_r^*}{\omega} \cos(x^*) + \frac{\omega_{ns}^2}{\omega^2} Z_r^* \sin(x^*)$$

(4.24)

This damped behavior equation is similar to the undamped equation in that it is a linear, second-order differential equation. However, the solution is a bit more involved.

The classic approach to solving Eq (4.24) is to assume a form of the homogeneous solution, which in this case is

$$z_{s,h}^* = C e^{\lambda x^*} \tag{4.25}$$

Substituting Eq (4.25) into the homogeneous part of Eq (4.24) generates the characteristic equation

$$\lambda^2 + \frac{b}{m_s} \frac{1}{\omega} \lambda + \frac{\omega_{ns}^2}{\omega^2} = 0 \tag{4.26}$$

The characteristic roots found from Eq (4.26) are

$$\lambda_{1,2} = \frac{-b}{2\,m_s\,\omega} \pm \frac{1}{2\,\omega} \sqrt{\left(\frac{b}{m_s}\right)^2 - 4\,\omega_{ns}^2} \tag{4.27}$$

In vibrational mechanics the phrase "critical damping" is used to define a condition for which the radical term of the characteristic roots is equal to zero, i.e.,

$$\left(\frac{b_{crit}}{m_s}\right)^2 - 4\,\omega_{ns}^2 = 0 \tag{4.28a}$$

or

$$b_{crit} = 2\,m_s\,\omega_{ns} \tag{4.28b}$$

The damping ratio, ζ, is defined as the ratio of the damping coefficient to it's critical value. Then

$$b = \zeta b_{crit} = 2\,m_s\,\zeta\,\omega_{ns} \tag{4.29}$$

For practical vehicle suspension simulation studies, the value of the damping ratio will be less than unity. Substituting Eq (4.29), with the provision that $\zeta < 1$, into Eq (4.27) results in

$$\lambda_{1,2} = -\frac{\zeta\omega_{ns}}{\omega} \pm j\,\frac{\omega_{ns}}{\omega}\sqrt{1 - \zeta^2} \tag{4.30}$$

where j is the complex operator $\sqrt{-1}$. Eq (4.30) can be slightly simplified by introducing the definition of "damped" natural frequency, i.e.,

$$\omega_{ds} \equiv \omega_{ns}\sqrt{1 - \zeta^2} \tag{4.31}$$

Then

$$\lambda_{1,2} = -\zeta \frac{\omega_{ns}}{\omega} \pm j \frac{\omega_{ds}}{\omega} \tag{4.32}$$

Using the information provided by Eq (4.32) the homogeneous solution becomes

$$z_{s,h}^* = \exp\left(-\zeta \frac{\omega_{ns}}{\omega} x^*\right)\left[C_0 \sin\left(\frac{\omega_{ds}}{\omega} x^*\right) + C_1 \cos\left(\frac{\omega_{ds}}{\omega} x^*\right)\right] \tag{4.33}$$

As before, the constants C_0 and C_1 are determined from the initial conditions.

To derive a particular solution we will find it more convenient to work with a variation of the nondimensional differential equation utilizing the definitions of damping ratio and frequency ratio, in which we rewrite Eq (4.24) as

$$\frac{d^2 z_s^*}{dx^{*2}} + \frac{2\zeta}{\omega_{rs}} \frac{dz^*}{dx^*} + \frac{1}{\omega_{rs}^2} z_s^* = \frac{2\zeta Z_r^*}{\omega_{rs}} \cos(x^*) + \frac{Z_r^*}{\omega_{rs}^2} \sin(x^*) \tag{4.34}$$

The particular solution of this equation is developed from an assumed form of the solution based on the forcing function.[2] From Eq (4.34) we assume that

$$z_{s,p}^* = Q_1 \cos(x^*) + Q_2 \sin(x^*) \tag{4.35}$$

Substituting Eq (4.35) into Eq (4.34) produces

$$\left[-Q_1 \cos(x^*) - Q_2 \sin(x^*)\right] + \frac{2\zeta}{\omega_{rs}}\left[-Q_1 \cos(x^*) + Q_2 \sin(x^*)\right]$$

$$+ \frac{1}{\omega_{rs}^2}\left[Q_1 \cos(x^*) + Q_2 \sin(x^*)\right] = \frac{2\zeta Z_r^*}{\omega_{rs}} \cos(x^*) + \frac{Z_r^*}{\omega_{rs}} \sin(x^*) \tag{4.36}$$

Equating the cosine and sine terms, respectively, creates the linear algebraic set of equations

$$\left(1 - \omega_{rs}^2\right) Q_1 + 2\zeta \omega_{rs} Q_2 = 2\zeta \omega_{rs} Z_r^* \tag{4.37}$$

$$-2\zeta \omega_{rs} Q_1 + \left(1 - \omega_{rs}^2\right) Q_2 = Z_r^* \tag{4.38}$$

---

[2] *Reference textbooks in mathematics refer to this approach as the "Method of Undetermined Coefficients."*

which can be simultaneously solved to yield

$$Q_1 = -\frac{2\zeta\omega_{rs}^3}{\left(1 - \omega_{rs}^2\right)^2 + \left(2\zeta\omega_{rs}\right)^2} Z_r^*$$

(4.39)

$$Q_2 = \frac{\left(1 - \omega_{rs}^2\right) + \left(2\zeta\omega_{rs}\right)^2}{\left(1 - \omega_{rs}^2\right)^2 + \left(2\zeta\omega_{rs}\right)^2} Z_r^*$$

(4.40)

The general solution to the original differential equation is found by adding the homogeneous solution, Eq (4.33), and the particular solution, Eq (4.35) with Eqs (4.39) and Eq (4.40):

$$z_s = \exp\left(-\zeta\omega_{ns}\frac{x}{V}\right)\left[C_0 \sin\left(\omega_{ds}\frac{x}{V}\right) + C_1 \cos\left(\omega_{ds}\frac{x}{V}\right)\right] -$$
$$\frac{2\zeta\omega_{rs}^3 Z_r}{\left(1 - \omega_{rs}^2\right)^2 + \left(2\zeta\omega_{rs}\right)^2}\cos\left(\frac{2\pi}{L}x\right) + \frac{\left[\left(1 - \omega_{rs}^2\right) + \left(2\zeta\omega_{rs}\right)^2\right]Z_r}{\left(1 - \omega_{rs}^2\right)^2 + \left(2\zeta\omega_{rs}\right)^2}\sin\left(\frac{2\pi}{L}x\right)$$

(4.41)

At this point in the analysis we could solve for the integration constants C_0 and C_1; however, these values are of little interest. The exponential term in Eq (4.41) decays towards zero as the traveled distance, x, increases. The combination of remaining terms, which do not vanish, is called the steady-state solution. Therefore, based on Eq (4.41) the damped sprung mass steady-state displacement is

$$\overline{z}_s = -\frac{2\zeta\omega_{rs}^3 Z_r}{\left(1 - \omega_{rs}^2\right)^2 + \left(2\zeta\omega_{rs}\right)^2}\cos\left(\frac{2\pi}{L}x\right)$$
$$+ \frac{\left[\left(1 - \omega_{rs}^2\right) + \left(2\zeta\omega_{rs}\right)^2\right]Z_r}{\left(1 - \omega_{rs}^2\right)^2 + \left(2\zeta\omega_{rs}\right)^2}\sin\left(\frac{2\pi}{L}x\right)$$

(4.42)

We also note that the steady-state solution is nothing more than the particular solution for the damped sprung mass displacement.

To study bounce characteristics it is more advantageous to express the steady-state solution in a compact form using a single sprung mass amplitude, Z_s, and a phase angle, Φ_s, as

$$\bar{z}_s = Z_s \sin\left(\frac{2\pi}{L} x + \Phi_s \right) \qquad (4.43)$$

With reference to Eq (4.35), when the steady-state (particular) solution exists as a summation of a cosine term plus a sine term the amplitude and phase angle are determined from

$$Z_s = \sqrt{Q_1^2 + Q_2^2} \qquad (4.44)$$

and

$$\Phi_s = \arctan\left(\frac{Q_2}{Q_1} \right) \qquad (4.45)$$

Evaluating Eqs (4.44) and (4.45) for the present analysis results in

$$Z_s = Z_r \sqrt{ \frac{1 + \left(2\zeta\omega_{rs}\right)^2}{\left(1 - \omega_{rs}^2\right)^2 + \left(2\zeta\omega_{rs}\right)^2} } \qquad (4.46)$$

and

$$\Phi_s = \arctan\left[-\frac{2\zeta\omega_{rs}^3}{\left(1 - \omega_{rs}^2\right) + \left(2\zeta\omega_{rs}\right)^2} \right] \qquad (4.47)$$

A bounce characteristic of particular interest in the "transmissibility ratio." The definition of which is identical to what is normally referred to as system gain in control theory, i.e., the steady-state system output amplitude divided by the input amplitude. From Eq (4.46) the transmissibility ratio of our simple damped vehicle model is

$$\frac{Z_s}{Z_r} = \sqrt{ \frac{1 + \left(2\zeta\omega_{rs}\right)^2}{\left(1 - \omega_{rs}^2\right)^2 + \left(2\zeta\omega_{rs}\right)^2} } \qquad (4.48)$$

A sprung mass frequency plot for several values of damping ratio is shown in Figure (4.9). Examining the results of the plots we draw the same conclusion as before. To control the bounce motion it is necessary to adjust the spring stiffness of the suspension system such that the sprung mass has a low natural frequency. We also notice that damping in the suspension system increases the sprung mass amplitude above the undamped amplitude for frequency rations greater than $\sqrt{2}$. This trend indicates that while damping is required to maintain control of the vehicle, too much

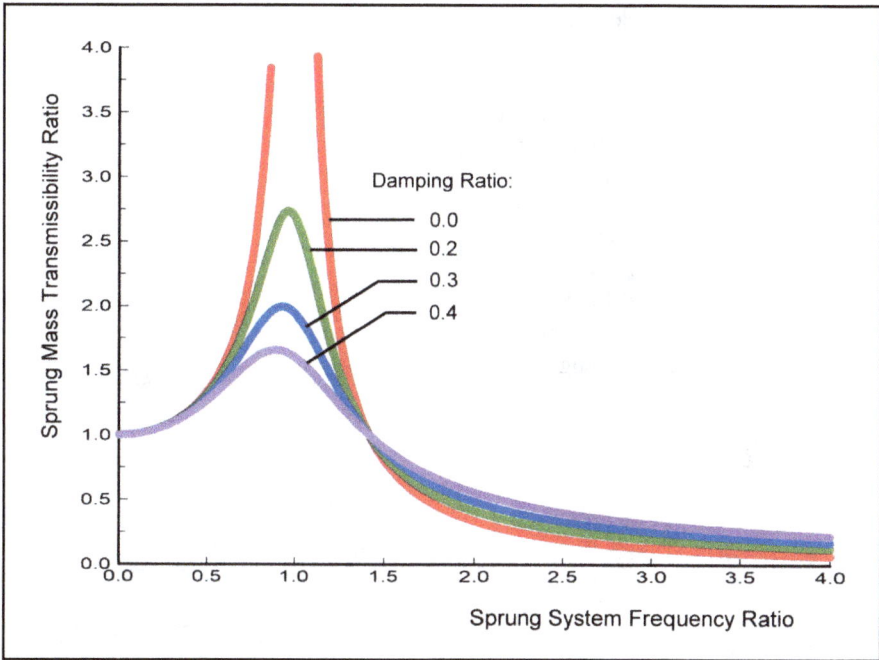

Figure (4.9): SPRUNG MASS FREQUENCY RESPONSE PLOT

damping results in a harsh ride. It is generally accepted that the damping ratio for a quality ride should range from about 0.25 to 0.38, depending on the amount of inherent friction (which also dampens movement) in the suspension linkage.

(4.2.2) NORMAL MODES OF BOUNCE AND PITCH

In this section we will examine two slightly more sophisticated vehicle models to further our understanding of vehicle ride. The first is bounce motion which includes unsprung mass and tire stiffness. The second model allows the sprung mass to possess both bounce and pitch motions, but ignores the effect of unsprung mass.

There is some commonality between the two models. Both of the models have two degrees of freedom. This means that there will be two natural frequencies which are characteristic of the vehicle system. In both vehicle models the mathematical approach taken to find these natural frequencies fall under the topic of eigenvalues[3]. For each eigenvalue there exists a unique natural frequency and an eigenvector, which describes a ratio of displacement amplitudes, corresponding to the different degrees of freedom. The amplitude ratio is called a "normal mode" [7]. In order to minimize the complexity of the mathematics involved in studying the two vehicle models, only free vibration without system damping will be considered.

---

[3] *The subject of eigenvalues is common to a variety of engineering fields, and can generally be found in any applied engineering mathematics textbook.*

(4.2.2A) Sprung and Unsprung Mass Bounce Model

The sprung and unsprung mass vehicle model shown in Figure (4.10) is an extension of the simple vehicle model studied in Section (4.2.1). The revised model includes two new features: (1) bounce movement of the unsprung mass, and (2) the combined tire stiffness. The governing equations of bounce motion for the sprung and unsprung mass bounce model are:

(i) sprung mass

$$m_s \frac{d^2 z_s}{dt^2} + k z_s = k z_u \tag{4.49}$$

(ii) unsprung mass

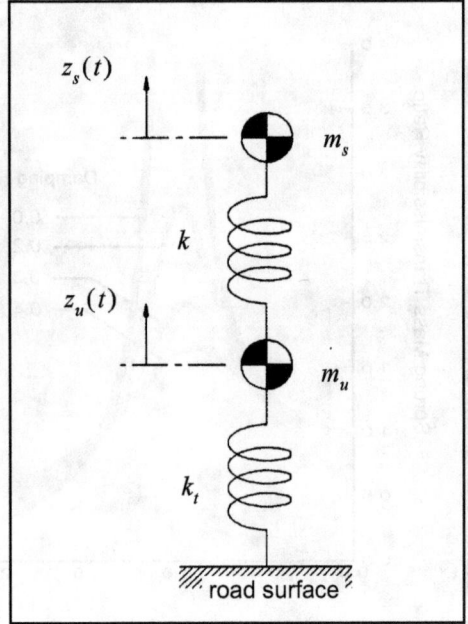

Figure (4.10): SPRUNG AND UNSPRUNG MASS BOUNCE MODEL

$$m_u \frac{d^2 z_u}{dt^2} + \left(k + k_t\right) z_u = k z_s \tag{4.50}$$

The previous equations can be written in matrix form as

$$\frac{d^2}{dt^2} \begin{Bmatrix} z_s \\ z_u \end{Bmatrix} = \begin{bmatrix} -\dfrac{k}{m_s} & \dfrac{k}{m_s} \\ \dfrac{k}{m_u} & -\dfrac{\left(k + k_t\right)}{m_u} \end{bmatrix} \begin{Bmatrix} z_s \\ z_u \end{Bmatrix} \tag{4.51}$$

or, in shorthand notation

$$\frac{d^2}{dt^2} \begin{Bmatrix} z_s \\ z_u \end{Bmatrix} = [K]\{Z\} \tag{4.52}$$

We have learned in the previous section that a periodic solution describes the sprung mass bounce motion. It seems reasonable, then, that a periodic solution should also apply to the present analysis; therefore, we will assume that the solution to Eq (4.52) is of the form

$$z_s = Z_s e^{j\omega_n t}$$

$$z_u = Z_u e^{j\omega_n t}$$

or, in matrix form

$$\left\{ \begin{matrix} z_s \\ z_u \end{matrix} \right\} = \{C\} e^{j\omega_n t} \; ; \qquad \{C\} = \left\{ \begin{matrix} Z_s \\ Z_u \end{matrix} \right\} \tag{4.53}$$

Inserting the assumed solution given by Eq (4.53) into Eq (4.52) produces the matrix equation

$$-\omega_n^2 \{C\} = [K]\{C\} \tag{4.54}$$

For convenience we define an unknown constant, call an eigenvalue, as $\lambda = \omega_n^2$. Rearranging Eq (4.54) results in

$$\big[[K] + \lambda [I]\big]\{C\} = 0 \tag{4.55}$$

where $[I]$ is the identity matrix.

Our attention is now focused on finding the eigenvalues which satisfy Eq (4.55). Since it is our desire to find all of the eigenvalues, we want a non-unique solution. This condition is met if

$$\det\big[[K] + \lambda [I]\big] = 0 \tag{4.56a}$$

or

$$\det \begin{bmatrix} -\dfrac{k}{m_s} + \lambda & \dfrac{k}{m_s} \\[3mm] -\dfrac{k}{m_u} & -\dfrac{(k + k_t)}{m_u} + \lambda \end{bmatrix} = 0 \tag{4.56b}$$

Eq (4.56b) evaluates as

$$\lambda^2 - \left[\frac{k}{m_s} + \frac{(k + k_t)}{m_u} \right] \lambda + \frac{k\,k_t}{m_s\,m_u} = 0 \tag{4.57}$$

Eq (4.57) is the "characteristic equation" for the sprung and unsprung mass system. The eigenvalues are the characteristic roots. Before we solve for the eigenvalues we will make one slight modification to Eq (4.57); and that is the introduction of the sprung to unsprung mass ratio, m_r. Using this definition Eq (4.57) can be rewritten as

$$\lambda^2 - \frac{(1 + m_r)k + m_r k_t}{m_s}\lambda + \frac{m_r k k_t}{m_s^2} = 0 \qquad (4.58)$$

The solution to Eq (4.58) is

$$\lambda_{1,2} = \frac{(1 + m_r)k + m_r k_t}{2 m_s} \pm \frac{1}{2 m_s}\left\{\left[(1 + m_r)k + m_r k_t\right]^2 - 4 m_r k k_t\right\}^{\frac{1}{2}} \qquad (4.59)$$

Subsequently, the two natural frequencies of the system are determined from

$$\omega_{n,1} = \sqrt{\lambda_1} \qquad (4.60a) \qquad \text{and} \qquad \omega_{n,2} = \sqrt{\lambda_2} \qquad (4.60b)$$

The physical significance of the natural frequencies as given by Eqs (4.59) and (4.60) is probably not readily apparent; however, we can develop two approximate natural frequencies which are easily interpreted. The approximations are based on the

Table (4.1): REPRESENTATIVE UNSPRUNG WEIGHT VALUES

Suspension Configuration	Unsprung Weight per Wheel [lb]
double A-arm front suspension with coil spring	80
double A-arm rear suspension with coil spring and inboard brake	65
double A-Arm rear suspension with coil spring and outboard brake	90
swing axle with coil spring	90
four-link solid rear axle with coil spring	110
solid rear axle with leaf spring	130

Note: (1) Data taken from Reference [8],
(2) Values based on a typical 2000 lb sports car.

concept that a well designed vehicle is one in which the unsprung mass is kept to a minimum. Representative values of unsprung weight (mass) for a 2000 lb sports car with various suspension configurations is given in Table (4.1). If we assume that the ratio of sprung to unsprung mass is much greater than one, then we can make the following approximations:

(i)

$$\left(1 + m_r\right)k + m_r k_t \approx m_r\left(k + k_t\right) \tag{4.61}$$

and (ii), expanding the square root term of Eq (4.59)

$$\left[\left(1 + m_r\right)k + m_r k_t\right]^2 - 4 m_r k k_t$$

$$= \left(1 + m_r\right)^2 k^2 + k k_t\left[m_r^2 + \left(1 - m_r\right)^2 - 1\right] + m_r^2 k_t^2$$

$$\approx m_r^2\left(k + k_t\right)^2$$

$$\tag{4.62}$$

Using the foregoing approximations, Eq (4.59) simplifies to

$$\lambda_1 = \frac{k + k_t}{m_u} \quad \text{(4.63a)} \qquad \text{and} \qquad \lambda_2 = 0 \quad \text{(4.63b)}$$

Recalling that the eigenvalue is defined as a natural frequency squared, we conclude from the first characteristic root, λ_1, that

$$\omega_{nu} = \sqrt{\frac{k + k_t}{m_u}} \tag{4.64}$$

which is the unsprung mass circular natural frequency. Essentially it describes the resonant frequency causing wheel hop.

The second root identified by Eq (4.63b) as equal to zero is not very useful; however, it does suggest that the remaining root has a small magnitude. If this is indeed the case, then by "order of magnitude" approximation the λ^2 term in Eq (4.58) can be neglected in comparison to the other terms. Rearranging the remaining terms of Eq (4.58), and using the approximation given by Eq (4.61), yields

$$\lambda_2 = \frac{k k_t}{m_s\left(k + k_t\right)} \tag{4.65}$$

From this root we conclude that

Chapter 4: RIDE CHARACTERISTICS

$$\omega_{ns} = \sqrt{\dfrac{\left(\dfrac{k\,k_t}{k + k_t}\right)}{m_s}} \qquad (4.66)$$

which is the sprung mass circular natural frequency. We should note that the algebraic combination of suspension spring stiffness and tire stiffness is the equivalent stiffness of two springs in a series arrangement. The equivalent stiffness of the suspension springs and tires appears as the terms in parentheses in Eq (4.66), and is sometimes referred to as the "ride rate"[9].

How do these approximate natural frequencies compare to exact values deduced from Eq (4.59)? This can best be answered by looking at a numeric example. Consider the following performance vehicle:

sprung weight:	**2000 lb**
unsprung weight:	**150 lb**
suspension spring stiffness:	**460 lb/in**
combined tire stiffness:	**4000 lb/in**

The eigenvalues obtained from Eq (4.59) are: $\lambda_1 = 11,489$ (rad/s)^2 and $\lambda_2 = 79.578$ (rad/s)^2. Next, the corresponding circular natural frequencies are calculated, i.e., $\omega_{n,1} = \sqrt{\lambda_1} = 107.2$ rad/s and $\omega_{n,2} = \sqrt{\lambda_2} = 8.921$ rad/s. The approximate unsprung mass natural frequency calculated from Eq (4.64) is $\omega_{nu} = 107.1$ rad/s and the approximate sprung mass natural frequency evaluated from Eq (4.66) is $\omega_{ns} = 8.924$ rad/s. Comparing the results, our approximate natural frequencies appear to be quite good. Indeed, this is generally the case for modern vehicle design.

Deriving a symbolic form for the eigenvectors, which describe the sprung mass amplitude relative to the unsprung mass amplitude, is not an easy task. Perhaps the best way to discuss this facet of our analysis is to continue the example above.

In general the eigenvectors are found by substituting the eigenvalues, one at a time, in Eq (4.55)[4]. In our case, since we have two eigenvalues, we have two eigenvectors to evaluate:

(i) for $\lambda_1 = 11,489$ (rad/s)^2, Eq (4.55) becomes (including appropriate units conversion):

$$\begin{bmatrix} 11,400 & 88.8 \\ 1,184 & 9.223 \end{bmatrix}_{\lambda_1} \begin{Bmatrix} Z_s \\ Z_u \end{Bmatrix}_{\lambda_1} = 0$$

---

[4] *The reader should be advised that eigenvalue problems in vehicle dynamics are very susceptible to round off error. Greater accuracy was used in the calculations than recorded in the text.*

If we decompose this matrix equation into two algebraic equations we find that they are linearly dependent, i.e., one is just a scaler multiplier of the other. Mathematically, the best we can reduce the equations to is

$$128.4 \, Z_s + Z_u = 0$$

If we arbitrarily select $Z_s = 1$, then from the previous equation $Z_u = -128.4$. The eigenvector is then

$$\left\{ \begin{array}{c} Z_s \\ Z_u \end{array} \right\}_{\lambda_1} = \left\{ \begin{array}{c} 1 \\ -128.4 \end{array} \right\}_{\lambda_1}$$

Slightly more useful is the amplitude ratio, which is found by rearranging the previous algebraic equation as

$$\left. \frac{Z_u}{Z_s} \right|_{\lambda_1} = -128.4$$

The value of the amplitude ratio tells us that large vertical movement of the unsprung mass occurs relative to the sprung mass. This condition suggests that the circular natural frequency derived from the present eigenvalue is the resonant frequency of the unsprung mass. This is also the conclusion we were able to derive from Eqs (4.63a) and (4.64).

(ii) for $\lambda_2 = 79.578 \ (\text{rad/s})^2$, we have

$$\left[\begin{array}{cc} -9.223 & 88.8 \\ 1{,}184 & -11{,}400 \end{array} \right]_{\lambda_2} \left\{ \begin{array}{c} Z_s \\ Z_u \end{array} \right\}_{\lambda_2} = 0$$

which reduces to the algebraic equation

$$Z_s - 9.628 \, Z_u = 0$$

The amplitude ratio follow from this equation as

$$\left. \frac{Z_u}{Z_s} \right|_{\lambda_2} = 0.1039$$

The amplitude ratio calculated from the present eigenvalue has just the opposite interpretation from the previously encountered amplitude ratio. The small magnitude of the amplitude ratio indicates a small amount of unsprung mass vertical movement relative to the sprung mass. Or, conversely, there is a large sprung mass movement. This means that the natural frequency corresponding to the second eigenvalue is the resonance frequency of the sprung mass, which is the same conclusion given by Eqs (4.63b) and (4.66).

For pure bounce motion the amplitude ratios are not nearly important as the natural frequencies in quantifying ride characteristics; and, for this purpose, the approximate natural frequency relationships expressed by Eqs (4.64) and (4.66) suffice. This is not the case, however, for bounce with pitch motion.

(4.2.2B) Sprung Mass Bounce with Pitch Mode

The consequence of having both a front and rear suspension on the movement of the vehicle sprung mass is two degrees of freedom. Therefore, we are required to use two coordinate variables to define the position of the sprung mass at any instant in time. One time dependent variable, which is familiar by now, is the vertical displacement, z_s. The second time dependent variable describing position is accomplished by introducing the pitch angle, θ, which describes the rotational position of the sprung mass about a lateral axis passing through the sprung mass center (or, center of gravity, CG). The vehicle model we will use to study sprung mass bounce with pitch motion is shown in Figure (4.11).

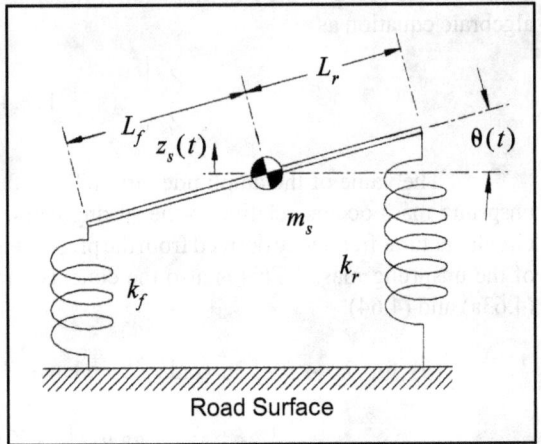

Figure (4.11): BOUNCE WITH PITCH VEHICLE RIDE MODEL

The fundamental equation which describes the bounce motion is

$$\left(L_f \sin\theta - z_s\right) k_f - \left(L_r \cos\theta + z_s\right) k_r = m_s \frac{d^2 z_s}{dt^2} \qquad (4.67)$$

But, before we can continue in analyzing bounce motion we need to further our understanding of pitch motion.

The pitch motion governing equation is developed from the basic relationship

$$\sum T_{CG} = I_s \frac{d^2\theta}{dt^2} \qquad (4.68)$$

The symbol I_s is the sprung mass moment of inertia about the center of gravity. It is mathematically defined as

$$I_s = \int r^2 \, dm = \rho^2 m_s \tag{4.69}$$

where ρ is the radius of gyration. The radius of gyration is a vehicle parameter dependent of the longitudinal distribution of sprung mass around the center of gravity. In some sense it is a design parameter because it's value changes as we add, subtract, or shift vehicle weight during the design process. For now we assume that it is a known quantity which provides a means to calculate the sprung mass moment of inertia. Continuing with the evaluation of Eq (4.68):

$$-k_f\left(L_f \sin\theta - z_s\right) L_f \cos\theta - k_r\left(L_r \cos\theta + z_s\right) L_r \cos\theta = I_s \frac{d^2\theta}{dt^2} \tag{4.70}$$

The maximum pitch angle for typical vehicles is usually under five degrees. This realization allows us to make a "small angle" approximation, i.e., $\sin\theta \approx \theta$ and $\cos\theta \approx 1$. Using this approximation we are able to generate the linear set of bounce with pitch governing equations:

$$\frac{d^2 z_s}{dt^2} = -\frac{k_f + k_r}{m_s} z_s - \frac{k_r L_r - k_f L_f}{m_s} \theta \tag{4.71}$$

and

$$\frac{d^2\theta}{dt^2} = -\frac{k_r L_r - k_f L_f}{I_s} z_s - \frac{k_f L_f^2 + k_r L_r^2}{I_s} \theta \tag{4.72}$$

The solution of Eqs (4.71) and (4.72) begins by expressing the equations in matrix form; that is

$$\frac{d^2}{dt^2}\begin{Bmatrix} z_s \\ \theta \end{Bmatrix} = \begin{bmatrix} -\dfrac{k_f + k_r}{m_s} & -\dfrac{k_r L_r - k_f L_f}{m_s} \\[3mm] -\dfrac{k_r L_r - k_f L_f}{I_s} & -\dfrac{k_f L_f^2 + k_r L_r^2}{I_s} \end{bmatrix} \begin{Bmatrix} z_s \\ \theta \end{Bmatrix} \tag{4.73a}$$

or

$$\frac{d^2}{dt^2}\begin{Bmatrix} z_s \\ \theta \end{Bmatrix} = [K]\begin{Bmatrix} z_s \\ \theta \end{Bmatrix} \tag{4.73b}$$

Based on our previous experience with pure bounce motion, we will assume that the solution for bounce with pitch motion should also be periodic in nature; therefore, let

$$z_s = Z_s e^{j\omega_n t} \qquad (4.74)$$

$$\theta = \Theta e^{j\omega_n t} \qquad (4.75)$$

or, in matrix form

$$\begin{Bmatrix} z_s \\ \theta \end{Bmatrix} = \{C\} e^{j\omega_n t} ; \qquad \{C\} = \begin{Bmatrix} Z_s \\ \Theta \end{Bmatrix} \qquad (4.76)$$

Placing Eq (4.76) into Eq (4.73b) results in

$$[[K] + \lambda[I]]\{C\} = 0 \qquad (4.77)$$

Similar to the evaluation process in the previous section, λ is an eigenvalue defined as ω_n^2 and is determined from

$$\det[[K] + \lambda[I]] = 0 \qquad (4.78)$$

Evaluating Eq (4.78) leads to the algebraic equation

$$\lambda^2 - (a + c)\lambda + \left[ac - \left(\frac{b}{\rho}\right)^2 \right] = 0 \qquad (4.79)$$

where the symbols a, b, and c are shorthand notations for

$$a = \frac{k_f + k_r}{m_s} \qquad b = \frac{k_r L_r - k_f L_f}{m_s} \qquad c = \frac{k_f L_f^2 + k_r L_r^2}{I_s}$$

The eigenvalues obtained by solving Eq (4.79) are

$$\lambda_{1,2} = \frac{1}{2}(a + c) \pm \sqrt{\frac{1}{4}(a - c)^2 + \left(\frac{b}{\rho}\right)^2} \qquad (4.80)$$

By definition, the two circular natural frequencies characteristic of the vehicle and suspension are then

$$\omega_{n1}, \omega_{n2} = \sqrt{\frac{1}{2}(a + c) \pm \sqrt{\frac{1}{4}(a - c)^2 + \left(\frac{b}{\rho}\right)^2}} \qquad (4.81)$$

Once again it is difficult to associate a physical significance to the natural frequencies as they exist in symbolic form. However the amplitude ratio, which is derived from the eigenvector and is dependent on the natural frequency, will help us understand the movement of the sprung mass. In order to evaluate the eigenvectors we rewrite Eq (4.77) as

$$\begin{bmatrix} -a + \omega_n^2 & -b \\ -\dfrac{b}{\rho^2} & -c + \omega_n^2 \end{bmatrix} \begin{Bmatrix} Z_s \\ \Theta \end{Bmatrix} = 0 \qquad (4.82)$$

Since the two algebraic equations contained in Eq (4.82) are linearly dependent, we are free to chose only one equation to examine. For this purpose we select:

$$\left(-a + \omega_n^2 \right) Z_s - b\Theta = 0 \qquad (4.83)$$

which can be written as

$$\frac{Z_s}{\Theta} = -\frac{b}{a - \omega_n^2} \qquad (4.84)$$

The left-hand-side of Eq (4.84) is a ratio of the bounce amplitude to the pitch amplitude. We can use this amplitude ratio to define an "oscillation center." An oscillation center essentially describes the center of rotation for combined bounce and pitch motion. Figure (4.12) illustrates this concept. The variable p is defined as the longitudinal distance, along a horizontal road surface, to the oscillation center from the

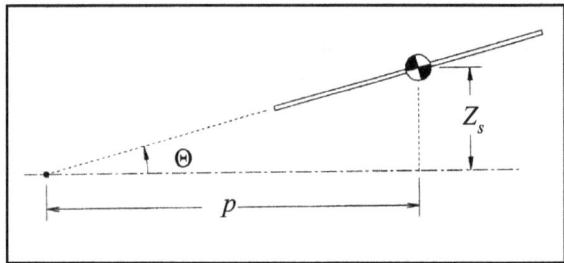

Figure (4.12): OSCILLATION CENTER GEOMETRIC DEFINITION

sprung mass center of gravity. The mathematical definition of p is

$$p = \frac{Z_s}{\tan\Theta} \qquad (4.85)$$

As before, we restrict out attention to small values of pitch angle. This allows us to make the approximation $\tan\Theta \approx \Theta$. Therefore Eq (4.85) can be approximated as

$$p \approx \frac{Z_s}{\Theta} \qquad (4.85)$$

Eq (4.84) indicates that two unique amplitude ratios exist corresponding to ω_{n1} and ω_{n2}. With a little bit of deductive reasoning we can show from Eq (4.81) that $\omega_{n1} > \sqrt{2}$ and $\omega_{n2} < \sqrt{2}$; therefore, one amplitude ratio is greater than zero, and the other is less than zero. This means one of the oscillation centers is located in front of the center of gravity, and the other oscillation center is behind. The location of the oscillation center governs the motion of the sprung mass.

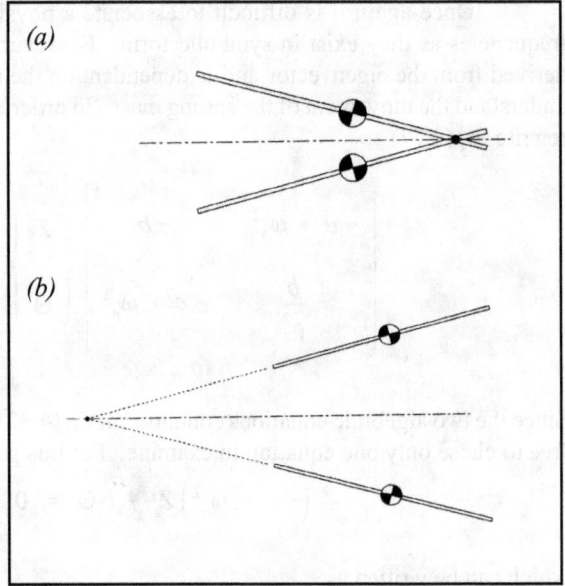

Figure (4.13): PRIMARY SPRUNG MASS MOTION: (a) PITCH and (b) BOUNCE

Referring to Figure (4.13), the oscillation center which lies within the wheelbase results primarily in pitch motion. Whereas the oscillation center outside the wheelbase produces predominantly bounce motion.

The location of the oscillation centers relative to the wheelbase is an important consideration in evaluating a vehicle suspension design. To illustrate this point lets examine two possible design specifications. The first specification under consideration is "$k_f L_f = k_r L_r$". In this case bounce and pitch motion are independent of each other because the equations of motion become uncoupled. We notice the term $\left(k_r L_r - k_f L_f \right)$ as a coefficient of θ and z_s in Eqs (4.71) and (4.72) becomes zero under this specification, and the equations of motion reduce to an independent set of equations:

$$\frac{d^2 z_s}{dt^2} + \frac{k_f + k_r}{m_s} z_s = 0 \qquad \text{and} \qquad \frac{d^2 \theta}{dt^2} + \frac{k_f L_f^2 + k_r L_r^2}{I_s} \theta = 0$$

These equations yield the two natural circular frequencies corresponding to bounce movement and pitch motion, respectively,

$$\omega_{ns} = \sqrt{\frac{k_f + k_r}{m_s}}$$

and

$$\omega_{n\theta} = \sqrt{\frac{k_f L_f^2 + k_r L_r^2}{I_s}}$$

The bounce oscillation center in this case is located at infinity, and the pitch oscillation center coincides with the sprung mass center of gravity.

Many people find pitch motion to be p a r t i c u l a r l y uncomfortable. So, when human perception of ride is considered, we endeavor to design the suspension such that the pitch oscillation center will be as far away from the center of gravity as possible. Therefore, the circumstances arising from the previous suspension specification result in an inferior ride.

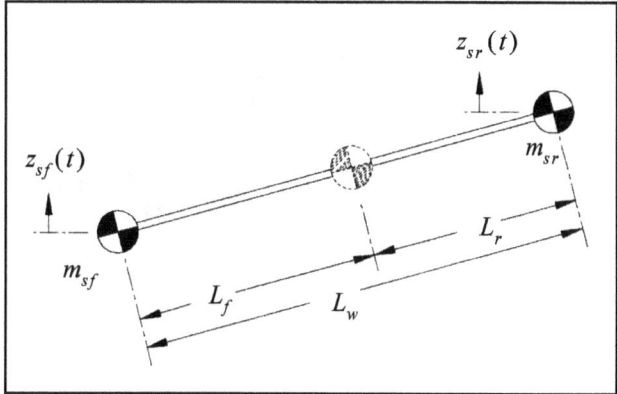

Figure (4.14): FRONT AND REAR AXLE PLANE COORDINATE SYSTEM

Before we examine the second design specification we will momentarily divert our attention to an alternate method of describing bounce and pitch motion. In the next section we will find it convenient to express the equations of motion, i.e., Eqs (4.71) and (4.72), in terms of (z_{sf}, z_{sr}) coordinates. As shown in Figure (4.14) these coordinates are the vertical movement of proportioned sprung mass at the front and rear axle planes.

The coordinate transformation to the (z_s, θ) system is derived from the geometric relationships

$$z_s = \frac{L_r z_{sf} + L_f z_{sr}}{L_w} \tag{4.87}$$

and

$$\sin\theta = \frac{z_{sr} - z_{sf}}{L_w}$$

or, using a small angle approximation

$$\theta = \frac{z_{sr} - z_{sf}}{L_w} \tag{4.88}$$

Expressions for the second derivative terms, which are required in the differential equations of motion, follow directly from the previous relationships, that is

$$\frac{d^2 z_s}{dt^2} = \frac{L_r}{L_w} \frac{d^2 z_{sf}}{dt^2} + \frac{L_f}{L_w} \frac{d^2 z_{sr}}{dt^2}$$

(4.89)

and

$$\frac{d^2 \theta}{dt^2} = \frac{1}{L_w} \frac{d^2 z_{sr}}{dt^2} - \frac{1}{L_w} \frac{d^2 z_{sf}}{dt^2}$$

(4.90)

Placing Eqs (4.87) through (4.90) into Eqs (4.71) and (4.72) results in

$$m_s \frac{L_r}{L_w} \frac{d^2 z_{sf}}{dt^2} + m_s \frac{L_f}{L_w} \frac{d^2 z_{sr}}{dt^2} + k_f z_{sf} + k_r z_{sr} = 0$$

(4.91)

and

$$\frac{I_s}{L_w} \frac{d^2 z_{sr}}{dt^2} - \frac{I_s}{L_w} \frac{d^2 z_{sf}}{dt^2} + k_r L_r z_{sr} - k_f L_f z_{sf} = 0$$

(4.92)

In order to complete the transformation of the differential equations it is necessary to express m_s and I_s in terms of sprung mass components which act at the front and rear axle planes. The basic relationships utilizing the sprung mass components are:

$$m_s = m_{sf} + m_{sr}$$

(4.93)

and

$$I_s = L_f^2 m_{sf} + L_r^2 m_{sr}$$

(4.94)

Some simplification can be achieved by introducing the "dynamic index", q, which we will discover has vehicle design implications. It is define as

$$q = \frac{\rho^2}{L_f L_r}$$

(4.95)

Substituting Eqs (4.95) and (4.69) into Eqs (4.93) and (4.94), along with a bit of algebraic manipulation, provides us with

$$m_s = \frac{L_w(L_f - L_r)}{L_r(qL_f - L_r)} m_{sf} \tag{4.96a}$$

$$= \frac{L_w(L_r - L_f)}{L_f(qL_r - L_f)} m_{sr} \tag{4.96b}$$

and

$$I_s = qL_r \frac{(L_r - L_f)}{qL_r - L_f} L_w m_{sr} \tag{4.97a}$$

$$= qL_f \frac{(L_f - L_r)}{qL_f - L_r} L_w m_{sf} \tag{4.97b}$$

Placing Eq (4.96) into Eq (4.91) yields

$$\left[\frac{L_r - L_f}{qL_r - L_f} m_{sr} \frac{d^2 z_{sr}}{dt^2} + k_r z_{sr} \right] + \left[\frac{L_f - L_r}{qL_f - L_r} m_{sf} \frac{d^2 z_{sf}}{dt^2} + k_f z_{sf} \right] = 0$$

$$\tag{4.98}$$

and

$$L_r \left[\frac{L_r - L_f}{qL_r - L_f} m_{sr} \frac{d^2 z_{sr}}{dt^2} + k_r z_{sr} \right] - L_f \left[\frac{L_f - L_r}{qL_f - L_r} m_{sf} \frac{d^2 z_{sf}}{dt^2} + k_f z_{sf} \right] = 0$$

$$\tag{4.99}$$

The transformation process is now complete; we have two equations with two unknown functions, i.e., the vertical movement at the front axle plane and the vertical movement at the rear axle plane.

The second suspension specification we wish to examine is q = 1. In this case Eqs (4.98) and (4.99) become uncoupled and reduce to the independent set of equations

$$m_{sf} \frac{d^2 z_{sf}}{dt^2} + k_f z_{sf} = 0 \tag{4.100}$$

and

$$m_{sr} \frac{d^2 z_{sr}}{dt^2} + k_r z_{sr} = 0 \tag{4.101}$$

We realize that since these differential equations are independent of each other, the response of the front suspension to a road input has no affect on the rear suspension, and vice versa. This condition was first recognized by Maurice Olley [10] as being a desirable design criterion to achieve a quality ride. The oscillation centers in this case coincide with the concentrated sprung mass points at the front and rear axle planes.

Actual dynamic index values range from around 0.8 for rear engine sports cars through 0.9 to 1.0 for modern production cars, to approximately 1.2 for some front wheel drive cars [6].

The differential equations do not mean, however, that bounce and pitch modes are uncoupled. Pitch motion can be controlled to a certain extent by specifying a rear suspension natural frequency value higher

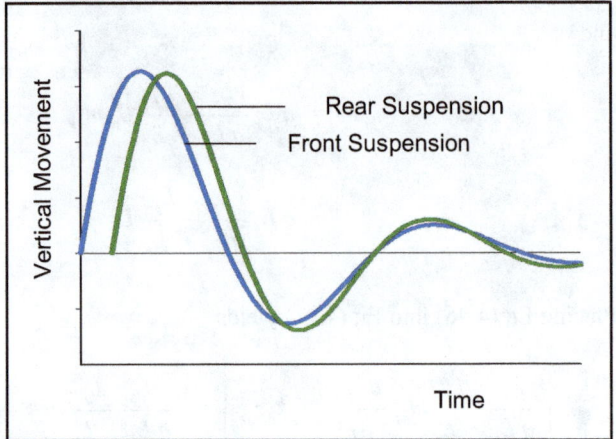

Figure (4.15): EXAMPLE SUSPENSION SYSTEM RESPONSE TO A ROAD STEP INPUT

than the adjoining front suspension. This point can be demonstrated by considering what happens to a vehicle encountering a sudden rise in the road pavement. The front and rear suspensions when modeled as simple damped second order systems have a response to a step input as shown in Figure (4.15). We notice: (1) that there is a short delay in the response of the rear suspension due to the time it takes for the rear suspension to travel to the rise in the road surface, and (2) that the front and rear of the vehicle move together after about one and a half oscillations. The general rule of thumb for suspension design is to have a rear suspension natural frequency of around 1.1 to 1.3 time greater than the front suspension natural frequency. Although, it should be realized that the actual required value is very dependent on the vehicle wheelbase and the forward design velocity.

(4.2.3) IMPEDANCE METHOD OF VEHICLE MODELING

The material presented Section (4.2.2) discussed modeling sprung and unsprung mass bounce motion, and also sprung mass bounce with pitch motion. We might ask ourselves if it is possible to develop a model which combines both sprung and unsprung mass vertical movement with pitch motion. Of course, since we have a section devoted to this subject, the answer is yes! An effective means to generate steady-state ride characteristics, without the use of differential equations, is impedance modeling.

The concept of impedance modeling and complex algebra is a well known tool in analyzing the output of RLC electronic circuits subjected to a sinusoidal input. The same logic can be applied to a vehicle suspension if the suspension is represented as an

Table (4.2): CROSS-REFERENCE OF VEHICLE AND ELECTRICAL ANALOG ELEMENTS

Vehicle System	Electric Analog	Impedance
mass, m:	inductor, $L = m$:	$j\omega m$
linear damper, b:	resistor, $R = b$:	b
linear spring, k:	capacitor, $C = \dfrac{1}{k}$:	$\dfrac{k}{j\omega}$
displacement, $z(t)$	current, $I(t)$	

analogous electrical network. The goal of impedance modeling is to develop a frequency dependent "gain" function which simulates the steady-state suspension response to the road surface vertical profile.

The first step in developing an impedance model is to construct an electric circuit analogous to the mechanical system representing the vehicle. The general scheme is to replace the sprung and unsprung masses, and suspension components with suitable electrical elements. However, the electrical elements must be arranged in a circuit such that the current flow through the circuit symbolizes displacement in the mechanical counterpart. This is important because there is generally not a one-to-one correspondence in the series and parallel arrangement of elements between the mechanical and electrical models. It should also be noted that a voltage drop across an electrical element represents a net applied force on the mechanical element. A cross-reference of vehicle components, electrical elements and their respective impedances is given in Table (4.2).

The road surface is treated as a current generator in the electric circuit. Furthermore, for the purpose of developing an impedance model, the road surface must be mathematically modeled as a sinusoid containing a single fundamental frequency. The road surface model is exactly the same as the model used in Section (4.2.1), i.e.,

$$z_r = Z_r \sin(\omega t) \tag{4.11}$$

However, the frequency and amplitude of the road surface are left as variables.

The last step in the impedance model development process is to create an impedance circuit. The impedance circuit is essentially the same as the electric circuit with a minor variation. The impedance circuit is obtained by exchanging each electrical component for the appropriate impedance listed in Table (4.2). The resulting circuit is by definition the impedance model of the vehicle.

Ride characteristics of the vehicle are obtained by analyzing the impedance circuit. The circuit analysis starts with finding a transfer function, which is mathematically defined as a ratio of the output current to the input current. The transfer function, $H(\omega)$, is frequency dependent due to the fact that the input to the circuit is sinusoidal. Since the impedance circuit is a representation of the vehicle, the transfer function is also a simulation of the sprung mass (or unsprung mass) movement caused by the road surface vertical profile. As it turns out, the nature of transfer functions are such that complex numbers are involved. Consequently, it is more useful to quantify the steady-state ride characteristics in terms of (1) "system gain," $G(\omega)$, or transmissibility ratio {refer to Section (4.2.1B)} which describes the frequency response, and (2) a phase shift angle, $\Phi(\omega)$. Both of these characteristics are derived from the real and imaginary parts of the transfer function:

$$G(\omega) \;=\; |H(\omega)| \;=\; \sqrt{\big[\,\mathbb{R}\{H(\omega)\}\,\big]^2 + \big[\,\mathbb{I}\{H(\omega)\}\,\big]^2} \tag{4.102}$$

and

$$\Phi(\omega) \;=\; \arctan\!\left(\frac{\mathbb{I}\{H(\omega)\}}{\mathbb{R}\{H(\omega)\}}\right) \tag{4.103}$$

The mathematical simulation of the vehicle vertical movement, as related to the road surface displacement, is given by the basic equation:

$$z(t) \;=\; Z_r\, G(\omega)\, \sin\big[\,\omega\, t + \Phi(\omega)\,\big] \tag{4.104}$$

However, from the suspension design point of view, it is usually sufficient to examine just the gain and phase shift characteristics to determine vehicle ride behavior.

As an example, suppose we wish to represent a particular vehicle as a single sprung mass affixed to a simple linear spring and shock absorber system. The unsprung suspension components are lumped together as a single mass supported on a single equivalent tire, i.e., a unicycle model. In an effort to obtain more accurate ride information a viscoelastic tire model [11] will be used. A mechanical schematic of the unicycle model is shown in Figure (4.16). The electrical system analogous to the mechanical system is shown in Figure (4.17). It bears repeating that the electric circuit is constructed such that displacement in the mechanical system is represented by current flow in the electrical system. For a moment, consider the suspension spring and the shock absorber. Although in Figure (4.16) they appear to be in a parallel arrangement, they must be placed in series for the electric analog. This is due to the fact that both respective ends of the suspension components experience the same displacement; therefore, the current flow through the corresponding electrical elements must be the same. This construction process is probably the most difficult aspect of impedance modeling, and care must be taken to ensure an accurate analog circuit. An impedance model representing the unicycle vehicle directly follows from the electrical circuit. As illustrated in Figure (4.17), each electrical component in the electrical circuit has been replaced by an appropriate labeled impedance block.

Before proceeding any further it is convenient to divide the impedance circuit schematic up into smaller, more manageable, functional blocks. This philosophy is used regardless of the mechanical system being used to model the vehicle. The advantage in separating out functional blocks is that each block can be independently analyzed. Thisgreatly simplifies the mathematical equations. Each functional block can have any number of spring and damper elements, but contains at most one mass element, designated as either sprung or unsprung. In our example the functional blocks are bounded by the dashed lines shown in Figure (4.17b).

The analysis concludes with a combination of results from each block, or system, to describe the ride characteristics of the vehicle, e.g.,

$$G(\omega) = G_s(\omega) G_u(\omega) \qquad (4.105)$$

and

$$\Phi(\omega) = \Phi_s(\omega) + \Phi_u(\omega) \qquad (4.106)$$

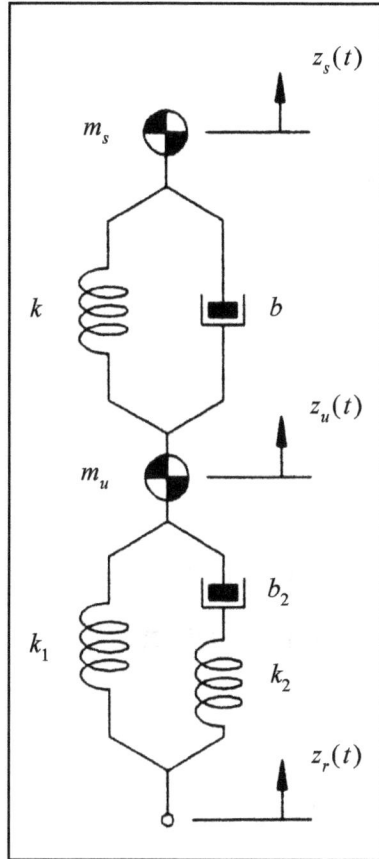

Figure (4.16): MECHANICAL SCHEMATIC OF VEHICLE AND VISCOELASTIC TIRE

The following analyses describe the specific steps required to develop sprung and unsprung gain equations for the unicycle model under consideration:

(i) Sprung Mass System

The output current from the sprung system functional block is found by applying a current divider relationship, in terms of impedances, to the unsprung system input current. The sprung system transfer function can then be found from:

$$H_s(\omega) = \frac{I_s}{I_u} = \frac{Z_b + Z_k}{Z_{m_s} + Z_b + Z_k} \qquad (4.107)$$

Substitution of the appropriate impedances from Table (4.2) into Eq (4.107) produces:

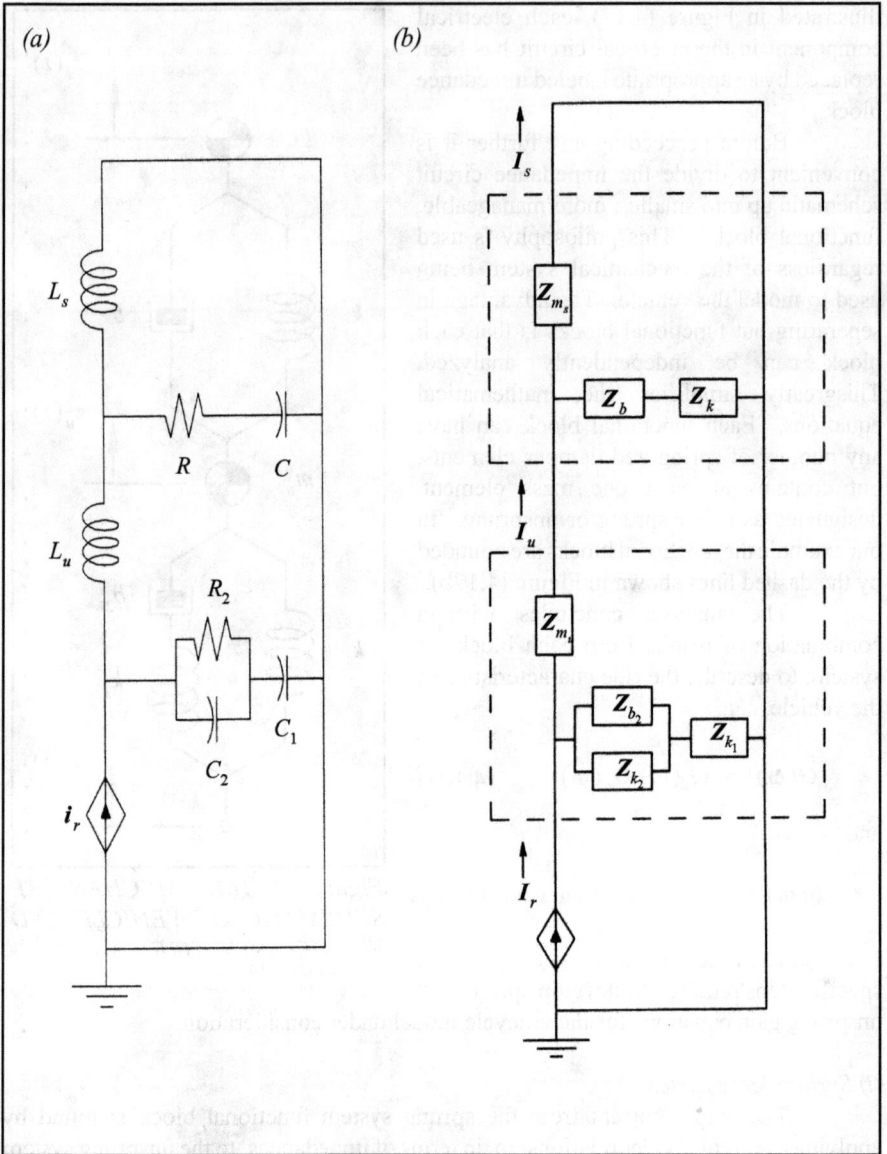

Figure (4.17): MECHANICAL SYSTEM ANALOG: (a) ELECTRIC CIRCUIT, and (b) IMPEDANCE CIRCUIT

$$H_s(\omega) = \frac{b + \dfrac{k}{j\omega}}{j\omega m_s + b + \dfrac{k}{j\omega}}$$ (4.108)

The previous equation can also be expressed in terms of a sprung system natural frequency, ω_{ns}, and a sprung damping ratio, ζ_s. Using these terms Eq (4.108) becomes

$$H_s(\omega) = \frac{j\omega \dfrac{2\zeta_s}{\omega_{ns}} + 1}{-\dfrac{\omega^2}{\omega_{ns}^2} + j\omega \dfrac{2\zeta_s}{\omega_{ns}} + 1} \qquad (4.109)$$

The transfer function relationship of Eq (4.109) can be slightly simplified by utilizing the dimensionless frequency ratio previously defined in Section (4.2.1A), i.e., $\omega_{rs} = \dfrac{\omega}{\omega_{ns}}$. The transfer function for the sprung mass system can then be rewritten as

$$H_s(\omega_{rs}) = \frac{1 + j2\zeta_s\omega_{rs}}{1 - \omega_{rs}^2 + j2\zeta_s\omega_{rs}} \qquad (4.110)$$

or, separated into real and imaginary parts as

$$H_s(\omega_{rs}) = \frac{\left(1 - \omega_{rs}^2\right) + \left(2\zeta_s\omega_{rs}\right)^2}{\left(1 - \omega_{rs}^2\right)^2 + \left(2\zeta_s\omega_{rs}\right)^2} + j\frac{-2\zeta_s\omega_{rs}^3}{\left(1 - \omega_{rs}^2\right)^2 + \left(2\zeta_s\omega_{rs}\right)^2}$$

$$(4.111)$$

The sprung system gain is determined from the absolute value of the transfer function as given by Eq (4.102):

$$G_s(\omega_{rs}) = \sqrt{\frac{1 + \left(2\zeta_s\omega_{rs}\right)^2}{\left(1 - \omega_{rs}^2\right)^2 + \left(2\zeta_s\omega_{rs}\right)^2}} \qquad (4.112)$$

At this point we should note that Eq (4.112) is identical to the transmissibility ratio equation, i.e., Eq (4.48), derived for the simple damped vehicle model used in Section (4.2.1B). Although different suspension modeling approaches were used, we expect the results to be the same since the mechanical model of a single spring and a single damper is the same in both instances.

The phase angle of the sprung system is found by dividing the imaginary part of the transfer function by the real part as indicated by Eq (4.103):

$$\Phi_s(\omega_{rs}) = \arctan\left[\frac{-2\zeta_s\omega_{rs}^3}{\left(1 - \omega_{rs}^2\right)^2 + \left(2\zeta_s\omega_{rs}\right)^2}\right] \qquad (4.113)$$

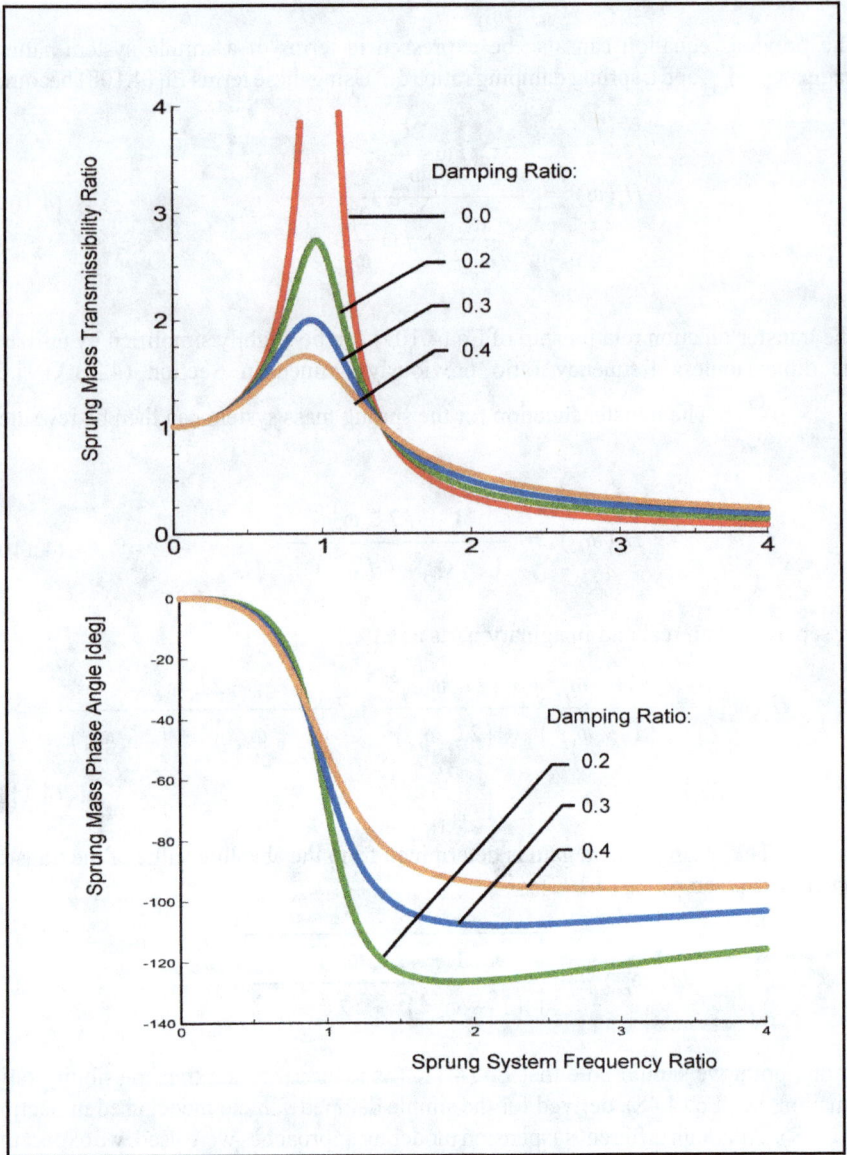

Figure (4.18): SPRUNG MASS SYSTEM RESPONSE

Again the result is the same as we found earlier. Graphic examples of Eqs (4.112) and (4.113) are shown in Figure (4.18).

(ii) Unsprung System Analysis

The same basic steps used in the foregoing analysis are followed to arrive at the unsprung system equations. The transfer function for the unsprung system is developed from

$$H_u(\omega) = \frac{I_u}{I_r} = \frac{Z_{k_1} + Z_{b_2} \parallel Z_{k_2}}{Z_{k_1} + Z_{b_2} \parallel Z_{k_2} + Z_{m_u}} \qquad (4.114)$$

Placing appropriate impedances from Table (4.2) into Eq (4.114) leads to the relationship

$$H_u(\omega_{ru}) = \frac{1 + k_r - \dfrac{j k_r}{2 \zeta_u \omega_{ru}}}{1 + k_r - \dfrac{j k_r}{2 \zeta_u \omega_{ru}} - \omega_{ru}^2 \left(1 - \dfrac{j k_r}{2 \zeta_u \omega_{ru}}\right)} \qquad (4.115)$$

where ω_{ru} is the unsprung mass system frequency ratio, ζ_u is the unsprung system damping ratio, and k_r is a tire stiffness ratio parameter. The mathematic definitions are

$$\omega_{ru} = \frac{\omega}{\omega_{nu}}, \quad \omega_{nu} = \sqrt{\frac{k_1}{m_u}}, \quad \zeta_u = \frac{b_2}{b_{crit,u}}, \quad b_{crit,u} = 2 m_u \omega_{nu}, \text{ and } k_r = \frac{k_2}{k_1}. \text{ After}$$

a bit of algebraic manipulation of Eq (4.115), the real and imaginary parts of the unsprung system transfer function can be found. They are, respectively,

$$\mathbb{R}\{H_u(\omega_{ru})\} = \frac{k_r^2\left(1 - \omega_{ru}^2\right) + \left(2 \zeta_u \omega_{ru}\right)^2 \left(1 + k_r\right)\left(1 + k_r - \omega_{ru}^2\right)}{k_r^2\left(1 - \omega_{ru}^2\right)^2 + \left(2 \zeta_u \omega_{ru}\right)^2 \left(1 + k_r - \omega_{ru}^2\right)^2} \qquad (4.116)$$

and

$$\mathbb{I}\{H_u(\omega_{ru})\} = \frac{-2 k_r^2 \zeta_u \omega_{ru}^3}{k_r^2\left(1 - \omega_{ru}^2\right)^2 + \left(2 \zeta_u \omega_{ru}\right)^2 \left(1 + k_r - \omega_{ru}^2\right)^2} \qquad (4.117)$$

We will not attempt to derive a concise equation for the unsprung system gain. Numeric values for the unsprung system gain are found by calculating the real and imaginary parts of the transfer function from Eqs (4.116) and (4.117), and then substituting these values into Eq (4.102). In addition, we can compute the unsprung mass system phase shift angle from the real and imaginary values via Eq (4.103). Sample unsprung system results are plotted in Figure (4.19).

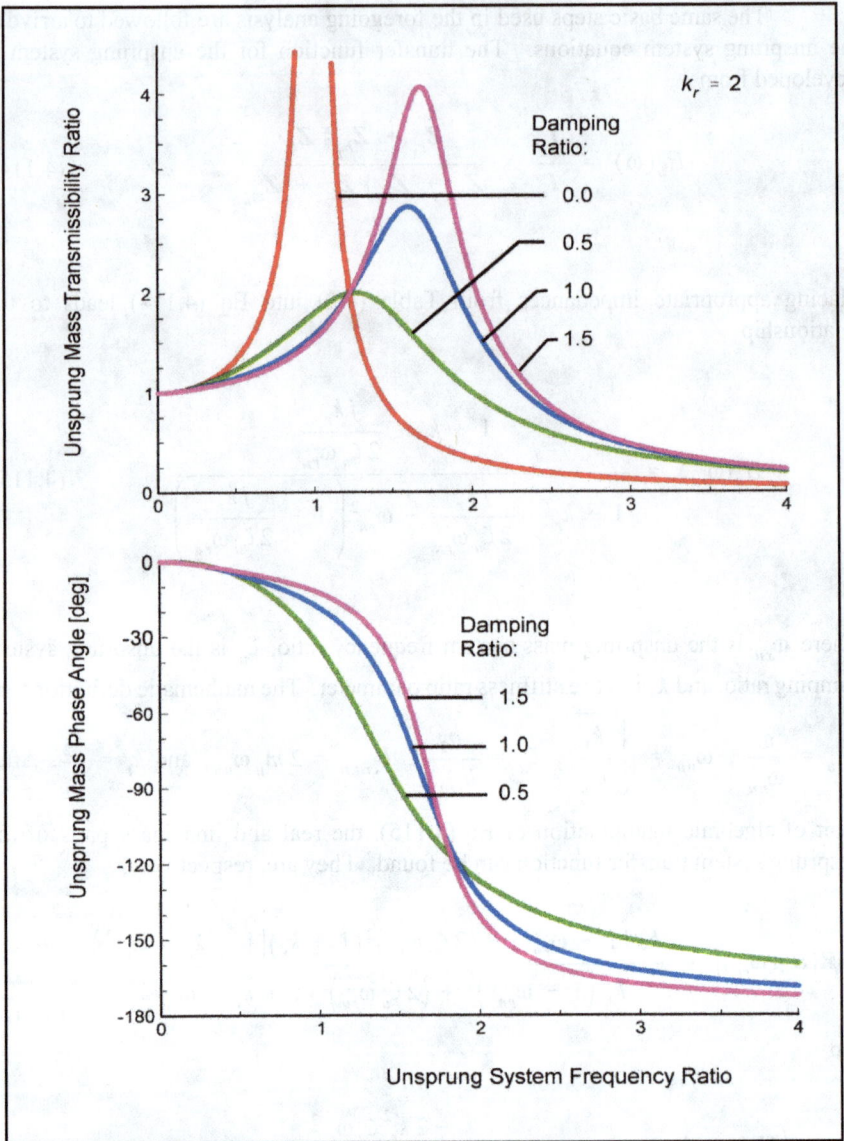

Figure (4.19): UNSPRUNG MASS SYSTEM RESPONSE

The unsprung system frequency response plot has an interesting characteristic. As the damping increases there is a shift in the natural frequency. This is to be expected because in the limit of infinite damping the unsprung system degenerates into two springs in parallel, which can be further reduced into a single equivalent spring. The original spring constant and the infinitely damped equivalent spring constant yield two distinct natural frequencies.

Summarizing the analysis thus far, we have been able to derive gain and phase relationships for the two functional blocks which comprise the unicycle vehicle model. As indicated by Eq (4.105), the sprung and unsprung system gains can be multiplied together to define the gain for the vehicle. However, traditionally the vehicle gain is calculated with respect to the sprung system frequency ratio. This requires a modification of the unsprung system independent variable. The unsprung system frequency ratio can be expressed in terms of the sprung system frequency ratio as follows:

$$\omega_{ru} = W \omega_{rs} \tag{4.118}$$

where

$$W = \frac{\omega_{ns}}{\omega_{nu}} = \sqrt{\frac{k}{k_1} \frac{1}{m_r}}$$

Combining the results of the sprung and unsprung system with Eq (4.118) produces the unicycle vehicle gain relationship, i.e.,

$$G(\omega_{rs}) = G_s(\omega_{rs}) G_u(\omega_{rs})$$

Similarly the unicycle vehicle system phase shift angle, Eq (4.106), becomes

$$\Phi(\omega_{rs}) = \Phi_s(\omega_{rs}) + \Phi_u(\omega_{rs})$$

A sample response plot in shown in Figure (4.20). The parametric conditions for which the plots were generated are also given in the figure.

A fairly obvious extension of the unicycle model is to apply the unicycle gain and phase relationships to the front and rear suspensions as two individual, coupled models. If we consider a front suspension application, the sprung mass displacement equation originally given by Eq (4.104) is modified to represent the vertical movement of the sprung mass component acting in the front axle plane, i.e.,

$$z_{sf}(t) = Z_r G_{sf}(\omega) G_{uf}(\omega) \sin\left[\omega t + \Phi_{sf}(\omega) + \Phi_{uf}(\omega)\right] \tag{4.119}$$

We can write a similar expression for the rear axle plane, except we include a time delay, t_d, relative to the front axle plane. The delay in the rear sprung mass response accounts for the time it takes for the rear wheels to traverse a section of road equivalent to the vehicle wheelbase to assume the former position of the front wheels. The modified equation describing rear sprung mass movement is

$$z_{sr}(t) = Z_r G_{sr}(\omega) G_{ur}(\omega) \sin\left[\omega (t - t_d) + \Phi_{sr}(\omega) + \Phi_{ur}(\omega)\right] \tag{4.120}$$

A mathematical combination of the front and rear sprung mass vertical motion enables us to generate a bicycle vehicle ride model. However, a requisite in obtaining

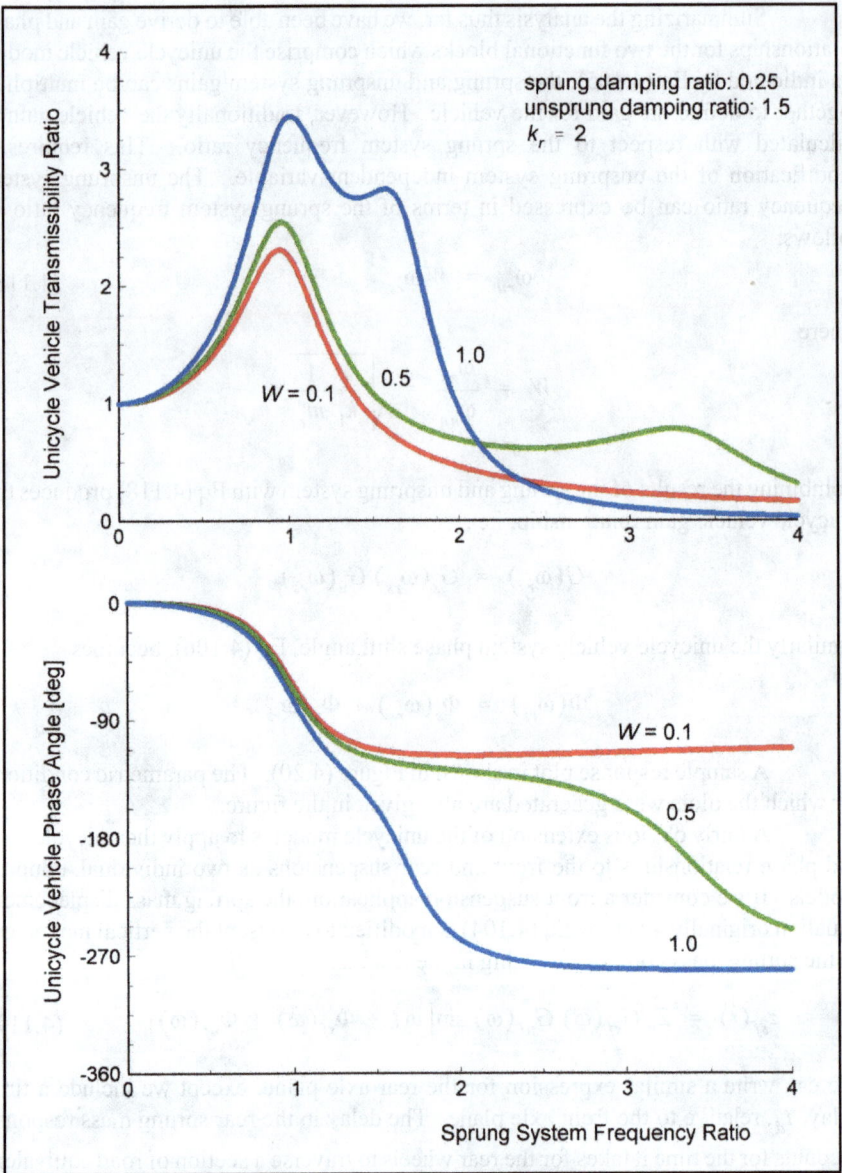

Figure (4.20): UNICYCLE VEHICLE SYSTEM RESPONSE

an accurate representation of bounce and pitch magnitudes is that the dynamic index must be close to unity. As you may recall, when the dynamic index is equal to one the rear suspension motion has no influence on the front, and vice versa. Since a mathematical mechanism does not exist to introduce effects of the opposite suspension into the unicycle impedance model, we are forced to place restrictions on the value of the dynamic index. Under this restriction the bounce motion of the vehicle sprung mass center of gravity, derived from Eq (4.87), is

$$z_s(t) = \frac{L_f}{L_w} z_{sf}(t) + \frac{L_r}{L_w} z_{sr}(t)$$

or

$$z_s(t) = Z_r \frac{L_f}{L_w} G_{sf}(\omega) G_{uf}(\omega) \sin\left[\omega t + \Phi_{sf}(\omega) + \Phi_{uf}(\omega)\right]$$

$$+ Z_r \frac{L_r}{L_w} G_{sr}(\omega) G_{ur}(\omega) \sin\left[\omega (t - t_d) + \Phi_{sr}(\omega) + \Phi_{ur}(\omega)\right]$$

$$(4.121)$$

and the pitch motion, derived from Eq (4.88), is

$$\theta(t) = \frac{1}{L_w} z_{sr}(t) - \frac{1}{L_w} z_{sf}(t)$$

or

$$\theta(t) = \frac{Z_r}{L_w} G_{sr}(\omega) G_{ur}(\omega) \sin\left[\omega (t - t_d) + \Phi_{sr}(\omega) + \Phi_{ur}(\omega)\right]$$

$$- \frac{Z_r}{L_w} G_{sf}(\omega) G_{uf}(\omega) \sin\left[\omega t + \Phi_{sf}(\omega) + \Phi_{uf}(\omega)\right]$$

$$(4.122)$$

In each of the previous equations the summation of the sine terms prevents us from developing a gain function. A typical approach to circumvent this problem is to examine the root-mean-square (RMS) value of bounce and pitch over a specified time interval. The RMS value of the bounce motion is determined from

$$\overline{Z}_s = \left\{ \frac{1}{\tau} \int_0^\tau z_s(t)^2 \, dt \right\}^{\frac{1}{2}} \tag{4.123}$$

Substituting Eq (4.121) into Eq (4.123) results in

$$\bar{Z}_s = \left\{ \frac{1}{\tau} \int_0^\tau Z_r^2 \left(\frac{L_f}{L_w} \right)^2 G_{sf}^2 \, G_{uf}^2 \, \sin^2 \left[\omega t + \Phi_{sf} + \Phi_{uf} \right] dt \right.$$

$$+ \frac{2}{\tau} \int_0^\tau Z_r^2 \frac{L_f L_r}{L_w^2} G_{sf} G_{uf} G_{sr} \, G_{ur} \, \sin \left[\omega t + \Phi_{sf} + \Phi_{uf} \right] \sin \left[\omega \left(t + t_d \right) + \Phi_{sr} + \Phi_{ur} \right] dt$$

$$\left. \frac{1}{\tau} \int_0^\tau Z_r^2 \left(\frac{L_r}{L_w} \right)^2 G_{sr}^2 \, G_{ur}^2 \, \sin^2 \left[\omega \left(t + t_d \right) + \Phi_{sr} + \Phi_{ur} \right] dt \right\}^{\frac{1}{2}}$$

$$= \left[\left| \frac{1}{\tau \omega} Z_r^2 \left(\frac{L_f}{L_w} \right)^2 G_{sf}^2 \, G_{uf}^2 \left\{ \frac{1}{2} \left(\omega t + \Phi_{sf} + \Phi_{uf} \right) - \frac{1}{4} \sin \left[2 \left(\omega t + \Phi_{sf} + \Phi_{uf} \right) \right] \right\} \right| _0^\tau \right.$$

$$+ \frac{2}{\tau} Z_r^2 \frac{L_f L_r}{L_w^2} G_{sf} G_{uf} G_{sr} G_{ur} \left\{ \frac{t}{2} \cos \left(\Phi_{sf} + \Phi_{uf} - \Phi_{sr} - \Phi_{ur} + \omega t_d \right) \right.$$

$$\left. \left. - \frac{1}{4\omega} \sin \left(2\omega t + \Phi_{sf} + \Phi_{uf} + \Phi_{sr} + \Phi_{ur} - \omega t_d \right) \right\} \right| _0^\tau$$

$$\frac{1}{\tau \omega} Z_r^2 \left(\frac{L_r}{L_w} \right)^2 G_{sr}^2 \, G_{ur}^2 \left\{ \frac{1}{2} \left(\omega t - \omega t_d + \Phi_{sr} + \Phi_{ur} \right) \right.$$

$$\left. \left. \left. - \frac{1}{4} \sin \left[2 \left(\omega t - \omega t_d + \Phi_{sr} + \Phi_{ur} \right) \right] \right\} \right| _0^\tau \right]^{\frac{1}{2}}$$

Or, after evaluating the limits of integration

$$
\frac{\bar{Z}_s}{Z_r} = \left\{ \left(\frac{L_f}{L_w} \right)^2 \frac{G_{sf}^2 \, G_{uf}^2}{2} \right.
$$

$$
+ \frac{L_f L_r}{L_w^2} G_{sf} G_{uf} G_{sr} G_{ur} \cos\left(\Phi_{sf} + \Phi_{uf} - \Phi_{sr} - \Phi_{ur} + \omega t_d \right)
$$

$$
\left. + \left(\frac{L_r}{L_w} \right)^2 \frac{G_{sr}^2 \, G_{ur}^2}{2} \right\}^{\frac{1}{2}}
$$

(4.124)

The RMS value of pitch rotational displacement can be found through a similar integration process to be:

$$
\frac{\bar{\Theta} L_w}{Z_r} = \left\{ \frac{G_{sr}^2 \, G_{ur}^2}{2} \right.
$$

$$
- G_{sr} G_{ur} G_{sf} G_{uf} \cos\left(\Phi_{sr} + \Phi_{ur} - \Phi_{sf} - \Phi_{uf} - \omega t_d \right)
$$

$$
\left. + \frac{G_{sf}^2 \, G_{uf}^2}{2} \right\}^{\frac{1}{2}}
$$

(4.125)

We can promote a better understanding of the application of a bicycle ride model in determining vehicle ride characteristics by working through an example. The vehicle in this case is summarized according to the specifications:

Vehicle Data
    2200 lb total weight (45% / 55% front to rear static distribution)
    108 in   wheelbase

    Independent Front Suspension:
            unsprung weight   176 lb
            combined effective spring stiffness 113.3 lb/in
            combined effective damping   81.6 lb/fps

    Solid Axle Rear Suspension:
            unsprung weight   286 lb
            combined effective spring stiffness 147.7 lb/in
            combined effective damping   135.3 lb/fps

Tire Parametric Data
      $k1 = 825$ lb/in
      $k2 = 1650$ lb/in

$$b2 = 495 \text{ lb/fps}$$

Forward Design Velocity
60 MPH (88 fps)

Before we begin we will need to know the location of the sprung mass center of gravity. Based on the static weight distribution, the sprung mass at the front and rear axle plane is, respectively

$$m_{sf} = (0.45)(2200) - 176 = 814 \text{ lb}$$

and

$$m_{sr} = (0.55)(2200) - 286 = 924 \text{ lb}$$

Then, the location of the sprung mass center of gravity from the front axle plane is

$$L_f = \left(\frac{m_{sr}}{m_{sf} + m_{sr}} \right) L_w = \left(\frac{924}{814 + 924} \right) (108) = 57.42 \text{ in}$$

Simple subtraction from the wheelbase yields the rear distance: $L_r = 50.28$ in. We will also need the time delay, which is easily calculated from

$$t_d = \frac{L_w}{V} = \frac{108/12}{88} = 0.1023 \text{ s}$$

Using the previous information, we can generate ride characteristics in terms of RMS bounce and pitch amplitudes as a function of road frequency from Eqs (4.124) and (4.125). The results are plotted in Figures (4.21) and (4.22). In an effort to make a compact graph, $\dfrac{\bar{Z}_s}{Z_r}$ is defined as the bounce transmissibility ratio, and $\dfrac{\bar{\Theta} L_w}{Z_r}$ is the pitch transmissibility ratio. Also, the bounce frequency ratio is arbitrarily defined as $\dfrac{\omega}{\omega_{ns,f}}$, and the pitch frequency ratio is $\dfrac{\omega}{\omega_{n\theta}}$. {Refer to Section

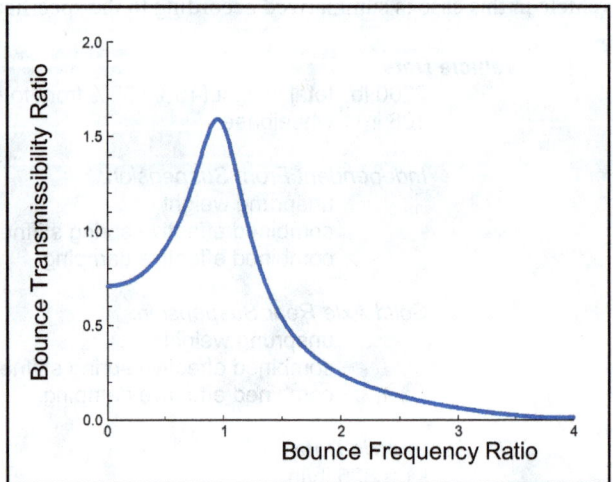

Figure (4.21): EXAMPLE RMS BOUNCE RESPONSE

(4.2.2B) for natural frequency expressions.} We have finally come to the point where we want to interpret the results. To satisfy our desire we must learn more about the vehicle suspension reaction to general road surface conditions.

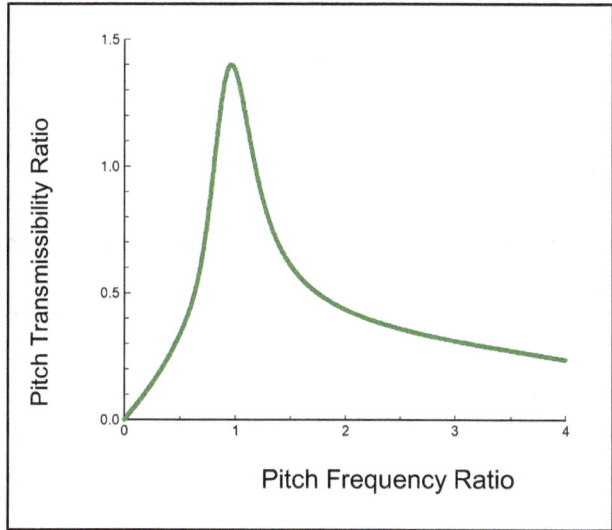

Figure (4.22): EXAMPLE RMS PITCH RESPONSE

(4.3) RIDE EVALUATION

Thus far in our study of ride characteristics we have invested a considerable amount of time developing vehicle ride models for various mass and suspension component configurations. From these models we have been able to ascertain the effect of design parameters, such as the sprung mass natural frequency, on vehicle bounce and pitch movement. However, the results of these models are limited due to the premise of modeling the vehicle as if it were traveling on a sinusoidal road surface of a single frequency. In addition, we have not been able to estimate the perception of ride quality. As most of you can imagine, a true sinusoidal surface exists only in theory, not reality. Rather than starting over with new models, we would obviously like to extend what we have already developed and learned to encompass more realistic road surface conditions. We can accomplish this goal in two phases. The first phase is to explore a semi-empirical road surface model based on random distribution of surface irregularities. The second phase is directed at combining previously developed steady-state ride models with the revised road model. As we shall see, the end result of the new ride model is one we can compare to ISO standards.

(4.3.1) MODELING OF ROAD SURFACES

The sinusoidal road profile model, previously used in developing vehicle ride models, does not provide a very accurate simulation of a true road surface. An actual road follows the general contour of the earth's surface and contains many small bumps and dips caused from the affects of thermal expansion and weathering. These irregularities in the road surface create a more or less random variation in vertical displacement, which we shall refer to as surface roughness. A sample road surface profile may appear as shown in Figure (4.23). Based on this information, a more preferable road simulation would appear to the vehicle suspension as a random vibrational input at the wheels. Our discussion suggests that in order to revamp the road profile model (pure sinusoidal profile), we need to quantify the surface roughness base on calculus of stochastic processes. Since many engineers are unfamiliar with the subject of stochastic processes, we will momentarily divert our attention from vehicle dynamics and introduce, or perhaps review, a few basic concepts.

The concept of modeling a road surface as random vibration presents an interesting challenge. The natural frequency of a random function is one that is nondeterministic. That is to say, in this case the road surface roughness at any particular location cannot be predicted; however, most random functions possess behavior repetition to some degree. To develop a random vibration road model from experimental evidence we make an assumption that the random distribution of bumps and dips in the actual road surface is a stationary, ergodic process. In essence this

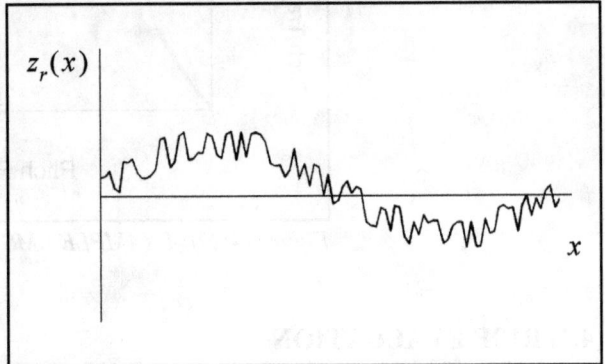

Figure (4.23): ROAD SURFACE SAMPLE

means that a small sample of road exhibits the same behavior as the entire length of road. For engineering purposes this is a good approximation. Although random functions are not periodic, they can be represented by a Fourier series in which the period approaches infinity. The Fourier series description of the sample road surface shown in Figure (4.23) is:

$$z_r(x) = \frac{A_0}{2} + \sum_{i=1}^{\infty} A_i \cos(\Omega_i x) + \sum_{i=1}^{\infty} B_i \sin(\Omega_i x)$$

where Ω_i is a discrete spatial frequency defined as $\dfrac{2\pi}{L_i}$. In this manner the random vibration becomes a compilation of individual wavelengths and amplitudes which approaches a continuous spectrum. We will soon find that there is considerable mathematic advantage in representing the Fourier series in terms of Euler's exponential function

$$e^{j\Omega_i x} = \cos(\Omega_i x) + j \sin(\Omega_i x)$$

which leads to the trigonometric identities

$$\cos(\Omega_i x) = \frac{1}{2}\left(e^{j\Omega_i x} + e^{-j\Omega_i x}\right) \quad \text{and} \quad \sin(\Omega_i x) = -\frac{1}{2}\left(e^{j\Omega_i x} - e^{-j\Omega_i x}\right)$$

An equivalent description of the road surface using these identities [7] is

$$z_r(x) = \mathbb{R}\left\{ \sum_{i=1}^{\infty} Z_{r,i}\, e^{\,j\,\Omega_i\, x} \right\}$$

$$= \frac{1}{2} \sum_{i=1}^{\infty} \left(Z_{r,i}\, e^{\,j\,\Omega_i\, x} + Z_{r,i}^{*}\, e^{\,-j\,\Omega_i\, x} \right)$$

The coefficient $Z_{r,i}$ is determined from $Z_{r,i} = \frac{1}{2}\left(A_i - j\,B_i \right)$, and $Z_{r,i}^{*}$ is it's complex conjugate.

The statistical analysis of vibration, and eventually ride perception, is concerned mainly with the average energy of the random function, which in our case is the road surface. The average energy is related to the mean square value of the random function defined by

$$\overline{Y}^2 = \lim_{X \to \infty} \left\{ \frac{1}{X} \int_{-\frac{X}{2}}^{\frac{X}{2}} [\,Y(x)\,]^2 \, dx \right\}$$

An expression for the road surface mean square value is found by substituting it's exponential description into the previous integral equation. The evaluation details of the integration process are given below. While deriving the mean square value we must remember that the exponential terms are actually trigonometric functions and, consequently, they are bounded. Thus, when we take the limit as X approaches infinity in the last step, the exponential terms vanish.

Step (i):

$$\overline{z}_r^2 = \lim_{X \to \infty} \left\{ \frac{1}{X} \int_{-\frac{X}{2}}^{\frac{X}{2}} \left[\frac{1}{2} \sum_{i=1}^{\infty} \left(Z_{r,i}\, e^{\,j\,\Omega_i\, x} + Z_{r,i}^{*}\, e^{\,-j\,\Omega_i\, x} \right) \right]^2 dx \right\}$$

Step (ii):

$$\overline{z}_r^2 = \lim_{X \to \infty} \left\{ \frac{1}{4} \sum_{i=1}^{\infty} \frac{1}{X} \int_{-\frac{X}{2}}^{\frac{X}{2}} Z_{r,i}^2\, e^{\,2j\,\Omega_i\, x}\, dx \right.$$

$$\left. + \frac{1}{4} \sum_{i=1}^{\infty} \frac{1}{X} \int_{-\frac{X}{2}}^{\frac{X}{2}} 2\, Z_{r,i}\, Z_{r,i}^{*}\, dx + \frac{1}{4} \sum_{i=1}^{\infty} \frac{1}{X} \int_{-\frac{X}{2}}^{\frac{X}{2}} Z_{r,i}^{*\,2}\, e^{\,-2j\,\Omega_i\, x}\, dx \right\}$$

and Step (iii):

$$\bar{z}_r^2 = \lim_{X \to \infty} \left\{ \frac{1}{4} \sum_{i=1}^{\infty} \left(\frac{Z_{r,i}^2 \, e^{2j\Omega_i x}}{2j\Omega_i X} + 2 Z_{r,i} Z_{r,i}^* + \frac{Z_{r,i}^{*\,2} \, e^{-2j\Omega_i x}}{-2j\Omega_i X} \right) \right\}$$

$$= \sum_{i=1}^{\infty} \frac{1}{2} Z_{r,i} Z_{r,i}^*$$

We should note that a complex function multiplied by it's complex conjugate produces the same result as the absolute value of the complex function squared; therefore, and equivalent expression for the road surface mean square value is

$$\bar{z}_r^2 = \sum_{i=1}^{\infty} \frac{1}{2} |Z_{r,i}|^2$$

We conclude from this equation that the mean square value of the road surface random function is one-half the sum of the amplitude squared of each frequency in the spectrum.

In reality it is at best difficult to evaluate the mean square value for an infinite number of frequencies. Thus, we consider only a finite number of wavelengths or frequencies that compose the random function. The power spectrum, $P(\Omega_i)$, is defined as the mean square value centered at a specific frequency, Ω_i, evaluated over a frequency interval of $\Delta\Omega_i$. For instance, if we select a certain range of frequencies, the power spectrum plotted against frequency for a sample road surface might look similar to the bar graph shown in Figure (4.24). Since the power spectrum is determined over a finite frequency interval, it's value is dependent on the frequency bandwidth, i.e., a larger bandwidth produces a larger power

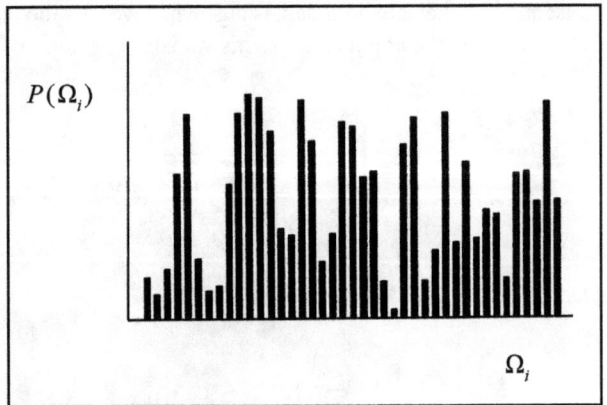

Figure (4.24): EXAMPLE POWER SPECTRUM

spectrum magnitude. We can alleviate this problem by dividing the power spectrum by the frequency interval. The result is defined as the discrete power spectrum density:

$$S(\Omega_i) = \frac{P(\Omega_i)}{\Delta\Omega}$$

If we combine definitions in reverse order the mean square value of the random function can be written in terms of the power spectrum, or spectral density, as

$$\overline{z}_r^2 = \sum_{i=1}^{\infty} S(\Omega_i) \Delta\Omega$$

When examining the limiting case of $\Delta\Omega \to 0$, a slightly different expression for the mean square value evolves

$$\overline{z}_r^2 = \int_0^{\infty} S(\Omega_i) \, d\Omega$$

The advantage of using power spectrum density to determine the mean square value is the relative ease by which experimental data can be procured from spectrum analyzers. Although the spectral density data experimentally obtained from each sample road surface is unique, all road surfaces tend to have the same general roughness characteristics. Investigators have been able to propose empirical correlations for road surface roughness from these trends. Some of the correlations which describe the spectral density is terms of a roughness coefficient, S_0, and wave number[5], N, are:

(*i*) Van Deusen [13]

$$S(N) = S_0 \, N^{-2.1}$$

where

$$N \, [\text{cycle/m}]$$

$$S_0 \, [\text{m}^2/\text{cycle/m}] = \begin{cases} 4.8 \times 10^{-7}, & \text{smooth highway} \\ 4.4 \times 10^{-6}, & \text{gravel road} \end{cases}$$

(*ii*) Dodds and Robson[6] [14]

---

[5] *The wave number is equivalent to the reciprocal of wavelength. It is almost universally used as a correlation parameter in empirically based road roughness spectral density relationships.*

[6] *The investigative work of Dodds and Robson was concerned mainly with road condition classification based on measured roughness spectral density. For example, a road surface with* $8 \le S_0 \times 10^6 \le 32 \, \text{m}^2/\text{cycle/m}$, $w_1 \approx 2.05$ *and* $w_2 \approx 1.440$ *would be classified as a "good principal road." They did not go so far as to make a design recommendation for vehicle ride.*

$$S(N) = \begin{cases} S_0 \left(\dfrac{N}{N_0} \right)^{-w_1} & ; N \leq N_0 \\[4mm] S_0 \left(\dfrac{N}{N_0} \right)^{-w_2} & ; N \geq N_0 \end{cases}$$

where $N_0 = \dfrac{1}{2\pi}$ cycle/m.

(*iii*) Gillespie [9]

$$S(N) = S_0 \frac{1 + \left(\dfrac{N_0}{N} \right)^2}{(2\pi N)^2}$$

where

N [cycle/ft]

$$S_0 \ [\text{ft}^2/\text{cycle/ft}] = \begin{cases} 1.25 \times 10^{-5} \text{ , rough surface} \\ 1.25 \times 10^{-6} \text{ , smooth surface} \end{cases}$$

$$N_0 \ [\text{cycle/ft}] = \begin{cases} 0.05 \text{ , asphalt road} \\ 0.02 \text{ , concrete pavement} \end{cases}$$

The foregoing correlations, along with a sample of roughness spectral data [15] for a "typical principle road" are plotted in Figure (4.25). It is somewhat difficult to draw a conclusion from these plots. One explanation for this dilemma is that researchers have attempted to develop a quantitative description for qualitative data. After all, how rough is rough? Nevertheless, we find ourselves in the position of having to select a correlation for later use in vehicle ride modeling. A logical approach is to develop a hybrid road roughness correlations based on a combination of features of the earlier correlations. We formulate a general road roughness correlation as

$$S(N) = S_0 \left(\frac{N}{N_0} \right)^{-2.05} \tag{4.126}$$

where, as before, $N_0 = \dfrac{1}{2\pi}$ cycle/m. It is beneficial to retain the N_0 term to nondimensionalize the wave number. This alleviates the necessity of having to interpret what happens when units of cycle/m are raised to a power of -2.05. The value of roughness coefficient is selected from a probable range, which is derived from the minimum and maximum roughness correlations in Figure (4.25), of

$4.2 \leq S_0 \times 10^6 \leq 209 \, \text{m}^2/\text{cycle/m}$. The average value of S_0 is approximately $32 \times 10^{-6} \, \text{m}^2/\text{cycle/m}$. It is worth noting that the range and average value of road roughness coefficient agrees quite well with Dodds and Robson's proposed road conditions of good (lower range limit), average (average value), and poor (upper range limit).

(4.3.2) VEHICLE RESPONSE TO RANDOM VIBRATION

Now that we have spent some time becoming acquainted with random vibration theory and empirically based road surface models, we are ready to re-examine our earlier vehicle ride models. Specifically, we would like to use the road surface roughness spectral density as an input to the vehicle suspension model to produce a truer simulation of vehicle ride. If the suspension response is expressed in terms of the root-mean-square

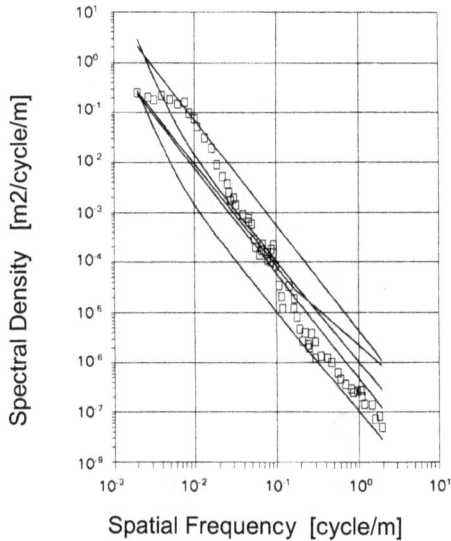

Figure (4.25): ROAD ROUGHNESS CORRELATIONS WITH SAMPLE MIRA [15] DATA

acceleration, then we can compare the response to ISO exposure time limits, detailed in Section (4.1), to estimate the quality of ride comfort.

An elementary ride model, for which we already have some familiarity, is the "simple damped vehicle model" found in Section (4.2.1B). Our analytical analysis of

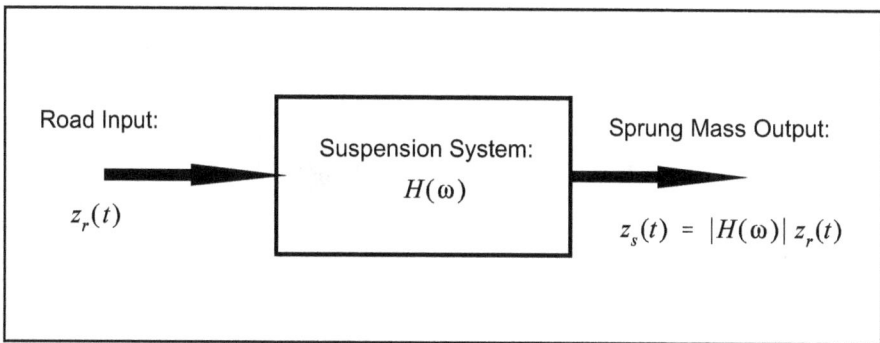

Figure (4.26): LINEAR SUSPENSION SYSTEM INPUT/OUTPUT BLOCK DIAGRAM

this model revealed that it is a linear system. We can make the general statement that in any linear system there is a direct relationship between the system input and the system output, as shown in the block diagram illustrated in Figure (4.26). This relationship holds true for multiple frequency random input/output functions as well.

The transfer function, $H(\omega)$, is probably more accurately referred to now as a frequency response function. There are two basic methods we learned in Section (4.2) to derive frequency response functions. When modeling simple suspension systems we used the steady-state solution of the sprung mass movement, and for more complex systems we used the impedance modeling technique. In either case, the frequency response function is a characteristic of a particular suspension system, which we can find without consideration of random road surface behavior.

The Euler exponential form of the road input found in the previous section is

$$z_r(x) = \mathbb{R}\left\{\sum_{i=1}^{\infty} Z_{r,i}\, e^{j\Omega_i x}\right\} \tag{4.127}$$

We want to convert the road input from a spatial independent variable to a time base so we can utilize the frequency response function to determine the suspension output. The is accomplished using the forward design velocity, V, and time, t, to eliminate x in the exponential term, i.e.,

$$j\Omega_i x = j\Omega_i V t = j\left(\frac{2\pi V}{L_i}\right)$$

Next, we define a discrete road frequency, ω_i, as ; $\dfrac{2\pi V}{L_i}$ then, $j\Omega_i x = j\omega_i t$. Finally, the road input becomes

$$z_r(t) = \mathbb{R}\left\{\sum_{i=1}^{\infty} Z_{r,i}\, e^{j\omega_i t}\right\} \tag{4.128}$$

The sprung mass movement can now be related to the road input, using the frequency response function, by the equation

$$z_s(t) = \mathbb{R}\left\{\sum_{i=1}^{\infty} Z_{r,i}\, H(\omega)\, e^{j\omega_i t}\right\}$$

$$= \sum_{i=1}^{\infty} \frac{Z_{r,i}}{2}\left[H(\omega_i)\, e^{j\omega_i t} + H^*(\omega_i)\, e^{j\omega_i t}\right] \tag{4.129}$$

The sprung mass displacement defined by Eq (4.129) is also a random function. The direction our analysis takes now is to derive an expression for the RMS acceleration of the sprung mass as a function of road frequency. Then, using the information we can make a direct comparison to ISO standards for exposure limits. The comparison will allow us to surmise an anticipated level of ride comfort. The first step in our quest is to find the mean square value of the sprung mass displacement via the auto-correlation.

The auto-correlation of $z_s(t)$ is found from the equation

$$R_s(\tau) = \left\langle z_s(t) \, z_s(t + \tau) \right\rangle$$

$$= \lim_{T \to \infty} \left\{ \frac{1}{2T} \int_{-T}^{T} z_s(t) \, z_s(t + \tau) \, dt \right\}$$

$$(4.130)$$

The $\langle \, \rangle$ brackets in the previous equation are customary symbols signifying a time average. They also represent the expected value of a random function. We proceed by placing Eq (4.129) into Eq (4.130):

$$R_s(\tau) = \lim_{T \to \infty} \left\{ \frac{1}{2T} \int_{-T}^{T} \left[\sum_{i=1}^{\infty} \frac{1}{2} \left(Z_{r,i} H e^{j\omega_i t} + Z_{r,i}^* H^* e^{-j\omega_i t} \right) \right] \right.$$

$$\left. \times \left[\sum_{i=1}^{\infty} \frac{1}{2} \left(Z_{r,i} H e^{j\omega_i (t+\tau)} + Z_{r,i}^* H^* e^{-j\omega_i (t+\tau)} \right) \right] dt \right\}$$

$$= \sum_{i=1}^{\infty} \frac{1}{4} \lim_{T \to \infty} \left\{ \frac{1}{2T} \int_{-T}^{T} \left[Z_{r,i}^2 H^2 e^{j\omega_i (2t+\tau)} \right. \right.$$

$$\left. \left. + Z_{r,i} Z_{r,i}^* H H^* \left(e^{-j\omega_i t} + e^{j\omega_i t} \right) + Z_{r,i}^{*2} H^{*2} e^{-j\omega_i (2t+\tau)} \right] dt \right\}$$

Next, we evaluate the integrals:

$$R_s(\tau) = \sum_{i=1}^{\infty} \frac{1}{4} \lim_{T \to \infty} \left\{ \frac{Z_{r,i}^2 H^2}{2 j \omega_i T} \left[e^{j\omega_i (T+\tau)} - e^{j\omega_i (-T+\tau)} \right] \right.$$

$$\left. + Z_{r,i} Z_{r,i}^* H H^* \left(e^{-j\omega_i \tau} + e^{j\omega_i \tau} \right) - \frac{Z_{r,i}^{*2} H^{*2}}{2 j \omega_i T} \left[e^{-j\omega_i (T+\tau)} - e^{-j\omega_i (-T+\tau)} \right] \right\}$$

Although not obvious at first, we must remember that the exponential terms in the above equation evolved from sine and cosine functions, so they are bounded. When we take the limit as $T \to \infty$, the first and last terms inside the brace brackets vanish to zero. This result simplifies the auto-correlation to

$$R_s(\tau) = \sum_{i=1}^{\infty} \frac{1}{4} |Z_{r,i}|^2 |H(\omega_i)|^2 \left(e^{-j\omega_i\tau} + e^{j\omega_i\tau} \right) \qquad (4.131)$$

The mean square value of $z_s(t)$ is derived from the auto-correlation, Eq (4.131), by setting $\tau = 0$; that is,

$$\overline{z}_s^2 = \sum_{i=1}^{\infty} \frac{1}{2} |Z_{r,i}|^2 |H(\omega_i)|^2 \qquad (4.132)$$

The mean square value of the sprung mass displacement described by Eq (4.132) is in terms of discrete road amplitudes and discrete road frequencies. On the other hand, the revised road surface models we have are empirically based roughness spectral density correlations which use the wave number as the independent variable. In order to incorporate the road surface model into Eq (4.132) we need to find, or select, just one variable which we can use as an independent variable in all of our relationships. Eventually we will want to compare the results from our new vehicle ride model to ISO time exposure standards. If you recall, the information presented in the ISO standards is based on temporal frequency, f. The logical choice for an independent variable is then the temporal frequency.

We begin the variable transformation process by recalling the relationship between the mean square value of road roughness and the spectral density:

$$\overline{z}_r^2 = \int_0^{\infty} S(\Omega)\, d\Omega = \int_0^{\infty} S(N)\, dN$$

The wave number can be related to the frequency by re-introducing the vehicle forward design velocity, i.e., $f = VN$. The previous equation then becomes

$$\overline{z}_r^2 = \int_0^{\infty} \frac{S(N)}{V}\, df$$

For convenience sake it is desirable to define a frequency based on spectral density function as $S(f) \equiv \dfrac{S(N)}{V}$. This allow us to slightly simplify the mean square value of road roughness to

$$\overline{z}_r^2 = \int_0^{\infty} S(f)\, df$$

In our earlier discussion concerning the mean square value of road roughness, we found that

$$\overline{z}_r^2 = \sum_{i=1}^{\infty} \frac{1}{2} |Z_{r,i}|^2$$

Since we have two expressions which describe the same mean square value, they must be equivalent; therefore,

$$\sum_{i=1}^{\infty} \frac{1}{2} |Z_{r,i}|^2 = \int_0^{\infty} S(f)\, df$$

Let us digress for a moment. If we evaluate the absolute value of the transfer function, i.e., $|H(\omega_i)|$, at a particular discrete frequency, we ultimately end up with a scalar constant. This means that the expression $|Z_{r,i}|^2 |H(f)|^2$ also describes a constant value, which we might define as a scaled discrete amplitude squared. An infinite summation of these discrete values can be equated to a scaled spectral density by virtue of the previous equation, that is,

$$\sum_{i=1}^{\infty} \frac{1}{2} |Z_{r,i}|^2 |H(f_i)|^2 = \int_0^{\infty} S(f) |H(f)|^2\, df \tag{4.133}$$

The last variable transformation we need is $2\pi f_i = \omega_i$. Now we see some similarity between Eqs (4.132) and (4.133), and in fact we can combine them to produce

$$\overline{z}_s^2 = \int_0^{\infty} S(f) |H(f)|^2\, df \tag{4.134}$$

We now have an expression for the mean square value of the sprung mass displacement which can be evaluated from the mathematical models for the road roughness spectral density and the suspension system. However, in order to make a comparison to ISO time exposure standards, we must make two more refinements. First we need the RMS value of sprung mass acceleration instead of the mean square displacement as given by Eq (4.134). We remember for sine and cosine functions, which is the fundamental foundation on which our road surface input models are based, that acceleration and displacement amplitudes are proportional by a value equivalent to the circular frequency squared. So, we introduce a factor of $(2\pi f)^2$ into the integrand of Eq (4.134) to obtain mean square acceleration. The RMS acceleration is found simply by taking the square root of the resulting integral. Performing these steps gives us

$$\overline{A}_s = \left\{ \int_0^{\infty} S(f) \left[(2\pi f)^2 |H(f)| \right]^2 df \right\}^{\frac{1}{2}} \tag{4.135}$$

Secondly, we must adjust the limits of integration to reflect an evaluation of the RMS acceleration over a one-third octave band centered about a frequency, f_c. A one-third octave band is defined as a range of frequencies in which the upper frequency is $\sqrt[3]{2}$ times greater than the lower. Then for a prescribed center frequency, the lower limit of integration becomes $f_c/\sqrt[6]{2}$ and the upper limit becomes $\sqrt[6]{2}\,f_c$. Making these adjustments to Eq (4.135) results in

$$\bar{A}_s = \left\{ \int_{f_c/\sqrt[6]{2}}^{\sqrt[6]{2}\,f_c} S(f)\left[(2\pi f)^2\,|H(f)|\right]^2 df \right\}^{\frac{1}{2}} \tag{4.136}$$

Eq (4.136) provides us with a means to calculate one RMS acceleration value corresponding to a selected value of center frequency. We will refer to this pair of values as a ride datum point. Obviously we can generate a set of theoretical ride data points by varying the center frequency. The next logical step is to plot these data points on a graph of RMS acceleration versus center frequency. Exposure limits specified by ISO standards can also be plotted on this graph. To ascertain the perception of ride we look for the exposure limit which lies completely above the ride data.

Better estimates of normal RMS acceleration are obtained from the "bicycle" vehicle ride model. The basic random function input/output block diagram for this model is shown in Figure (4.27). The bicycle ride model is not that much more involved than the simple, single sprung mass model we just analyzed. But, there are two differences we should be aware of. First, the vehicle movement we are interested in is the vertical displacement of the sprung mass center of gravity. If the dynamic index is close to unity {refer to Section (4.2.2B)}, then the center of gravity displacement can be expressed as a geometric combination of the sprung mass displacements at the front and rear axle planes. Secondly, the rear axle suspension system experiences a time delay in the road input, relative to the front axle plane, which must be accounted for in the rear sprung mass output.

The linear input/output relationship between the road input and the sprung mass output is fundamentally the same as before. At the front axle plane the relationship is

$$z_{sf}(t) = \mathbb{R}\left\{ \sum_{i=1}^{\infty} Z_{r,i}\, H_{uf}(\omega_i)\, H_{sf}(\omega_i)\, e^{j\omega_i t} \right\}$$
$$+ \sum_{i=1}^{\infty} \frac{1}{2} Z_{r,i}\left[H_f(\omega_i)\, e^{j\omega_i t} + H_f^*(\omega_i)\, e^{-j\omega_i t} \right] \tag{4.137}$$

where

$$H_f(\omega) = \left[\mathbb{R}\{H_{uf}(\omega)\}\, \mathbb{R}\{H_{sf}(\omega)\} - \mathbb{I}\{H_{uf}(\omega)\}\, \mathbb{I}\{H_{sf}(\omega)\} \right]$$
$$+ j\left[\mathbb{R}\{H_{uf}(\omega)\}\, \mathbb{I}\{H_{sf}(\omega)\} + \mathbb{I}\{H_{uf}(\omega)\}\, \mathbb{R}\{H_{sf}(\omega)\} \right] \tag{4.138}$$

The relationship for the rear sprung mass output is similar to Eq (4.137), except we must include a time delay in the exponential terms:

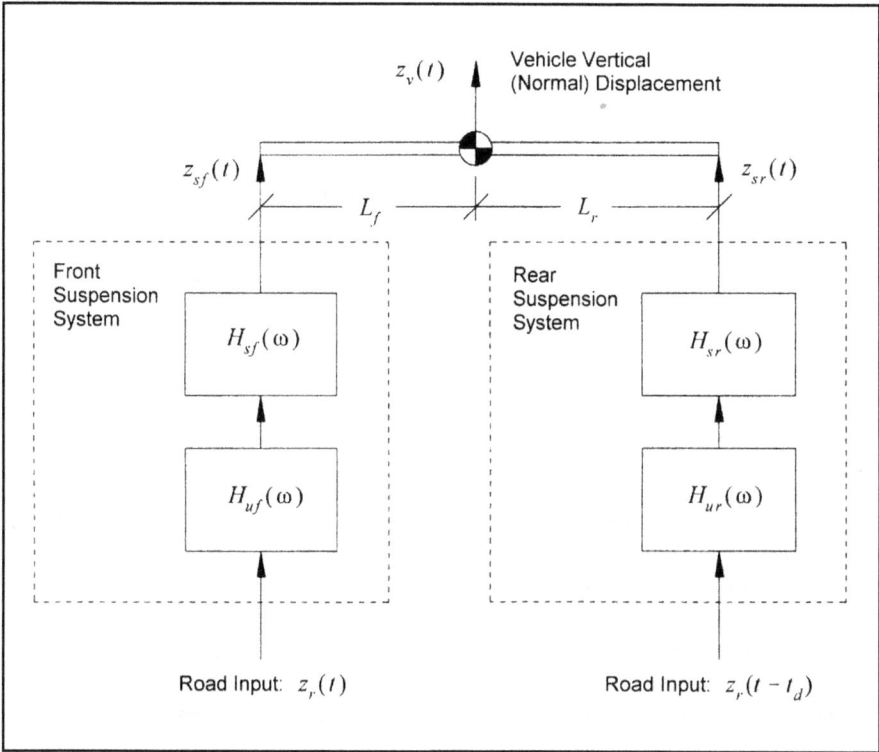

Figure (4.27): "BICYCLE" RIDE MODEL INPUT/OUTPUT BLOCK DIAGRAM

$$z_{sr}(t) = \sum_{i=1}^{\infty} \frac{1}{2} Z_{r,i} \left[H_r(\omega_i) e^{j\omega_i(t-t_d)} + H_r^*(\omega_i) e^{-j\omega_i(t-t_d)} \right] \qquad (4.139)$$

The expression for the transfer function $H_r(\omega)$ follows the same format as indicated by Eq (4.138), just change the "f" subscript to "r".

We start our new analysis with an expression for the random vertical, or normal, motion of the vehicle in terms of the front and rear sprung mass displacements, i.e.,

$$z_v(t) = \frac{L_r}{L_w} z_{sf}(t) + \frac{L_f}{L_w} z_{sr}(t) \qquad (4.140)$$

The derivation of the RMS acceleration now follows the same procedure as before. The fundamental equation describing the auto-correlation of $z_v(t)$ is

$$R_v(\tau) = \left\langle z_v(t) z_v(t+\tau) \right\rangle$$

Placing Eq (4.140) into the previous equation results in

$$R_v(\tau) = \left\langle \left[\frac{L_r}{L_w} z_{sf}(t) + \frac{L_f}{L_w} z_{sr}(t) \right] \left[\frac{L_r}{L_w} z_{sf}(t+\tau) + \frac{L_f}{L_w} z_{sr}(t+\tau) \right] \right\rangle$$

$$= \left(\frac{L_r}{L_w} \right)^2 \left\langle z_{sf}(t) z_{sf}(t+\tau) \right\rangle + \frac{L_f L_r}{L_w^2} \left\langle z_{sf}(t) z_{sr}(t+\tau) \right\rangle$$

$$+ \frac{L_f L_r}{L_w^2} \left\langle z_{sr}(t) z_{sf}(t+\tau) \right\rangle + \left(\frac{L_f}{L_w} \right)^2 \left\langle z_{sr}(t) z_{sr}(t+\tau) \right\rangle$$

$$(4.141)$$

The most expedient manner to evaluate Eq (4.141) is to analyze each of the time averaged terms individually. The first and last time average terms we recognize as the auto-correlation of the sprung mass displacements at the front and rear axle planes, respectively. The two middle terms are called cross-correlations. Mathematical expressions for the auto-correlation and cross-correlation terms are derived from the evaluation of the basic integral of time averaged values. Instead of going through all of the integration details this time, we will just summarize the results:

(*i*) The auto-correlation of the front axle sprung mass displacement is

$$R_{sf}(\tau) \equiv \left\langle z_{sf}(t) z_{sf}(t+\tau) \right\rangle$$

$$= \lim_{T \to \infty} \left\{ \frac{1}{2T} \int_{-T}^{T} z_{sf}(t) z_{sf}(t+\tau) \, dt \right\}$$

or, using Eq (4.137)

$$R_{sf}(\tau) = \sum_{i=1}^{\infty} \frac{1}{4} Z_{r,i}^2 H_f(\omega_i) H_f^*(\omega_i) \left[e^{j\omega_i \tau} + e^{-j\omega_i \tau} \right] \qquad (4.142)$$

(*ii*) The cross-correlation of the front axle to rear axle sprung mass displacement is defined as

$$R_{fr}(\tau) \equiv \left\langle z_{sf}(t) z_{sr}(t+\tau) \right\rangle$$

$$= \lim_{T \to \infty} \left\{ \frac{1}{2T} \int_{-T}^{T} z_{sf}(t) z_{sr}(t+\tau) \, dt \right\}$$

which, using Eqs (4.137) and (4.139), becomes

$$R_{fr}(\tau) = \sum_{i=1}^{\infty} \frac{1}{4} Z_{r,i}^2 \left[H_f(\omega_i) H_r^*(\omega_i) e^{j\omega_i(t_d - \tau)} \right.$$

$$\left. + H_f^*(\omega_i) H_r(\omega_i) e^{-j\omega_i(t_d - \tau)} \right] \qquad (4.143)$$

(*iii*) The cross-correlation of the rear to front axle sprung mass displacement (which is generally not the same as the previous front to rear cross-correlation) is

$$R_{rf}(\tau) \equiv \left\langle z_{sr}(t) z_{sf}(t + \tau) \right\rangle$$

$$= \lim_{T \to \infty} \left\{ \frac{1}{2T} \int_{-T}^{T} z_{sr}(t) z_{sf}(t + \tau) \, dt \right\}$$

which evaluates as

$$R_{rf}(\tau) = \sum_{i=1}^{\infty} \frac{1}{4} Z_{r,i}^2 \left[H_f(\omega_i) H_r^*(\omega_i) e^{j\omega_i(t_d + \tau)} \right.$$

$$\left. + H_f^*(\omega_i) H_r(\omega_i) e^{-j\omega_i(t_d + \tau)} \right] \qquad (4.144)$$

(*iv*) The auto-correlation of the rear axle plane sprung mass is

$$R_{sr}(\tau) \equiv \left\langle z_{sr}(t) z_{sr}(t + \tau) \right\rangle$$

$$= \lim_{T \to \infty} \left\{ \frac{1}{2T} \int_{-T}^{T} z_{sr}(t) z_{sr}(t + \tau) \, dt \right\}$$

using Eq (4.139) this definition becomes

$$R_{sr}(\tau) = \sum_{i=1}^{\infty} \frac{1}{4} Z_{r,i}^2 H_r(\omega_i) H_r^*(\omega_i) \left[e^{j\omega_i(\tau - t_d)} + e^{-j\omega_i(\tau - t_d)} \right] \qquad (4.145)$$

As in the previous analysis, the mean square value of vehicle displacement is found by evaluating the vehicle auto-correlations and cross-correlations at $\tau = 0$. From Eq (4.141)

$$\bar{z}_v^2 = \left(\frac{L_r}{L_w} \right)^2 R_{sf}(0) + \frac{L_f L_r}{L_w^2} \left[R_{fr}(0) + R_{rf}(0) \right] + \left(\frac{L_f}{L_w} \right)^2 R_{sr}(0) \qquad (4.146)$$

The front and rear axle sprung mass auto-correlations are evaluated from Eqs (4.142) and (4.145), respectively,

$$R_{sf}(0) = \sum_{i=1}^{\infty} \frac{1}{2} Z_{r,i}^2 |H_f(\omega_i)|^2 \qquad (4.147)$$

and

$$R_{sr}(0) = \sum_{i=1}^{\infty} \frac{1}{4} Z_{r,i}^2 |H_r(\omega_i)|^2 \left[e^{j\omega_i t_d} + e^{-j\omega_i t_d} \right]$$

The latter equation can be simplified by noting that $\cos(\omega_i t_d) \equiv \frac{1}{2} \left[e^{j\omega_i t_d} + e^{-j\omega_i t_d} \right]$, then

$$R_{sr}(0) = \sum_{i=1}^{\infty} \frac{1}{2} Z_{r,i}^2 |H_r(\omega_i)|^2 \cos(\omega_i t_d) \qquad (4.148)$$

Let's examine the summation of the cross-correlation terms a little more closely. We note from Eqs (4.143) and (4.144) that

$$R_{fr}(0) + R_{rf}(0) = \sum_{i=1}^{\infty} \frac{1}{2} Z_{r,i}^2 \left[H_f(\omega_i) H_r^*(\omega_i) e^{j\omega_i t_d} + H_f^*(\omega_i) H_r(\omega_i) e^{-j\omega_i t_d} \right]$$

However, using Euler's identities the exponential terms can be replaced with sine and cosine functions. An equivalent expression for the previous equation is then

$$R_{fr}(0) + R_{rf}(0) = \sum_{i=1}^{\infty} \frac{1}{2} Z_{ri}^2 \left\{ \left[H_f H_r^* + H_f^* H_r \right] \cos(\omega_i t_d) \right.$$

$$\left. + j \left[H_f H_r^* - H_f^* H_r \right] \sin(\omega_i t_d) \right\} \qquad (4.149)$$

The transfer function terms in square brackets can be more easily evaluated by expanding them as functions of real and imaginary components, that is,

(*i*)

$$H_f H_r^* + H_f^* H_r = 2 \left(\mathbb{R}\{H_f\} \mathbb{R}\{H_r\} + \mathbb{I}\{H_f\} \mathbb{I}\{H_r\} \right) \qquad (4.150)$$

and (*ii*)

$$H_f H_r^* - H_f^* H_r = -2j \left(\mathbb{R}\{H_f\} \mathbb{I}\{H_r\} - \mathbb{I}\{H_f\} \mathbb{R}\{H_r\} \right) \qquad (4.151)$$

Substituting Eqs (4.150) and (4.151) into Eq (4.149) results in

$$R_{fr}(0) + R_{rf}(0) = \sum_{i=1}^{\infty} Z_{ri}^2 \left\{ \left[\mathbb{R}\{H_f\}\,\mathbb{R}\{H_r\} + \mathbb{I}\{H_f\}\,\mathbb{I}\{H_r\} \right] \cos\left(\omega_i t_d\right) \right.$$

$$\left. + \left[\mathbb{R}\{H_f\}\,\mathbb{I}\{H_r\} - \mathbb{I}\{H_f\}\,\mathbb{R}\{H_r\} \right] \sin\left(\omega_i t_d\right) \right\}$$

(4.152)

Referring to Eq (4.152), we will define the lengthy term in brace brackets as a cross-spectral transfer function, and identify it by the symbol $|H_{fr}(\omega_i, t_d)|^2$. We can then save time and space by writing the summation of cross-spectral terms as

$$R_{fr}(0) + R_{rf}(0) = \sum_{i=1}^{\infty} Z_{ri}^2 \, |H_{fr}(\omega_i, t_d)|^2$$

(4.153)

Now that we have expressions for the auto-correlations and cross-correlation at $\tau = 0$, we are ready to resume the derivation of the mean square value of vehicle displacement. Inserting the correlations into Eq (4.136) yields

$$\overline{z_v^2} = \sum_{i=1}^{\infty} \frac{1}{2} Z_{ri}^2 \left[\left(\frac{L_r}{L_w} \right)^2 |H_f(\omega_i)|^2 + 2\frac{L_f L_r}{L_w^2} |H_{fr}(\omega_i, t_d)|^2 \right.$$

$$\left. + \left(\frac{L_f}{L_w} \right)^2 |H_r(\omega_i)|^2 \cos\left(\omega_i t_d\right) \right]$$

(4.154)

Although Eq (4.154) provides us with a means to evaluate the mean square vehicle displacement, we must remember that our goal is to derive an expression for the normal RMS acceleration. To achieve our objective the final steps employed in developing the RMS acceleration for the previous simple, unitary suspension vehicle ride model are duplicated. Basically, without a great deal of effort, we replace the summation of road surface amplitudes in Eq (4.154) with an integral expression incorporating the road roughness spectral density together with the transfer functions; and then manipulate the resulting equation to describe acceleration. The normal RMS acceleration for our bicycle ride model finally evolves as

$$\bar{A}_v = \left\{ \int_{f_c/\sqrt[6]{2}}^{\sqrt[6]{2} f_c} S(f) \left[\left(\frac{L_r}{L_w} \right)^2 |(2\pi f)^2 H_f(f)|^2 + 2\frac{L_f L_r}{L_w^2} |(2\pi f)^2 H_{fr}(f, t_d)|^2 \right. \right.$$

$$\left. \left. + \left(\frac{L_f}{L_w} \right)^2 |(2\pi f)^2 H_r(f)|^2 \cos\left(2\pi f t_d\right) df \right] \right\}^{\frac{1}{2}}$$

(4.155)

Earlier in Section (4.2.3) we worked through an example problem in which the vehicle suspension was simulated as a bicycle using impedance modeling techniques. We were able to determine the bounce response and the pitch response; but, we were not able to draw any conclusion pertaining to the perception of ride quality. The transfer functions, obtained from the bicycle ride model, together with Eq

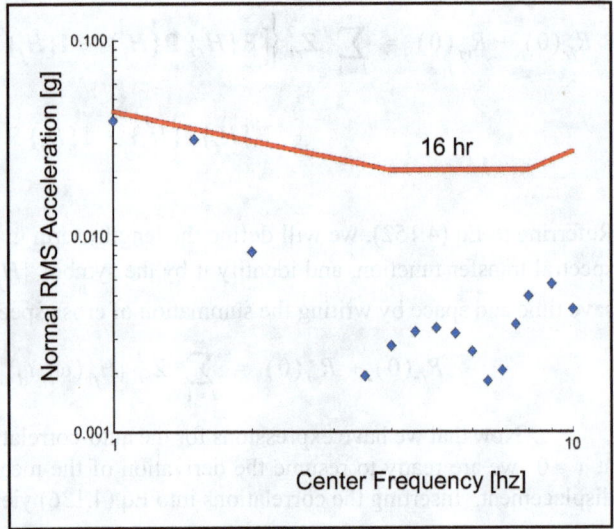

Figure (4.28): EXAMPLE ANTICIPATED RMS ACCELERATION PLOT

(4.155) enables us to calculate the RMS acceleration as a function of center frequency for various road surface roughness and vehicle velocities. The final aspect of the ride analysis is to overlay a graph of RMS acceleration versus center frequency on top of an appropriate ISO boundary, e.g., the "fatigued-decreased proficiency boundary" shown in Figure (4.4). The ISO time limit which is completely above the plotted acceleration points is an estimated time that an average driver will be able to operate the vehicle without experiencing discomfort.

Let's continue our earlier bicycle suspension analysis. The focus is to calculate the normal RMS acceleration as expressed by Eq (4.155); however, the integral does not lend itself to a closed form solution and thus a numerical integration process must be used in either a computer algorithm or spreadsheet application. Several calculations for the normal RMS acceleration, as along with the previous bicycle transfer functions found in Section (4.2.3) and the road spectral density for an "average" road surface, as modeled in Section (4.3.1), were performed at various center frequencies for a vehicle velocity of 60 MPH. The results are plotted in Figure (4.28) as discrete data points along with an overlaid of the 16 hour ISO "fatigued-decreased proficiency boundary." Since all of the anticipated data points are below this boundary, we make the subjective conclusion that this vehicle and suspension should provide a comfortable ride under these conditions. Different driving conditions can be simulated by varying the vehicle velocity and the type of road surface. In each case we compare the results to ISO exposure limits to gain a better insight into the subjective perception of ride quality.

Naturally, we could expand our analysis from a "bicycle" model into a more sophisticated "four wheel" model. We would start by placing the unicycle suspension blocks at each of the four corners of the vehicle. The combined output of the four suspension blocks is used as the input to the vehicle sprung mass {similar to the block diagram in Figure (4.27)}. We could also study the effects of the drivers seat on driver comfort as well. We would do this by including an additional functional block describing the frequency response characteristics of the seat. The output of the vehicle sprung mass is then used as the input to the seat. As you might expect, the amount of mathematical manipulation, although not difficult, is quite involved. However, the improvement to be gained in estimating driver comfort from this type of vehicle model are not great enough to include here.

(4.4) CLOSING REMARKS

Looking back over the contents of this chapter we have accumulated a variety of vehicle ride models and modeling techniques. All of which would be very useful if we had a pre-existing suspension system to analyze. However, the analytical models and experimental data we have studied do not provide substantial guidance in specifying spring constants, or rates, and damping coefficients, also referred to as rates, required in the initial suspension system design. So, we must rely on a few general "rules of thumb" which, for the most part, are accepted by suspension designers:

(1) A dynamic index close to unity is specified to minimize the influence of the front suspension motion on the rear suspension, and vice versa.

(2) Many designers believe an optimum ride should be based on a 60 cpm sprung mass bounce natural frequency at the front axle plane.

(3) A rear suspension bounce frequency around 1.1 to 1.3 times greater than the front is specified to control pitch motion.

These criteria originate mainly from physiological considerations and the results of suspension frequency response plots. Using the bounce frequency specification the initial spring rate is calculated from the basic definition of natural frequency {Eq (4.16)} and an estimate of the vehicle sprung mass. Beware that the calculated spring rate is the effective spring rate at the wheel. If necessary, we must adjust the actual spring rate to compensate for any mechanical advantage (or disadvantage) due to the placement of the spring in the suspension linkage. We should note that the vehicle static deflection resulting from a design specification of 60 cpm is relatively large, which must be accommodated by a long spring length. Incorporating such a spring into a compact vehicle suspension is a difficult design problem.

The majority of drivers and passengers would a vehicle suspension designed in the foregoing manner to be a very comfortable ride, which one expects of a "luxury" automobile. The sports car enthusiast, on the other hand, would probably prefer a more firm ride, and would find the amount of body roll of this car objectionable during cornering. For sports car applications a natural frequency of 70 to 90 cpm at the front axle is a good starting point, with slightly higher frequency for the rear. As one might imagine, spring rates for racing cars are not specified with driver comfort in mind. The general trend is to select spring rates which will provide about one inch of static deflection. This is a very stiff suspension. Even with these high spring rates, an anti-roll bar may still be required to limit excessive body roll. We will examine the application of anti-roll bars, or torsion bars, more closely in the next chapter. As far as ride performance is concerned, the roll frequency should be about the same as the bounce frequency.

The initial effective damping rate for the shock absorber is calculated from the damping ratio and natural frequency {Eq (4.29)}. The damping rate is generally in the range of 0.25 to 0.38 of the critical value. The actual damping rate depends on: (1) the mechanical advantage of the suspension linkage, and (2) the amount of inherent friction in the suspension system, such as the friction in the linkage pivot connections at the chassis. In addition for suspension systems using leaf springs, there is a great deal of interfacial friction between the spring leafs.

Chapter 5:
CORNERING PERFORMANCE

Common drivers generally define the performance characteristics of a vehicle executing a cornering maneuver at high speed as "handling." The meaning of vehicle handling or cornering performance to engineers and suspension designers, on the other hand, is somewhat different. We are primarily concerned with the response of the vehicle, which we think of as a mechanical system, to steering commands by the driver. Another aspect of vehicle handling is the affect of external forces, such as aerodynamic forces and road surface irregularities, on the direction of travel. Of equal importance is the study of cornering stability. That is to say, circumstances which result in a loss of control, e.g., spin out.

Our goal in this chapter is to gain a fundamental understanding of cornering and the design parameters which effect cornering performance. We will accomplish our goal by taking a look at basic steering geometry, and by analyzing two different vehicle handling models.

(5.1) ACKERMANN GEOMETRY

An idealized geometric description of a vehicle executing a cornering maneuver is show in Figure (5.1), and is referred to as Ackermann geometry. This ideal geometric model assumes that none of the tires experience slip, and that the steering system is designed to allow all of the wheels to move in an arc about a unique point. The steer angles required to accomplish this task can be expressed in terms of the location of the vehicle center of gravity, front track width[1], and turning radius:

$$\delta_i = \arctan\left[\frac{L_f + L_r}{\sqrt{R^2 - L_r^2} - \dfrac{T_f}{2}}\right] \tag{5.1}$$

---

[1] *The distance separating the front wheels is the Ackermann geometry is not actually the front track width, but rather the front track width less the sum of inboard scrub radii; however, the scrub radius of modern front suspension systems is so small compared to the track width that this correction is not usually necessary.*

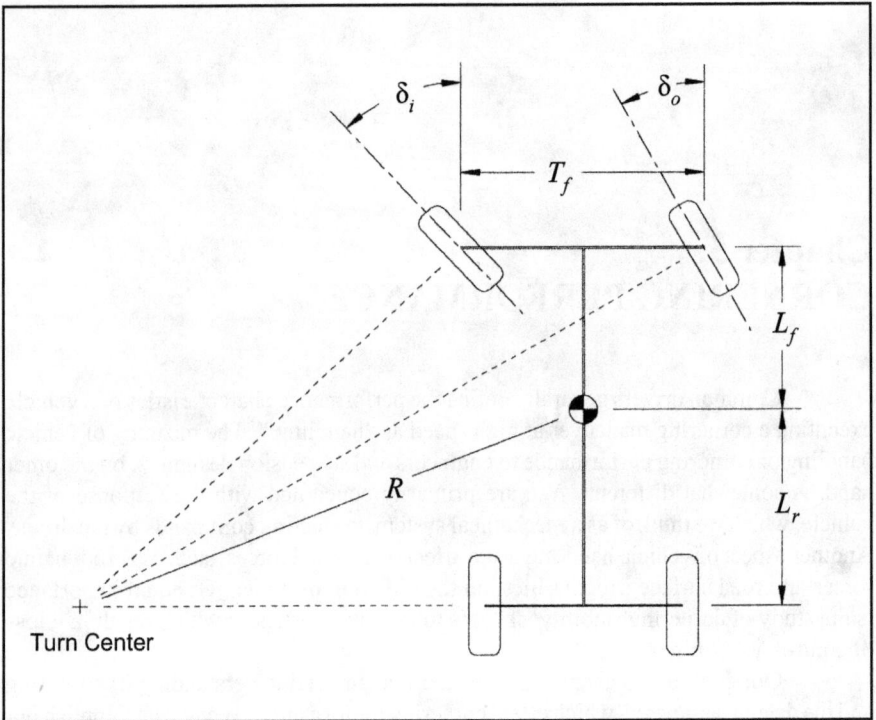

Figure (5.1): ACKERMANN GEOMETRY

and

$$\delta_o = \arctan \left[\frac{L_f + L_r}{\sqrt{R^2 - L_r^2} + \frac{T_f}{2}} \right] \qquad (5.2)$$

      A fundamental approach to understanding cornering behavior of vehicles is the "long radius turn." In this case we consider the order of magnitude of the turning radius to be much greater than either the front track width or the wheelbase. The denominators in the previous arc tangent arguments resulting from this consideration become:

$$\sqrt{R^2 - L_r^2} \pm \frac{T_f}{2} \approx R$$

Also implied by the long radius assumption is a required small angle of steer, which indicates that $\tan(\delta_o) \approx \delta_o$ and $\tan(\delta_i) \approx \delta_i$. Therefore, Eqs (5.1) and (5.2) can be approximated by

$$\delta_i = \delta_o \approx \frac{L_f + L_r}{R} = \frac{L_w}{R} \qquad (5.3)$$

Chapter 5: CORNERING PERFORMANCE

for long radius turns. This relationship for inside/outside steer angle is known as the Ackermann steer angle.

(5.2) TWO-WHEEL CORNERING MODEL

The Ackermann steering model provides a suitable foundation for establishing the basics of steering geometry. However, this model neglects the effect of the tires on the vehicle motion as it travels through a corner. The elastic behavior of pneumatic tires allows a twisting deformation of the tire caused indirectly by the centrifugal forces encountered during cornering. Since there is a twisting deformation the direction the wheel rim points and the direction the tire is pointing, where it makes contact with the road surface, need not be the same. The angle between these two different directions is the tire slip angle. As a result, the tire deformation that occurs in the rear wheels will also steer the vehicle. Thus, it is the combination of front wheel steering, front tire slip angles, and rear tire slip angles which actually determines the cornering motion of the vehicle.

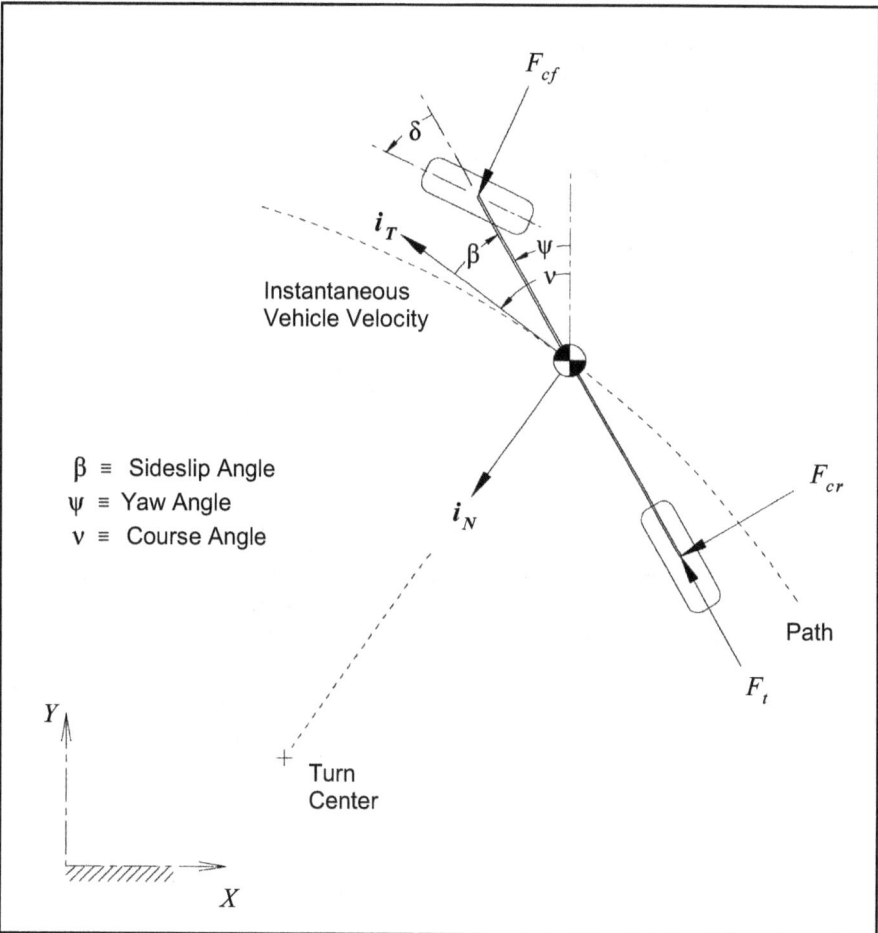

Figure (5.2): TWO-WHEEL VEHICLE CORNERING SCHEMATIC

The classic approach to understanding the dynamic principles of vehicle cornering is the "two-wheel" cornering model. In the previous section we found that the inside and outside steer angles are approximately the same for long radius turns. Under this condition the symmetry of the left and right sides of the vehicle suggests that a simple cornering model can be obtained by squeezing the sides of the car together to form a bicycle of sorts. Although the two-wheel cornering model is elementary, it does describe basic concepts and design considerations involved in vehicle cornering. The major drawback of this model is inability to study the importance of the suspension system, in particular the effects of roll stiffness and track width, on cornering performance.

(5.2.1) EQUATIONS OF MOTION

A schematic of a two-wheeled vehicle executing a cornering maneuver is shown in Figure (5.2). For our first look at this simple cornering model the only externally applied forces we will consider are the cornering forces, i.e., the lateral forces of the road surface acting on the tires, and the tractive force. Using a curvilinear coordinate system[2], the fundamental equations governing the vehicle motion are:

(*i*) $\sum F_T = m a_T$, or

$$-F_{cf}\sin(\delta - \beta) + F_{cr}\sin(\beta) + F_t\cos(\beta) = m\dot{V} \qquad (5.4)$$

(*ii*) $\sum F_N = m a_N$, or

$$F_{cf}\cos(\delta - \beta) + F_{cr}\cos(\beta) + F_t\sin(\beta) = m\frac{V^2}{R} \qquad (5.5)$$

and (*iii*) $\sum M_{CG} = I\ddot{\psi}$, or

$$F_{cf}L_f\cos(\delta) - F_{cr}L_r = I\ddot{\psi} \qquad (5.6)$$

In order to make the equations of motion more manageable we will restrict our analysis to long radius cornering maneuvers at constant forward (tangential) speed. The implication of a long turning radius is that the steer and sideslip angles will be small. In this event we are able to linearize the sine and cosine functions. The infinite series representation of these trigonometric functions are

$$\sin(x) = \sum_{n=1}^{\infty} \frac{(-1)^{n-1}x^{2n-1}}{(2n-1)!} = x - \frac{x^3}{6} + \cdots$$

and

$$\cos(x) = \sum_{n=0}^{\infty} \frac{(-1)^n x^{2n}}{(2n)!} = 1 - \frac{x^2}{2} + \cdots$$

---

[2] *A brief review of pertinent dynamics material is provided in Appendix C.*

For an angle of six degrees ($\pi/30$ radians), or less, the series representations converge very rapidly and can be approximated by $\sin(x) \approx x$ and $\cos(x) \approx 1$ within an error of 0.5%.

The long radius, constant forward velocity restriction together with the small angle approximations allows us to simplify the foregoing set of equations of motion to:

$$-F_{cf}(\delta - \beta) + F_{cr}\beta + F_t = 0 \qquad (5.7)$$

$$F_{cf} + F_{cr} + F_t\beta = m\frac{V^2}{R} \qquad (5.8)$$

$$F_{cf}L_f - F_{cr}L_r = I\ddot{\psi} \qquad (5.9)$$

The tractive force can be eliminated from further consideration by combing Eqs (5.7) and (5.8), which results in which results in

$$F_{cf}(1 + \delta\beta - \beta^2)$$
$$+ F_{cr}(1 - \beta^2) = m\frac{V^2}{R} \qquad (5.10)$$

The orders of magnitude involved in the small angle assumption allows us to further simplify Eq (5.10) by noting that

$$1 + \delta\beta - \beta^2 \approx 1$$

and

$$1 - \beta^2 \approx 1$$

Using these approximations the previous equation reduces to:

$$F_{cf} + F_{cr} = m\frac{V^2}{R} \qquad (5.11)$$

At first glance the insight into vehicle handling as provided by Eq (5.11) is somewhat uninspiring. The equation only indicates that combination of cornering forces generated by the front and rear wheels must balance the inertia force created as a result of cornering. It also appears that any combination of cornering

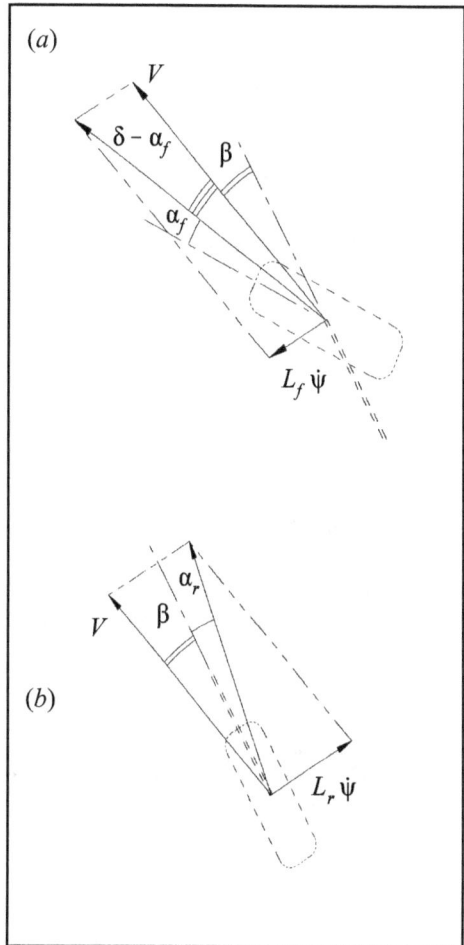

Figure (5.3): VELOCITY DIAGRAM OF TWO-WHEEL CORNERING MODEL AT (a) FRONT AXLE and (b) REAR AXLE

Chapter 5: CORNERING PERFORMANCE

forces would suffice; however, this is not the case. The cornering forces which appear in the equations of motion cannot have arbitrary values. A slip angle is generated at each tire by application of a lateral force. However, under non-skid conditions, the slip angles have geometric constraints which are defined by the velocity vector diagrams shown in Figure (5.3). In addition to the vehicle CG velocity, at each axle there is a velocity perpendicular to the chassis longitudinal centerline due to the angular speed of rotation of the vehicle about the center of gravity. The instantaneous velocity of the tire/rim combination is the vector sum of these two velocities. The direction of these velocities gives rise to the following geometric relationships which involve the tire slip angle:

(*i*) at the front axle

$$\tan(\delta - \alpha_f) = \frac{L_f \dot{\psi} + V\sin(\beta)}{V\cos(\beta)} \tag{5.12}$$

and (*ii*) at the rear axle

$$\tan(\alpha_r) = \frac{L_r \dot{\psi} - V\sin(\beta)}{V\cos(\beta)} \tag{5.13}$$

Once again using the small angle approximation, the slip angle relationships simplify to

$$\alpha_f = \delta - \frac{L_f \dot{\psi}}{V} - \beta \tag{5.14}$$

and

$$\alpha_r = \frac{L_r \dot{\psi}}{V} - \beta \tag{5.15}$$

Eqs (5.14) and (5.15) place additional constraints on the vehicle motion. The vehicle must achieve and angular orientation which allows the front and rear cornering forces to generate slip angles which satisfy these requirements.

The relationship between lateral force and slip angle for a particular tire is actually quite complex. It is dependent on parameters such as normal load, camber angle, inflation pressure, and temperature. Under the condition of long radius turns, the slip angles are generally small in magnitude. If we examine lateral force versus slip angle tire characteristics at small angles of slip (typically less than five degrees), we find that the effect of normal load, camber angle, etc., is slight. Furthermore, we can describe the lateral force as a linear function of slip angle by the relationship $F_c = C_\alpha \alpha$ where C_α is a constant value called the tire cornering stiffness. The cornering stiffness is empirically evaluated as $\left. \frac{\partial F_c}{\partial \alpha} \right|_{\alpha=0}$ from tire performance data, as discussed in Chapter 2. Consequently, using Eq (5.14), the front axle cornering force can be expressed as

$$F_{cf} = 2\,C_{af}\,\alpha_f$$

$$= 2\,C_{a_f}\left(\delta - \frac{L_r\dot{\psi}}{V} - \beta\right)$$

$$(5.16)$$

and, using Eq (5.15), the expression for the rear axle cornering force is

$$F_{cr} = 2\,C_{ar}\,\alpha_f$$

$$= 2\,C_{ar}\left(\frac{L_r\dot{\psi}}{V} - \beta\right)$$

$$(5.17)$$

It should be noted that the multiplication factor of two in the cornering force equations is required to adjust the tire properties of the two-wheel vehicle model to reflect that in a real four-wheel vehicle that are actually two tires at each axle.

Placing Eqs (5.16) and (5.17) into the first simplified equation of motion, i.e., Eq (5.8), results in

$$m\,\frac{V^2}{R} - \frac{2}{V}\left(C_{ar}L_r - C_{af}L_f\right)\dot{\psi} + 2\left(C_{ar} + C_{af}\right)\beta = 2\,C_{af}\delta \qquad (5.18)$$

The turning radius, R, can be eliminated in the previous equation by recalling that $V = R\dot{v}$. By virtue of the angular definitions depicted in Figure (5.2), $v = \psi + \beta$. Correspondingly, the angular speeds are related by $\dot{v} = \dot{\psi} + \dot{\beta}$. From here it follows that $\dfrac{V}{R} = \dot{\psi} + \dot{\beta}$. Eq (5.18) can then be written as

$$\left[m\,V - \frac{2}{V}\left(C_{ar}L_r - C_{af}L_f\right)\right]\dot{\psi} + m\,V\dot{\beta} + 2\left(C_{ar} + C_{af}\right)\beta = 2\,C_{af}\delta \quad (5.19)$$

Finally, we arrive at the second governing equation of motion by inserting Eqs (5.16) and (5.17) into Eq (5.9), and that is

$$I\ddot{\psi} + \frac{2}{V}\left(C_{ar}L_r^2 + C_{af}L_f^2\right)\dot{\psi} - 2\left(C_{ar}L_r - C_{af}L_f\right)\beta = 2\,C_{af}L_f\delta \qquad (5.20)$$

(5.2.2) CORNERING STABILITY

The steer angle, which appears as a forcing function on the right-hand-side of Eqs (5.19) and (5.20), is generally an unknown function of time, dependent on the reaction of the driver while cornering. Consequently, solutions to the equations of motion are available for only a few special cases. Nevertheless, we can learn much about the cornering behavior of a vehicle without actually having to solve these equations.

The first item we notice is that the yaw angle, ψ, does not appear in either Eq (5.19) or (5.20), but rather it's first derivative with respect to time. This time derivative is referred to as the yaw velocity, or yaw rate, Ω. Using the definition of yaw velocity, the simplified equations of motion become linear, first order differential equations. The classical mathematical approach to solving these equations is to assume that the homogeneous part of the solution is of the form

$$\Omega = A e^{\lambda t} \tag{5.21}$$

$$\beta = B e^{\lambda t} \tag{5.22}$$

Placing the assumed solution into the homogeneous equations of motion results in

$$\left[m V - \frac{2}{V} \left(C_{\alpha r} L_r - C_{\alpha f} L_f \right) \right] A + m V \lambda B + 2 \left(C_{\alpha r} + C_{\alpha f} \right) B = 0$$

$$I \lambda A + \frac{2}{V} \left(C_{\alpha r} L_r^2 + C_{\alpha f} L_f^2 \right) A - 2 \left(C_{\alpha r} L_r - C_{\alpha f} L_f \right) B = 0$$

or, in matrix form

$$\begin{bmatrix} m V - \dfrac{2}{V} \left(C_{\alpha r} L_r - C_{\alpha f} L_f \right) & m V \lambda + 2 \left(C_{\alpha r} + C_{\alpha f} \right) \\[2ex] I \lambda + \dfrac{2}{V} \left(C_{\alpha r} L_r^2 + C_{\alpha f} L_f^2 \right) & - 2 \left(C_{\alpha r} L_r - C_{\alpha f} L_f \right) \end{bmatrix} \begin{Bmatrix} A \\[1ex] B \end{Bmatrix} = \begin{Bmatrix} 0 \\[1ex] 0 \end{Bmatrix}$$

$$\tag{5.23}$$

To obtain non-trivial values for constants A and B the determinant of the coefficient matrix in Eq (5.23) must be zero, i.e.,

$$
\det \begin{bmatrix} mV - \dfrac{2}{V}\left(C_{ar}L_r - C_{af}L_f\right) & mV\lambda + 2\left(C_{ar} + C_{af}\right) \\[2em] I\lambda + \dfrac{2}{V}\left(C_{ar}L_r^2 + C_{af}L_f^2\right) & -2\left(C_{ar}L_r - C_{af}L_f\right) \end{bmatrix}
$$

$$
= \left[mV - \dfrac{2}{V}\left(C_{ar}L_r - C_{af}L_f\right)\right]\left[-2\left(C_{ar}L_r - C_{af}L_f\right)\right]
$$

$$
- \left[mV\lambda + 2\left(C_{ar} + C_{af}\right)\right]\left[I\lambda + \dfrac{2}{V}\left(C_{ar}L_r^2 + C_{af}L_f^2\right)\right] = 0
$$

or, with a little bit of algebra

$$
mVI\lambda^2 + \left[2m\left(C_{ar}L_r^2 + C_{af}L_f^2\right) + 2\left(C_{ar} + C_{af}\right)I\right]\lambda
$$

$$
+ 2mV\left(C_{ar}L_r - C_{af}L_f\right) + \dfrac{4}{V}C_{ar}C_{af}\left(L_r + L_f\right)^2 = 0
\tag{5.24}
$$

The previous equation is the characteristic equation associated with the equations of motion. You probably recognize the generic form of this equation as the quadratic equation. It follows that the characteristic roots would then be determined from

$$
\lambda_{1,2} = \left\{ -\left[m\left(C_{ar}L_r^2 + C_{af}L_f^2\right) + \left(C_{ar} + C_{af}\right)I\right] \right.
$$

$$
\pm \sqrt{\left[m\left(C_{ar}L_r^2 + C_{af}L_f^2\right) + \left(C_{ar} + C_{af}\right)I\right]^2}
$$

$$
\left. \overline{ -mVI\left[mV\left(C_{ar}L_r - C_{af}L_f\right) + \dfrac{2}{V}C_{ar}C_{af}\left(L_r + L_f\right)^2\right] } \right\} \Big/ (mVI)
$$

$$
\tag{5.25}
$$

The conditions under which the cornering maneuver will be stable are surmised by closer examination of the characteristic roots. The concept of cornering stability is mathematically founded on the criteria that the yaw velocity and the sideslip angle

$$
mVI\left[mV\left(C_{ar}L_r - C_{af}L_f\right) + \dfrac{2}{V}C_{ar}C_{af}\left(L_r + L_f\right)^2\right] \geq 0
$$

cannot become infinitely large, and both must be able to attain a steady-state solution. Based on the assumed linear solutions, Eqs (5.21) and (5.22), the possible values of λ must either be real and negative, or complex with a negative real part in order to satisfy the stability requirement. The leading term in Eq (5.25), i.e., $m\left(C_{\alpha r}L_r^2 + C_{\alpha f}L_f^2\right) + \left(C_{\alpha r} + C_{\alpha f}\right)I$, is always greater than zero. Therefore, cornering stability requires that which implies that

$$m\,V^2\left(C_{\alpha r}L_r - C_{\alpha f}L_f\right) + 2\,C_{\alpha r}C_{\alpha f}\left(L_r + L_f\right)^2 \geq 0 \tag{5.26}$$

We conclude from Eq (5.26) that if $C_{\alpha r}L_r \geq C_{\alpha f}L_f$, then the inequality always holds true and the vehicle retains cornering stability for all forward speeds. We also conclude that if $C_{\alpha r}L_r < C_{\alpha f}L_f$ it still possible for the car to remain stable provided that the forward velocity does not exceed a critical value of

$$V_{crit} =$$

$$\sqrt{-\frac{2\,C_{\alpha r}C_{\alpha f}\left(L_r + L_f\right)^2}{m\left(C_{\alpha r}L_r - C_{\alpha f}L_f\right)}} \tag{5.27}$$

(5.2.2A) Static Margin

An alternate method of measuring vehicle stability is static margin. This method is particularly useful in analyzing the stability of more complex vehicle models, such as the four-wheel model introduced later in this chapter.

Static margin is derived from the concept of static stability[3], which quantitatively describes a vehicle's ability to overcome the effects of

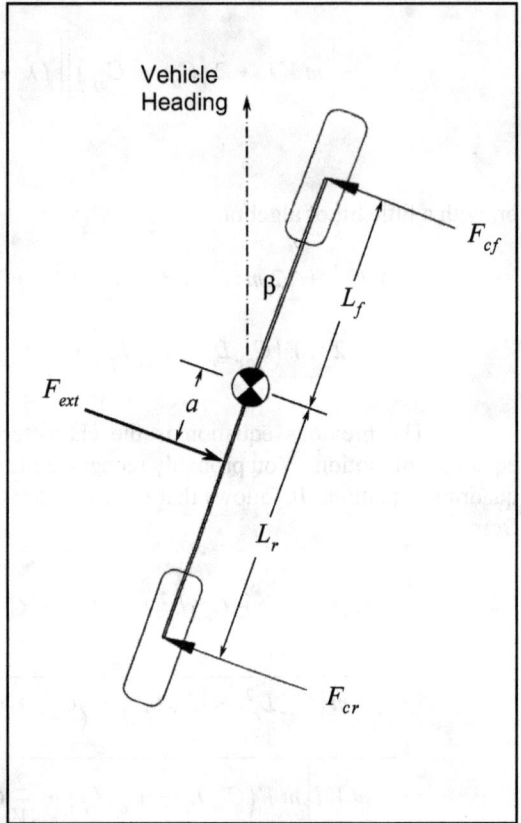

Figure (5.4): TWO-WHEEL VEHICLE EXPERIENCING AN EXTERNAL DISTURBING FORCE

---

[3] *Static stability is a well established technique used in designing aircraft. It was first used to study the stability of aircraft in wind tunnel tests.*

an external disturbance. To develop this concept, let's consider a two-wheel vehicle that is initially traveling in a straight line (zero steer angle). The vehicle is then disturbed by an external force. The external force is undefined but could be the result of a cross wind, aerodynamic drag, or perhaps a gravitational body force due to a crown in the road surface. The exact cause of the external force is unimportant in studying the static stability of the car. In any event, we will consider that the line of action of the external force is perpendicular to the vehicle longitudinal axis. The application of this force to the car causes lateral forces to be developed at both the front and rear tires, analogous to the cornering forces discussed earlier. As we have learned from tire behavior, each lateral force will generate a slip angle at their respective tires. The tire slip angles, in turn, cause vehicle sideslip to occur as illustrated in Figure (5.4).

If we ignore the fact that our vehicle is moving we can use statics to investigate the forces and moments acting on the car. Two conditions must be met if static equilibrium is to be maintained:

(*i*)

$$\sum F = 0 = F_{ext} - F_{cr} - F_{cf} \tag{5.28}$$

and (*ii*)

$$\sum M_{CG} = 0 = F_{ext} a - F_{cr} L_r + F_{cf} L_f + M_{ext} \tag{5.29}$$

You will notice that we have added an external yaw moment, M_{ext}, to Eq (5.29) that is not shown in Figure (5.4). At this point the distance a is unspecified, which means the application of the forces themselves will not necessarily result in static equilibrium. Therefore, it may be necessary for us to apply an external yawing moment to regain equilibrium.

Using Eq (5.28), the external force term is related to the tire slip angles by

$$F_{ext} = 2 C_{ar} \alpha_r + 2 C_{af} \alpha_f \tag{5.30}$$

Since the steer angle is zero, the tire slip angles and the vehicle sideslip have the same value, i.e., $\alpha_r = \alpha_f = \beta$. Then from Eq (5.30) the external force can be written in terms of the sideslip as

$$F_{ext} = 2(C_{ar} + C_{af})\beta \tag{5.31}$$

Also, from Eq (5.29) we have

$$F_{ext} a = 2(C_{ar} L_r - C_{af} L_f)\beta - M_{ext} \tag{5.32}$$

Substituting Eq (5.31) into (5.32) effectively eliminates the external disturbing force from any further consideration, i.e.,

$$2\left(C_{\alpha r} + C_{\alpha f}\right)\beta\, a = 2\left(C_{\alpha r} L_r - C_{\alpha f} L_f\right)\beta - M_{ext} \tag{5.33}$$

Of particular interest to suspension designers is the unique location along the vehicle longitudinal axis giving a distance a where an external yawing moment is not required for static equilibrium. This point is called the *Neutral Steer Point* (NSP). From Eq (5.33):

$$a_{NSP} = \frac{C_{\alpha r} L_r - C_{\alpha f} L_f}{C_{\alpha r} + C_{\alpha f}} \tag{5.34}$$

The *Static Margin* (SM) is a non-dimensional variant of the neutral steer point, defined as

$$SM \equiv \frac{a_{NSP}}{L_w} = \frac{C_{\alpha r} L_r - C_{\alpha f} L_f}{\left(C_{\alpha r} + C_{\alpha f}\right) L_w} \tag{5.35}$$

In the previous section we found that when $C_{\alpha r} L_r \geq C_{\alpha f} L_f$ the vehicle is always stable; therefore, based on Eq (5.35), when $SM \geq 0$, which results when the neutral steer point is at or aft of the center of gravity, is also an indication of a stable vehicle. On the other hand, if the neutral steer point is forward of the center of gravity, or $SM < 0$, the vehicle will become unstable for vehicle velocities in excess of the critical velocity {Eq (5.27)}.

The results of the stability analysis are very important for the suspension designers. From the results we conclude that the tire cornering stiffness properties and the vehicle center of gravity determine the stability of the vehicle.

(5.2.3) STEADY-STATE SOLUTION

The steady-state solutions to the equations of motion are easy to find. First, we recognize that derivatives with respect to time are zero under steady-state conditions. In this case Eqs (5.19) and (5.20) become

$$\left[mV - \frac{2}{V}\left(C_{\alpha r} L_r - C_{\alpha f} L_f\right)\right]\Omega_{SS} + 2\left(C_{\alpha r} + C_{\alpha f}\right)\beta_{SS} = 2C_{\alpha f}\delta \tag{5.36}$$

and

$$\frac{2}{V}\left(C_{\alpha r} L_r^2 + C_{\alpha f} L_f^2\right)\Omega_{SS} - 2\left(C_{\alpha r} L_r - C_{\alpha f} L_f\right)\beta_{SS} = 2C_{\alpha f} L_f\delta \tag{5.37}$$

Simultaneously solving Eqs (5.36) and (5.37) for the yaw velocity produces the result

$$\frac{\Omega_{SS}}{\delta} = \frac{V}{\left(L_r + L_f\right) + \dfrac{C_{ar}L_r - C_{af}L_f}{2\,C_{ar}\,C_{af}\left(L_r + L_f\right)}\,m\,V^2} \tag{5.38}$$

We can write the previous equation in a more compact form by noting that

$$\frac{W_f}{g} = \left(\frac{L_r}{L_r + L_f}\right)m \quad \text{and} \quad \frac{W_r}{g} = \left(\frac{L_f}{L_r + L_f}\right)m \; , \text{where } W_f \text{ and } W_r \text{ are the}$$

vehicle weights[4] at the front and rear axle planes, respectively. In addition, the sum $L_r + L_f$ can be replaced by the vehicle wheelbase L_w . Using these simplifications, Eq (5.38) then becomes

$$\frac{\Omega_{SS}}{\delta} = \frac{V}{L_w + \left(\dfrac{W_f}{C_{af}} - \dfrac{W_r}{C_{ar}}\right)\dfrac{V^2}{2\,g}} \tag{5.39}$$

The term on the left-hand-side of the previous equation is call the yaw velocity gain or yaw velocity response. Symbolically, $G_{yaw} \equiv \dfrac{\Omega_{SS}}{\delta}$. In addition, it is customary to define an understeer coefficient, K_{us}, (expressed in non-dimensional units of radians) as

$$K_{us} \equiv \frac{W_f}{C_{af}} - \frac{W_r}{C_{ar}} \tag{5.40}$$

We can, therefore, rewrite Eq (5.39) in it's simplest form as

$$G_{yaw} = \frac{V}{L_w + K_{us}\dfrac{V^2}{2\,g}} \tag{5.41}$$

In a similar fashion, the steady-state solution for the sideslip gain is found to be

---

[4] *The relationships for W_f and W_r are equal and opposite to the reaction forces acting at the front and rear wheels. The reactions are derived from a static analysis in the usual manner, i.e., (1) summation of forces {in a normal or vertical direction} equal to zero, and (2) the summation of moments {taken about the vehicle lateral axis passing through the center of gravity equal to zero.*

Chapter 5: CORNERING PERFORMANCE

$$G_{slip} \equiv \frac{\beta_{SS}}{\delta} = \frac{L_r - \dfrac{W_r}{C_{\alpha r}} \dfrac{V^2}{2g}}{L_w + K_{us} \dfrac{V^2}{2g}} \tag{5.42}$$

We notice that the denominators in Eqs (5.41) and (5.42) are exactly the same, and are dependent on the value of the understeer coefficient. This raises new questions. What is understeer, and what is its practical significance in the design of vehicle suspensions?

(5.2.3A) Understeer, Neutral Steer, and Oversteer

The origin of understeer and oversteer is attributed to Maurice Olley {circa 1940}. He used this concept to describe the behavior of a vehicle traveling in a forward direction when acted upon by a lateral force. In modern practice, suspension designers use these words to describe certain cornering characteristics of vehicles.

It is probably easiest to understand the modern concept of understeer, neutral steer, and oversteer by examining the geometric behavior of a vehicle as described by the curvature gain or response, defined as $G_{curv} \equiv \dfrac{1/R}{\delta}$. For steady-state cornering we also note that $V = \Omega_{SS} R$. Substituting these relationships into Eq (5.41) gives us

$$G_{curv} = \frac{1}{L_w + K_{us} \dfrac{V^2}{2g}} \tag{5.43}$$

Let's take a moment to consider what happens to a car if we hold the steer angle at some fixed position, and then increase the forward velocity ever so slightly. {The latter specification is required so that we maintain at least a quasi steady-state condition.} Figure (5.5) demonstrates that there are three different characteristics which the car could exhibit depending on the value of the understeer coefficient:

(i) understeer {$K_{us} > 0$}

A vehicle which understeers tends to move away from the turn center. The driver must, therefore, increase the steer angle while accelerating through a constant radius corner. A steer angle relationship which describes this point is obtained by rearranging the curvature gain equation, i.e.,

$$\delta = \frac{L_w}{R} + K_{us} \frac{V^2}{2gR} \tag{5.44}$$

A graph of Eq (5.44) for a constant radius turn is shown in Figure (5.6). We can see that the steer angle increases parabolically with the forward velocity, and that the vehicle is stable for all forward velocities.

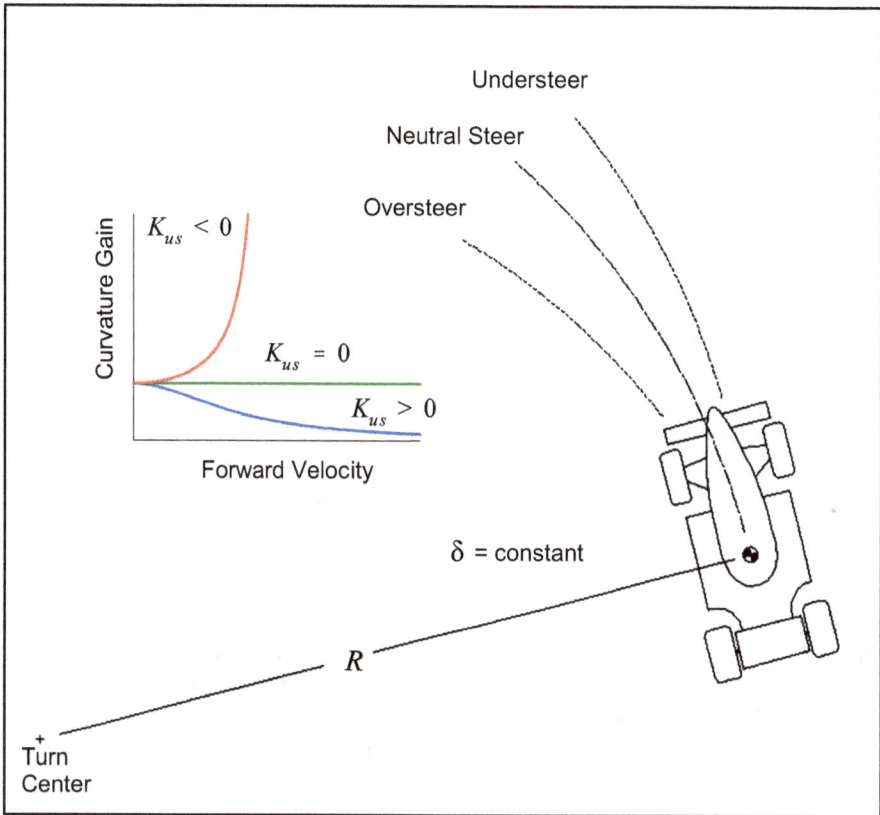

Figure (5.5): CURVATURE GAIN OF VEHICLES AT FIXED STEER ANGLES

The understeering phenomenon can also be explained in terms of the tire slip angles. When a lateral force, or centrifugal inertia force, acts at the vehicle CG the slip angles generated at the front tires are greater than the slip angles of the rear tires. The difference in the slip angles causes a yaw motion, which rotates the front of the vehicle away from the lateral force.

(ii) neutral steer $\{K_{us} = 0\}$

A vehicle which exhibits no change in curvature with velocity is said to have neutral steer characteristics. It is interesting to note that if we substitute $K_{us} = 0$ into

Eq (5.44) we obtain $\delta = \dfrac{L_w}{R}$, the Ackermann steer angle.

(iii) oversteer $\{K_{us} < 0\}$

As you might expect, oversteer is just the opposite of understeer. A vehicle which has this characteristic moves toward the turn center. In this case the application of a lateral side force at the vehicle CG creates larger tire slip angles at the rear than the front. The slip angle difference causes a yaw motion which rotates the front of the vehicle toward the lateral force. To compensate for the yaw affect the driver must decrease the steer angle, or steer out of the corner, while accelerating through a constant radius turn. The driver may continue to accelerate through the turn until the forward velocity reaches it's critical value {Eq (5.27)[5]}, at which time the vehicle becomes directionally unstable. We conclude that, while oversteer affects are opposite those of understeer, there is a major difference in vehicle stability between these two characteristics. Now that we have a better understanding of understeer, neutral steer, and oversteer characteristics, let's return to the yaw velocity gain and sideslip gain equations. Typical plots of both of these relationships are shown in Figures (5.7) and (5.8).

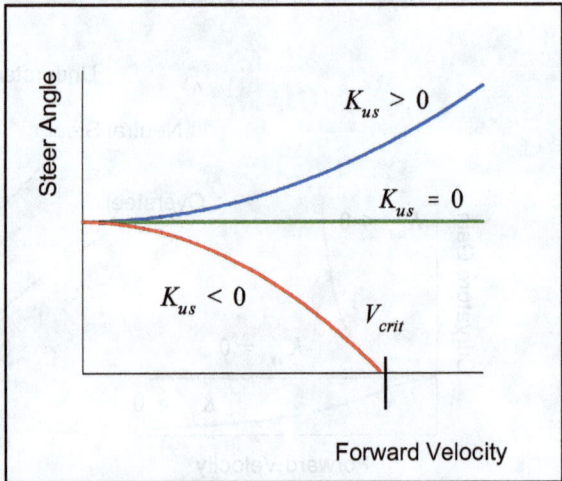

Figure (5.6): STEER ANGLE AS A FUNCTION OF FORWARD VELOCITY FOR A CONSTANT RADIUS TURN

The graph in Figure (5.7) demonstrates that the yaw velocity gain for an understeering vehicle has a maximum value. The forward velocity at which the yaw velocity reaches this value is called the characteristic velocity. We can derive an expression for the characteristic velocity by first taking the derivative of yaw velocity gain with respect to forward velocity. Secondly, this derivative is set equal to zero:

$$\frac{\partial G_{yaw}}{\partial V} = \frac{L_w - K_{us} \dfrac{V^2}{2g}}{\left[L_w - K_{us} \dfrac{V^2}{2g} \right]^2} = 0$$

therefore,

---

[5] Using the definition of understeer coefficient, we can write a slightly different expression for the critical forward velocity, i.e., $V_{crit} = \sqrt{-\dfrac{2g L_w}{K_{us}}}$

$$V_{char} = \sqrt{\frac{2\,g\,L_w}{K_{us}}} \qquad (5.45)$$

Astute readers will observe that, with the exception of the minus sign, the expression for the critical velocity is the same as the characteristic velocity. As we have learned, however, the physical significance of these two velocities is quite different.

There is one other curvature relationship worthy of our attention. The relationship is a variation of the steering angle equation (5.44), i.e.,

$$\frac{1}{R} = \frac{\delta}{L_w} - \frac{K_{us}}{2\,L_w}\,a_{lat}$$
$$(5.46)$$

where a_{lat} is the lateral acceleration {in dimensionless units of g's} defined as $\dfrac{V^2/R}{g}$. A typical graph of curvature versus lateral acceleration for fixed angles of steer is shown in Figure (5.9). The actual numbers for curvature are not nearly important as the slope, or tangent line, of the function. As we can see, if the slope is positive the vehicle has oversteer characteristics. Likewise, a zero slope indicates neutral steer and a negative slope indicates understeer. We will find these results particularly useful when we develop a four-wheel cornering model.

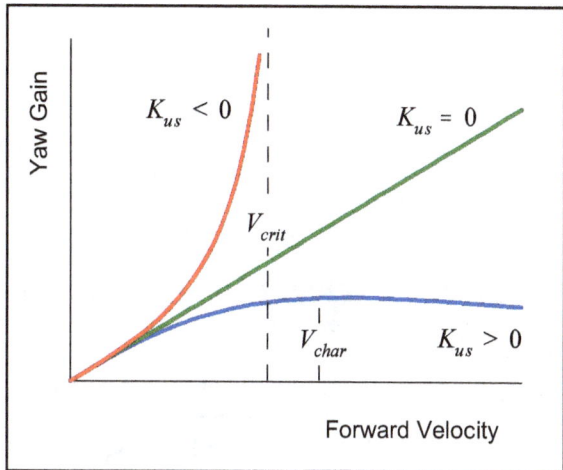

Figure (5.7): YAW VELOCITY GAIN AS A FUNCTION OF FORWARD VELOCITY

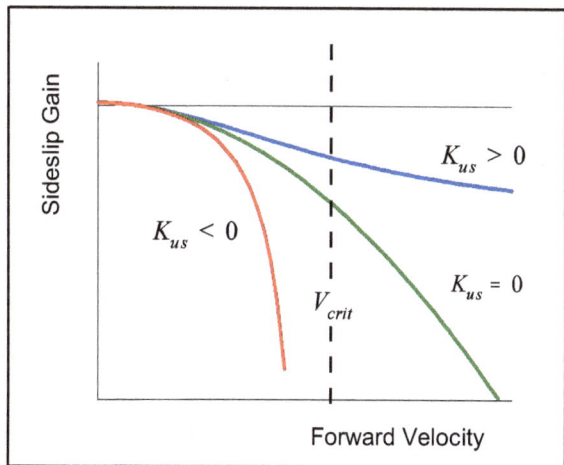

Figure (5.8): SIDESLIP GAIN AS A FUNCTION OF FORWARD VELOCITY

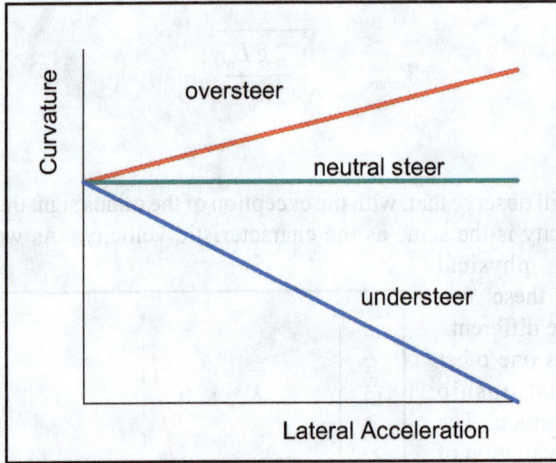

Figure (5.9): CURVATURE AS A FUNCTION OF LATERAL ACCELERATION

(5.2.4) TRANSIENT RESPONSE TO STEP CHANGE IN STEER ANGLE

Earlier we mentioned that solutions to the simple linear equations of motion {Eqs (5.19) and (5.20)} are available for only a few special cases, since the steer angle forcing function is generally an unknown function of time. One specific example we will examine is the solution, or cornering response, to a step change in steer angle. This will help us gain a better understanding of vehicle transient behavior.

Much of the analysis is the foregoing sections involves aspects of the homogeneous solution to the simple equations of motion. A natural extension of the previous investigation is to complete the homogeneous solution. Two characteristic roots are identified by Eq (5.25). Depending on whether the roots are real, complex, or double the homogeneous solution for the yaw velocity and the sideslip angle can have three different forms:

(i) real distinct roots

$$\Omega_h = A_1 e^{\lambda_1 t} + A_2 e^{\lambda_2 t} \tag{5.47}$$

$$\beta_h = B_1 e^{\lambda_1 t} + B_2 e^{\lambda_2 t} \tag{5.48}$$

(ii) complex roots

$$\Omega_h = e^{\lambda t}\left[A_1 \cos(\omega t) + A_2 \sin(\omega t)\right] \tag{5.49}$$

$$\beta_h = e^{\lambda t}\left[B_1 \cos(\omega t) + B_2 \sin(\omega t)\right] \tag{5.50}$$

where $\lambda = -\dfrac{b}{2a}$ and $\omega = \dfrac{1}{2a}\sqrt{|b^2 - 4ac|}$. Referring to Eq (5.24), the symbols a, b, and c are shorthand notations for:

$$a = m V I$$

$$b = 2 m \left(C_{\alpha r} L_r^2 + C_{\alpha f} L_f^2 \right) + 2 \left(C_{\alpha r} + C_{\alpha f} \right) I$$

$$c = 2 m V \left(C_{\alpha r} L_r - C_{\alpha f} L_f \right) + \frac{4}{V} C_{\alpha r} C_{\alpha f} \left(L_r + L_f \right)^2$$

(iii) double roots

$$\Omega_h = \left(A_1 + A_2 t \right) e^{\lambda t} \qquad (5.51)$$

$$\beta_h = \left(B_1 + B_2 t \right) e^{\lambda t} \qquad (5.52)$$

where $\lambda = -\dfrac{b}{2 a}$.

In all cases, the particular part of the solution could be found by employing the "method of undetermined coefficients." However, the particular solutions have a special significance in initial value problems, such as we currently have, they represent the steady-state solutions. The steady-state yaw velocity, Ω_{SS}, has already been derived, and is given by Eq (5.41). Likewise, the sideslip steady-state solution, β_{SS}, is delineated by Eq (5.42).

Since the simple equations of motion are linear differential equations, the general solutions for the yaw velocity and the sideslip angle are obtained by adding their respective homogeneous and particular parts together. All that remains is to evaluate the constants A_1 and A_2 in the yaw velocity general solution; and, similarly, constants B_1 and B_2 found in the sideslip angle solution. For this we need to specify the initial conditions. If our car was initially traveling in a straight line, then (i) $\Omega(0) = \dot{\Omega}(0) = 0$ and (ii) $\beta(0) = \dot{\beta}(0) = 0$. Using these initial conditions three different forms of the final solutions for the yaw velocity and the sideslip angle are obtained, again depending on the type of characteristic roots:

(i) real distinct roots

$$\Omega(t) = \Omega_{SS} \left[1 - \frac{\lambda_1 e^{\lambda_2 t} - \lambda_2 e^{\lambda_1 t}}{\lambda_1 - \lambda_2} \right] \qquad (5.53)$$

$$\beta(t) = \beta_{SS} \left[1 - \frac{\lambda_1 e^{\lambda_2 t} - \lambda_2 e^{\lambda_1 t}}{\lambda_1 - \lambda_2} \right] \qquad (5.54)$$

(ii) complex roots

$$\Omega(t) = \Omega_{SS}\left\{1 + e^{\lambda t}\left[\frac{\lambda}{\omega}\sin(\omega t) - \cos(\omega t)\right]\right\} \qquad (5.55)$$

$$\beta(t) = \beta_{SS}\left\{1 + e^{\lambda t}\left[\frac{\lambda}{\omega}\sin(\omega t) - \cos(\omega t)\right]\right\} \qquad (5.56)$$

and (iii) double roots

$$\Omega(t) = \Omega_{SS}\left[1 + (\lambda t - 1)e^{\lambda t}\right] \qquad (5.57)$$

$$\beta(t) = \beta_{SS}\left[1 + (\lambda t - 1)e^{\lambda t}\right] \qquad (5.58)$$

In all three instances the yaw angle, if desired, is determined from the integral

$$\psi(t) = \int_0^t \Omega(\tau)\,d\tau \qquad (5.59)$$

The yaw velocity and sideslip solutions, Eqs (5.53) through (5.58), provides us with a means to simulate the transient cornering behavior of a two-wheel vehicle model. These are simple equations which can be programmed into a spreadsheet or computer program to execute multiple calculations. Let's look at a specific example. Suppose we have a car which weighs 3188 lbs and has a wheelbase of 108 in. The front tires on the car have a cornering stiffness of 157.5 lb/deg, and the stiffness of the rear tires are 191.5 lb/deg. The present car has the engine mounted in the front producing a weight distribution of 57.6% on the front axle plane and 42.4% on the rear. With this information, the understeer coefficient is calculated {Eq (5.40)} to be 0.08029 radians, indicating that the vehicle exhibits understeer characteristics. Since the transient response is being sought, we need a value for the moment of inertia; which can be estimated[6] from

$$I = \frac{W_f}{g}L_f^2 + \frac{W_r}{g}L_r^2 = L_f L_r \frac{W}{g}$$

The transient cornering response to a three degree steer angle in shown in Figure (5.10A). We notice that the maximum yaw velocity, which is a measure of how fast our vehicle can execute a change in course heading, is around 15 deg/s. We also notice that under all forward speeds we are able to attain stable, steady-state characteristics.

Now suppose we redesign our vehicle and move the engine to the rear without making any further modifications. In this instance the weight distribution may become 35.0% front and 65.0% rear. Our redesigned car now oversteers. Figure (5.10B)

---

[6] *Details can be found in Appendix B.*

illustrates the cornering behavior of the redesigned car under the same conditions as the original car. In general the yaw velocity has greatly increased, and we conclude that the redesigned car can react to a cornering maneuver much faster. However, we should bear in mind two critical items. The first is that the car is not stable for all forward speeds. We are not able to attain steady-state cornering if the forward velocity exceeds the critical velocity of 64.25 MPH. Secondly, at high forward speeds the vehicle may respond faster to a course change that the driver's reaction time. This is, of course, assuming that the tires can develop sufficient lateral cornering forces to keep the car from spinning out. Under these conditions the driver would find the vehicle difficult to control, at best.

The last modification we will examine is, at least from a practical point of view, only theoretically possible. If we use identical tires on the front and the rear, and if we redesign our car to have a 50/50 weight distribution (perhaps by moving the engine once again), then we will have a neutral steering car. The neutral steer transient cornering results are given in Figure (5.10C). Neutral steer characteristics are a good compromise between high yaw velocity and vehicle stability. However, at high speeds it may be more advantageous to have a vehicle which slightly understeers to limit the amount of sideslip.

(5.2.4A) Moment Diagram

The vehicle characteristics presented in Figures (5.10) are valid only for one value of steer angle. Obviously, designing a vehicle suspension dictates that we consider a variety of steer angles. To minimize the quantity of tabular data, graphs, etc., it is beneficial to us to find another way to correlate the results. A convenient correlation method is the modified moment diagram.

The moment diagram originally conceived by Milliken [1] is a technique used to correlate steady-state empirical data gathered from full scale wind tunnel testing. During the tests, the vehicle is supported by a moving, flat belt; but, is constrained in both longitudinal and lateral directions, as well as yaw. Force moments are taken at various sideslip (or drift) angles and steer angles. The results are then plotted on a moment diagram, which is a plot of yawing moment coefficient, C_{yaw} (dimensionless), versus lateral acceleration a_{lat} (g's). The yawing moment is defined as

$$C_{yaw} = \frac{M_{yaw}}{W L_w} \tag{5.60}$$

where M_{yaw} is the yawing moment required to maintain equilibrium.

Our present interest, however, is correlating analytical results for transient and steady-state cornering behavior. We can still use the concept of moment diagram, but to meet our current needs we must alter the yawing moment coefficient definition to:

$$C_{yaw} = \frac{I \dot{\Omega}}{W L_w} \tag{5.61}$$

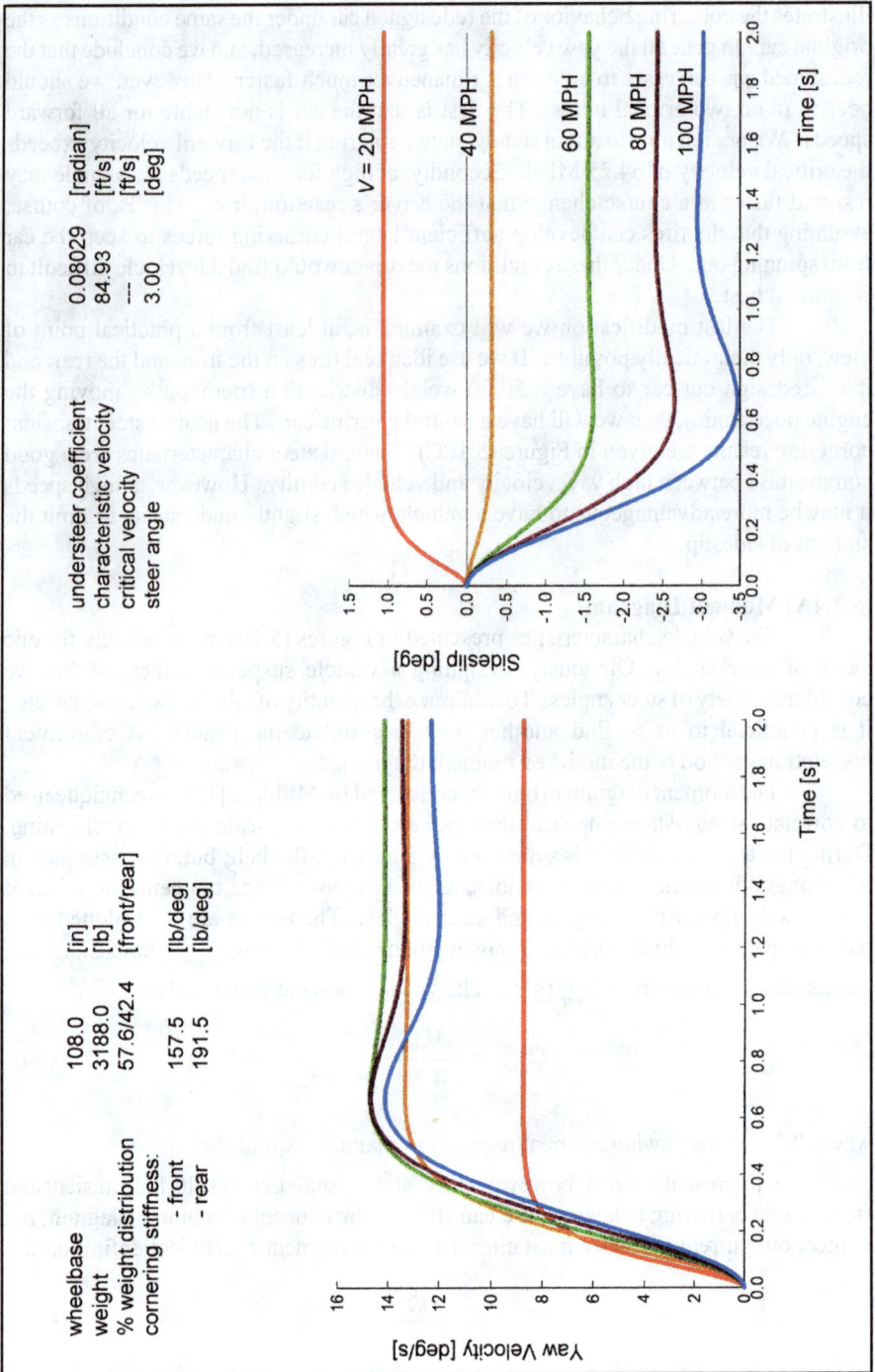

wheelbase	108.0	[in]
weight	3188.0	[lb]
% weight distribution	57.6/42.4	[front/rear]
cornering stiffness:		
- front	157.5	[lb/deg]
- rear	191.5	[lb/deg]

understeer coefficient	0.08029	[radian]
characteristic velocity	84.93	[ft/s]
critical velocity	----	[ft/s]
steer angle	3.00	[deg]

Figure (5.10A): TRANSIENT CORNERING CHARACTERISTICS OF AN UNDERSTEERING TWO-WHEEL VEHICLE

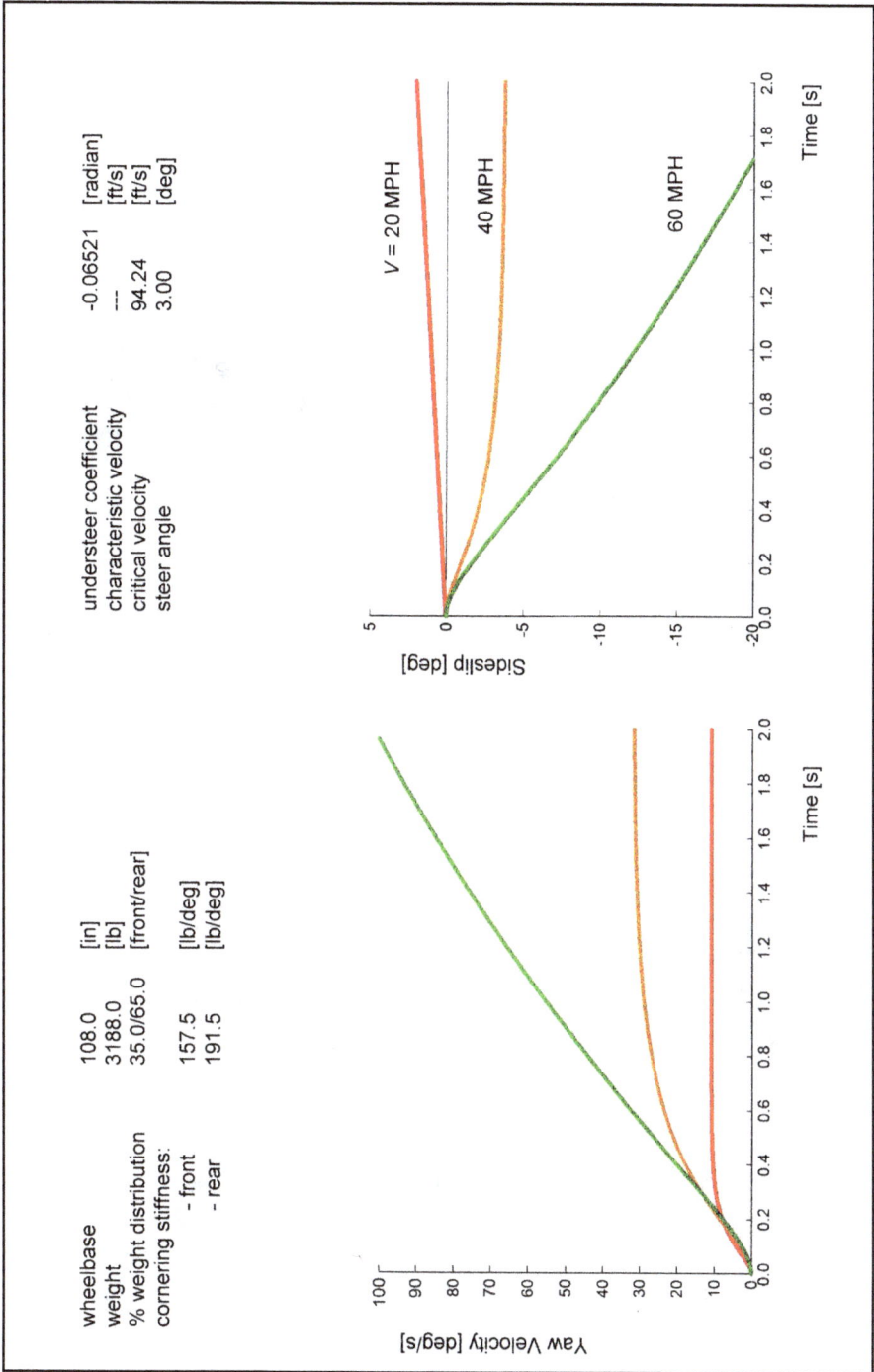

Figure (5.10B): TRANSIENT CORNERING CHARACTERISTICS OF AN OVERSTEERING TWO-WHEEL VEHICLE

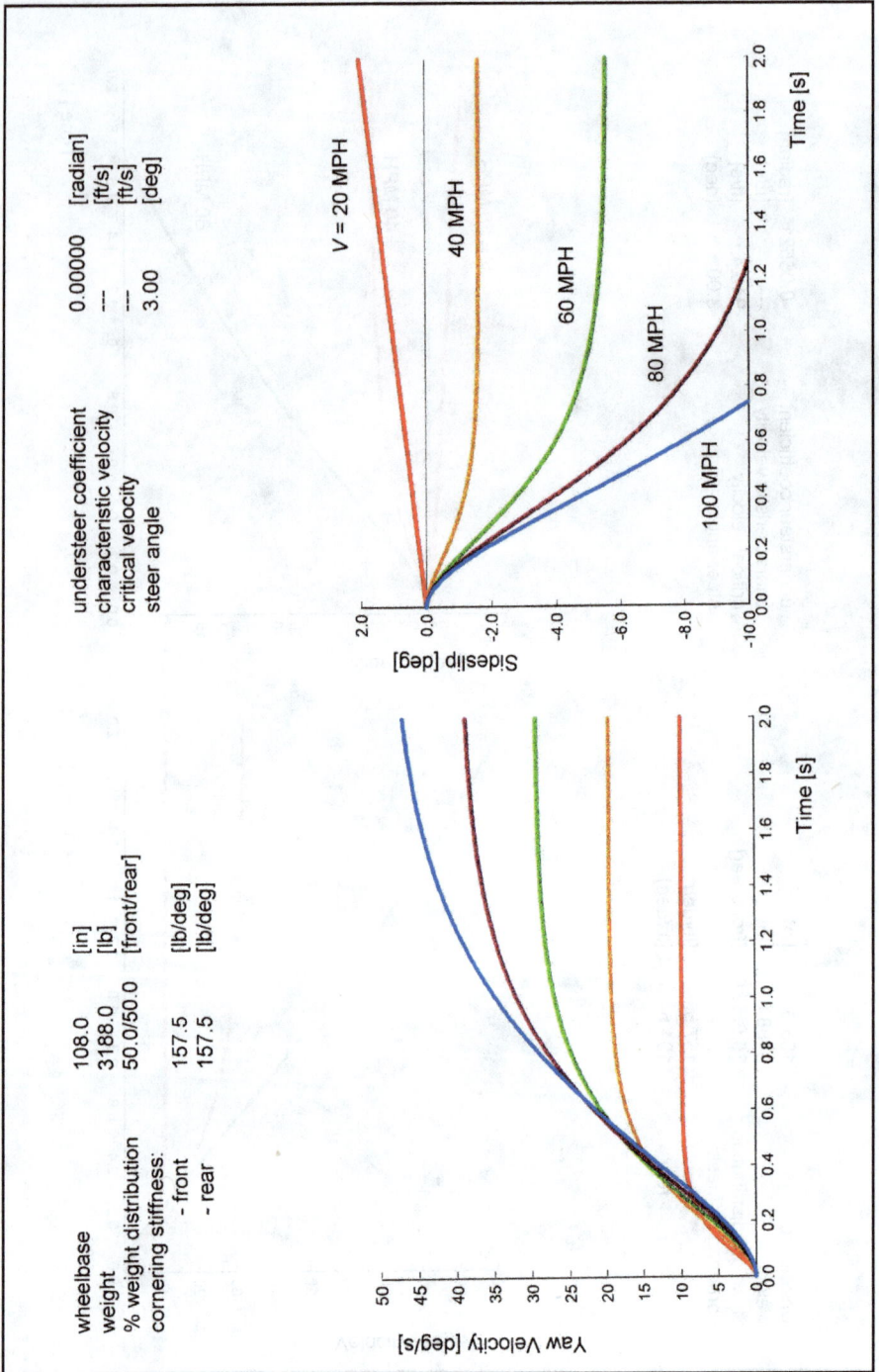

Figure (5.10C): TRANSIENT CORNERING CHARACTERISTICS OF A NEUTRAL STEERING TWO-WHEEL VEHICLE

Chapter 5: CORNERING PERFORMANCE   235

In this case the yaw acceleration, $\dot{\Omega}$, is determined by analytical means. For instance, the yaw acceleration based on a steering step input to the two-wheel cornering model is found by differentiating the yaw velocity solution:

(i) real distinct roots

$$\dot{\Omega}(t) = \Omega_{SS}\left[\frac{\lambda_1\lambda_2}{\lambda_1 - \lambda_2}\left(e^{\lambda_2 t} - e^{\lambda_1 t} \right) \right] \qquad (5.62)$$

(ii) complex roots

$$\dot{\Omega}(t) = \Omega_{SS}\left[e^{\lambda t}\left(\frac{\lambda^2}{\omega} + \omega \right) \sin(\omega t) \right] \qquad (5.63)$$

(iii) double roots

$$\dot{\Omega}(t) = \Omega_{SS}\left[\lambda^2 t\, e^{\lambda t} \right] \qquad (5.64)$$

For the curvilinear coordinate system {see Section (5.2.1)}, the lateral acceleration can be expressed as

$$a_{lat} = \frac{V^2/R}{g} = \frac{V}{g}\left(\Omega + \dot{\beta} \right) \qquad (5.65)$$

The yawing moment coefficient and lateral acceleration are calculated by specifying a steer angle, a forward velocity, followed by calculating yaw velocity, yaw acceleration and sideslip velocity for a particular time. A moment diagram is developed by varying the time from an initial value of zero to a final time determined by one of two conditions. Normally the final time is that which is required to reach steady-state conditions, but there are instances where the lateral acceleration overcomes the grip capacity of the tires. (A maximum value of 1.5 g's is a practical upper limit.) In this case the real vehicle would lose traction and either plough out or spin out. The simple two-wheel cornering model does not recognize the limit of tire adhesion, so we arbitrarily limit the maximum lateral acceleration to 1.5 g's.

An example moment diagram which illustrates the cornering behavior of the understeering and oversteering vehicles of Figures (5.10A) and (5.10B) is shown in Figure (5.11). Again we see that the understeering and oversteering vehicles exhibit entirely different cornering characteristics.

(5.3) SUSPENSION MECHANICS IN ROLL

One of the limitations of the two-wheel cornering model is its inability to account for the effects of suspension movement and body roll on cornering performance. Before we develop a more refined cornering model, however, we must introduce three new concepts: roll center, roll axis, and roll stiffness. These are required to quantify the suspension characteristics in roll, and the sprung mass roll behavior.

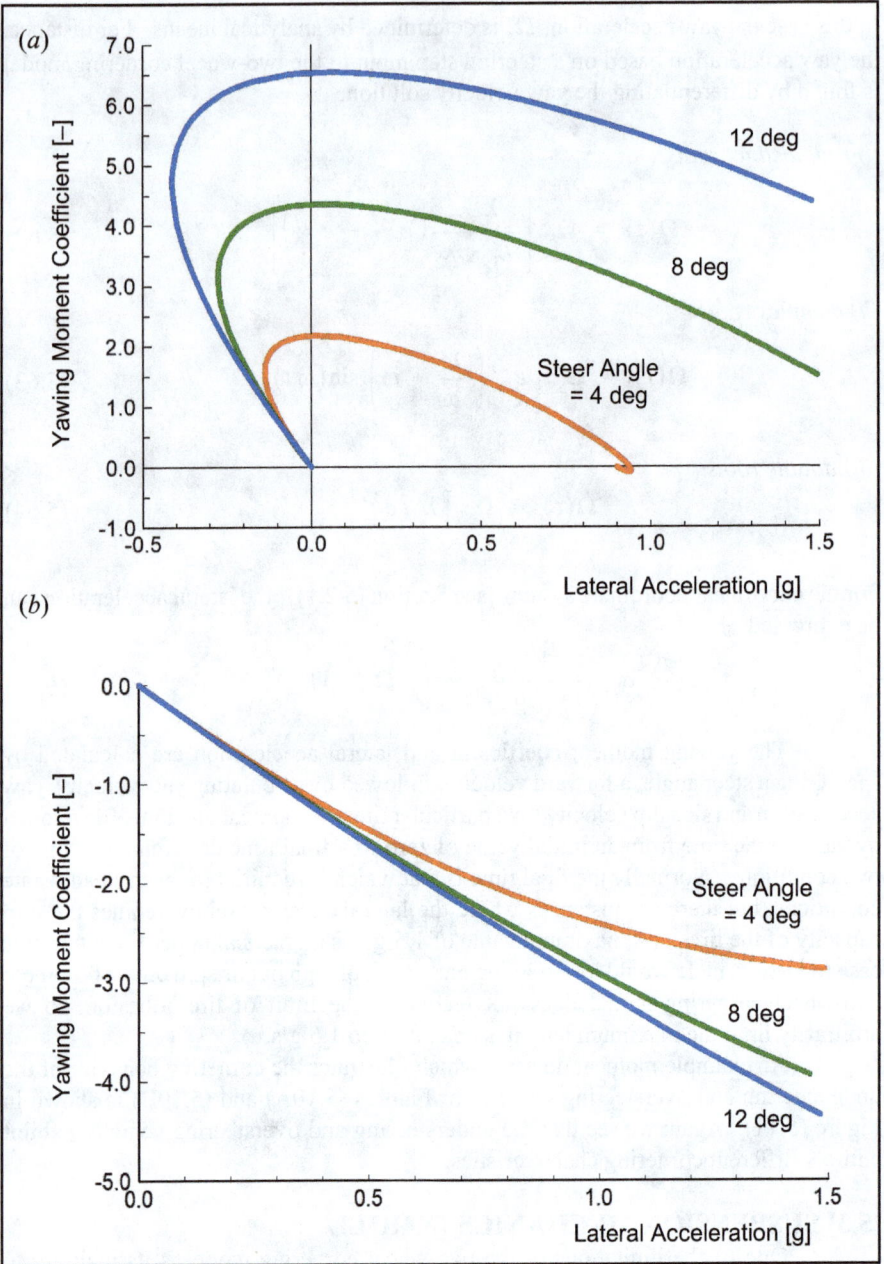

Figure (5.11): MOMENT DIAGRAM OF TRANSIENT CORNERING BEHAVIOR FOR (a) UNDERSTEERING CAR and (b) OVERSTEERING CAR

(5.3.1) ROLL CENTER

The roll center is an important characteristic of suspension geometry. It is a real or imagined point located in the suspension axle plane through which the lateral forces acting on the suspension can be transmitted to the sprung mass without causing body roll. If we think of the suspension as a system of mechanical links between the tire and chassis, then we can view the roll center as being similar to the concept of instant centers used in kinematics. Even though the concept of instant center applies only to bodies in plane motion, in many suspension the geometric techniques used to find suspension roll centers. A pictorial summary of the methodology required to locate roll centers, for commonly encountered suspensions, is provided in Figures (5.12) through (5.18). In all cases, the location of the roll center depends solely on the type of suspension and the suspension geometry.

When analyzing suspension geometry it must be kept in mind that any movement in the suspension control arms generally causes a shift in the location of the roll center. Therefore, a permanent location cannot be identified as the suspension roll center. For example, consider the double a-arm suspension system shown in Figure (5.19). At zero degrees roll (0° tire camber) the location of the roll center aligns with the chassis centerline. But, if we allow the body to roll to an extent that the right tire cambers in, say 5°, then the location of the roll center shifts away from the chassis centerline.

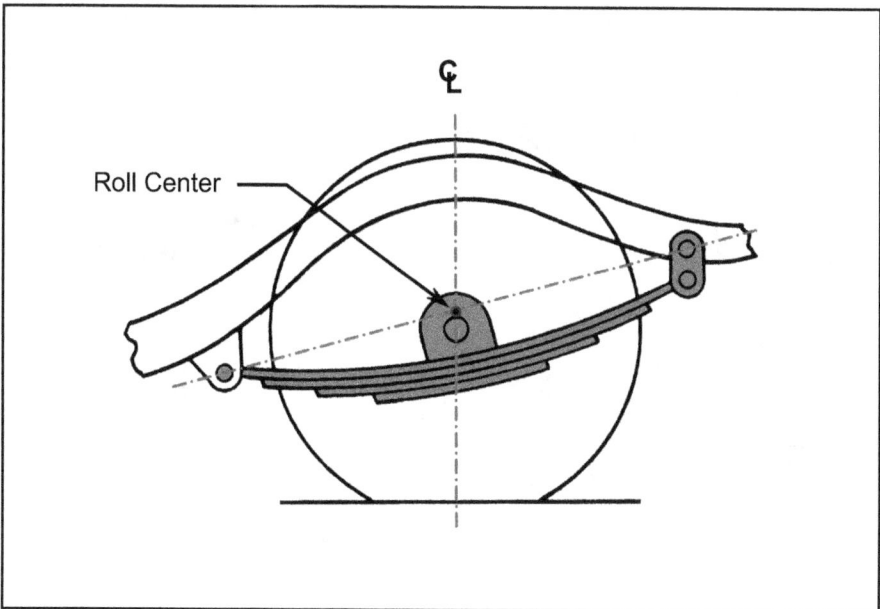

Figure (5.12): ROLL CENTER LOCATION FOR HOTCHKISS SUSPENSION

Chapter 5: CORNERING PERFORMANCE

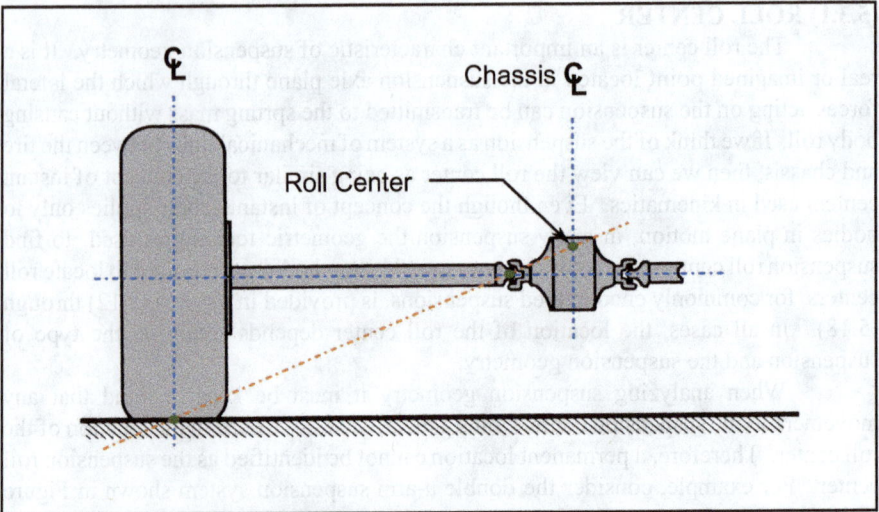

Figure (5.13): ROLL CENTER LOCATION FOR SWING AXLE SUSPENSION

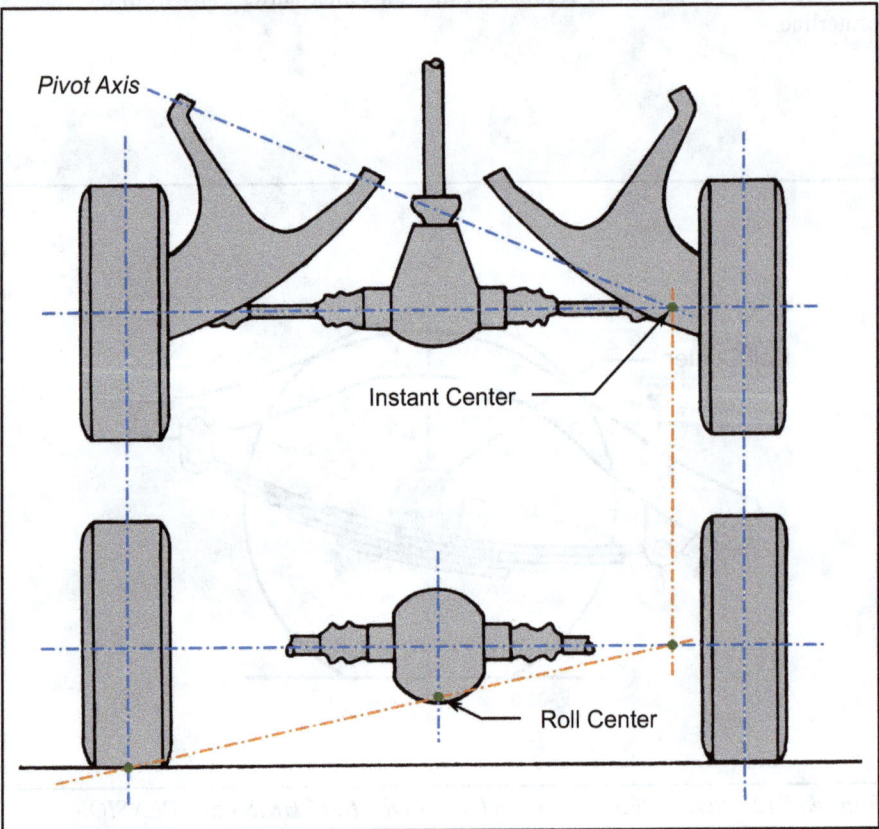

Figure (5.14): ROLL CENTER LOCATION FOR SEMI-TRAILING ARM SUSPENSION

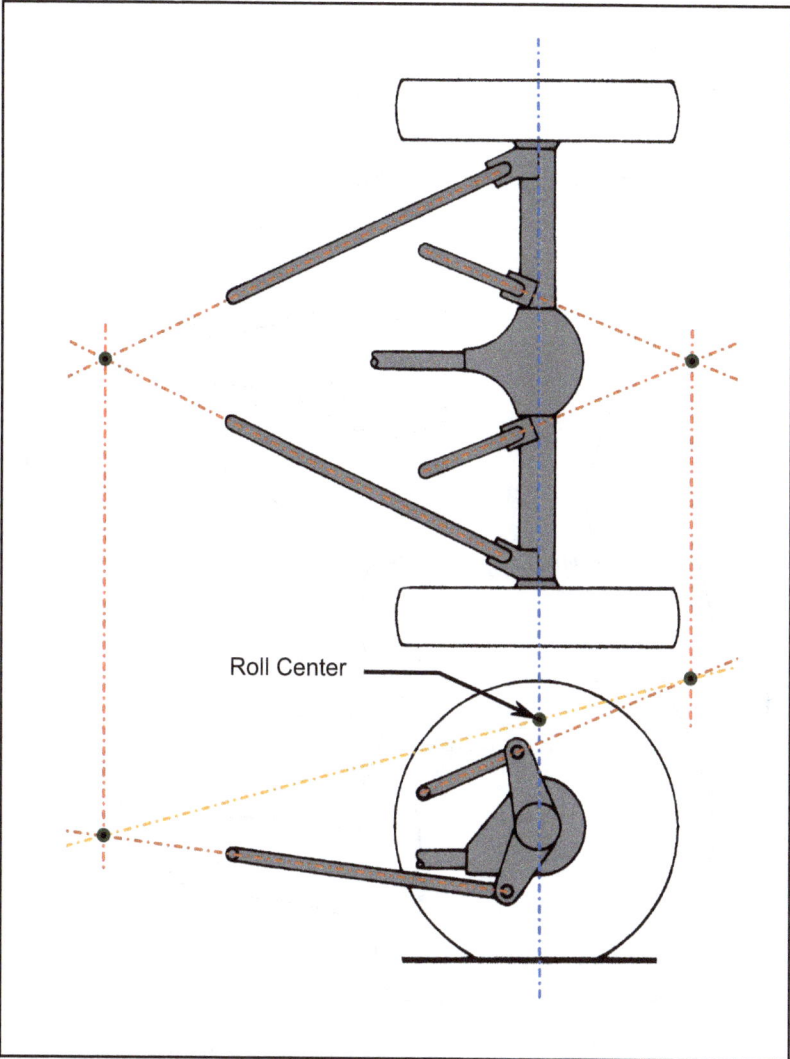

Figure (5.15): ROLL CENTER LOCATION FOR FOUR LINK SUSPENSION WITH ANGLED LOWER CONTROL ARMS.

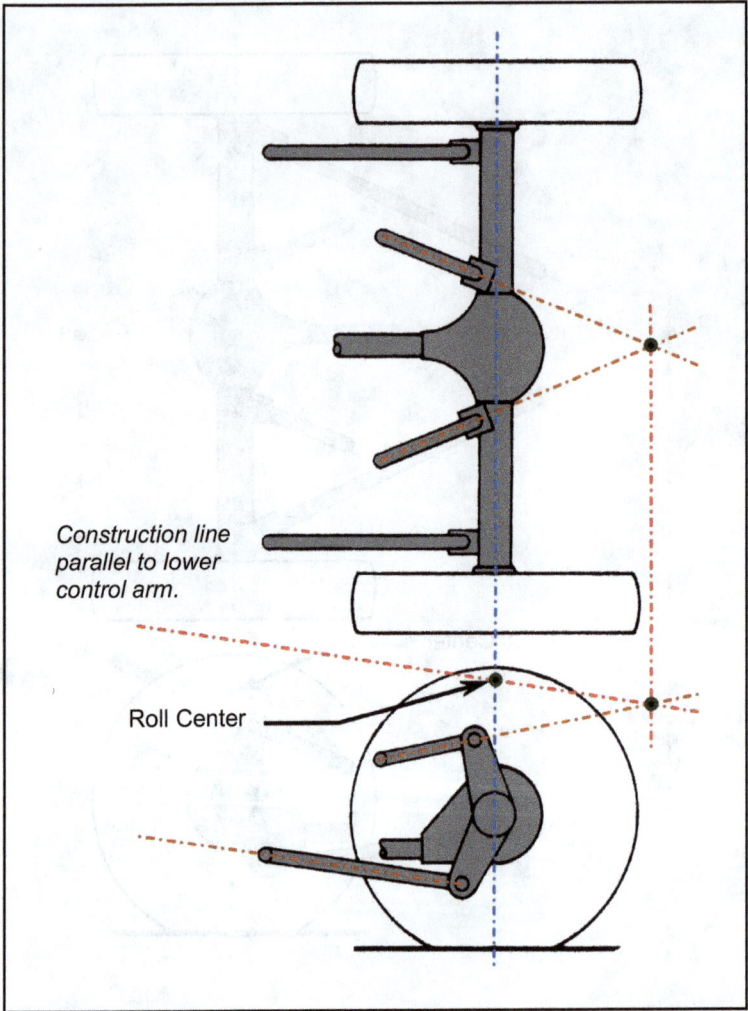

Figure (5.16): ROLL CENTER LOCATION FOR FOUR LINK SUSPENSION WITH PARALLEL LOWER CONTROL ARMS

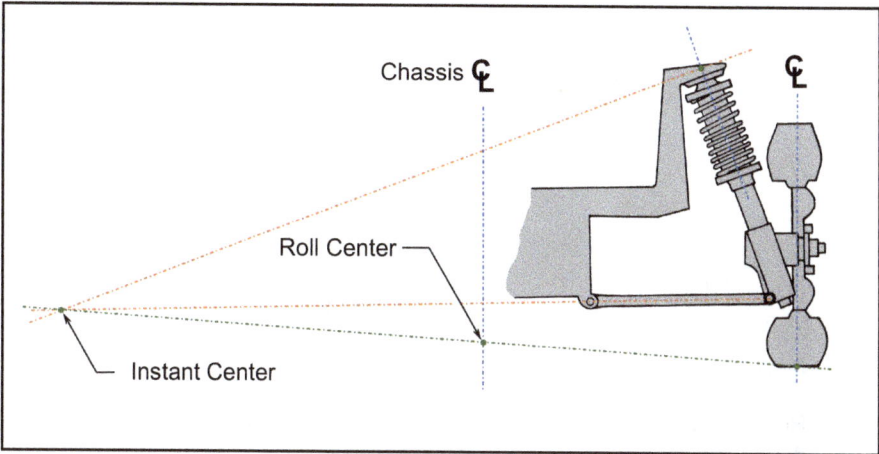

Figure (5.17): ROLL CENTER LOCATION FOR MacPHERSON/CHAPMAN STRUT SUSPENSION

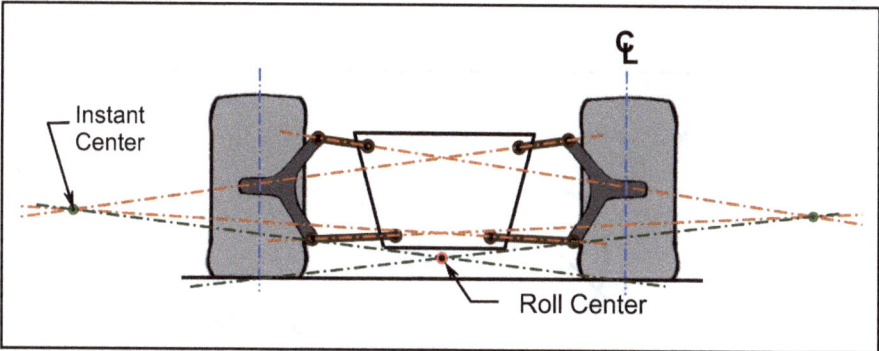

Figure (5.18): ROLL CENTER LOCATION FOR DOUBLE A-ARM SUSPENSION

(5.3.2) ROLL AXIS

The roll axis is an imagined line that passes through the front and rear suspension roll centers. The vehicle sprung mass rolls about this axis when acted upon by an external force, such as the inertia force encountered during cornering. Using the concept of roll axis we can devise a simple, generic four-wheel vehicle as shown in Figure (5.20).

There are some limitations of the simple four-wheel model which we should be aware of. As illustrated in the figure, the suspension systems have been replaced by rigid axles of fixed track width, even though the original suspension may have been an independent design. We have previously discovered that the roll center can, as a result of the suspension geometry, migrate away from it's location at zero body roll. Consequently, we may find it necessary to adjust the symmetrical rigid axle, about the roll center, to an asymmetrical representation according to the amount of roll. Also, resulting from independent suspension geometry is the possibility of track width variation as the suspension accommodates body roll. Essentially what happens is that

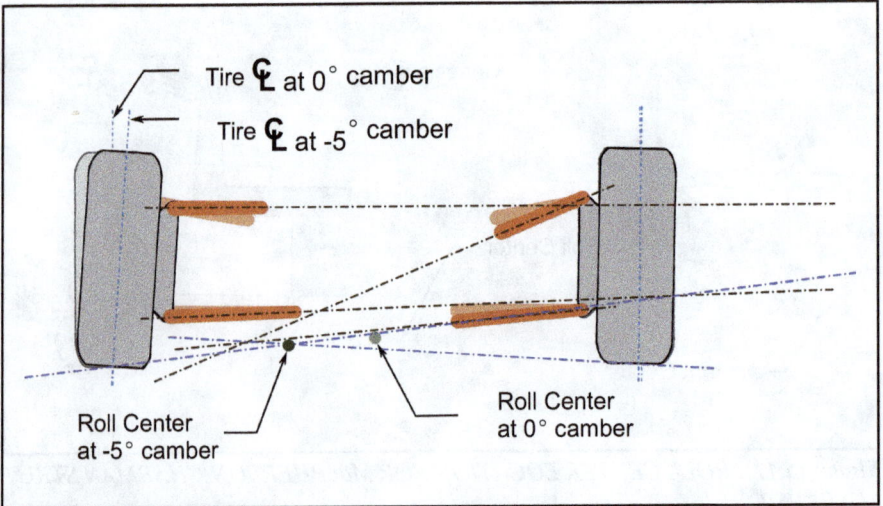

Figure (5.19): EXAMPLE OF VARIATION IN ROLL CENTER LOCATION WITH BODY ROLL

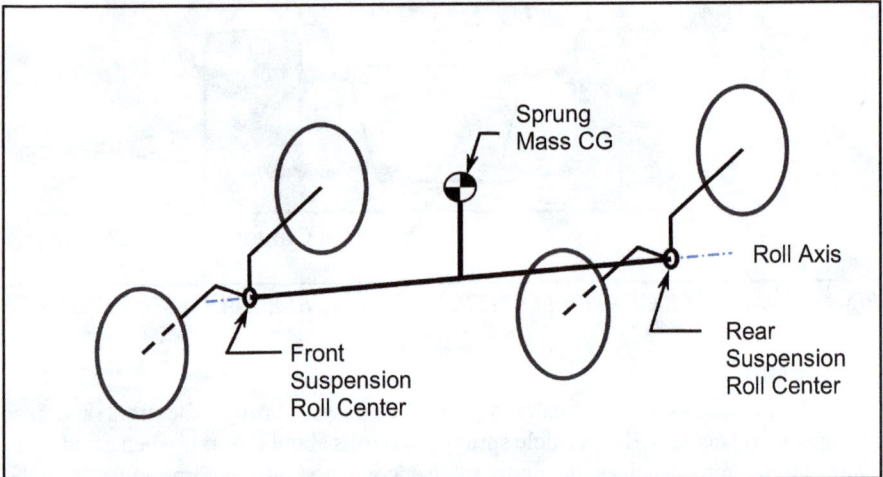

Figure (5.20): FOUR-WHEEL VEHICLE CORNERING MODEL

the wheels move in, or out, relative to the chassis longitudinal axis, as the wheel moves in jounce or rebound.

(5.3.3) ROLL STIFFNESS

The function of suspension springs is not only to control the bounce motion of the vehicle sprung mass, but also offer resistance to roll motion. In some cases where the conventional coil or leaf spring does not provide enough resistance, a torsion bar (or anti-roll bar) is incorporated into the suspension design. Roll stiffness defines the amount of roll motion resistance provided by the suspension system.

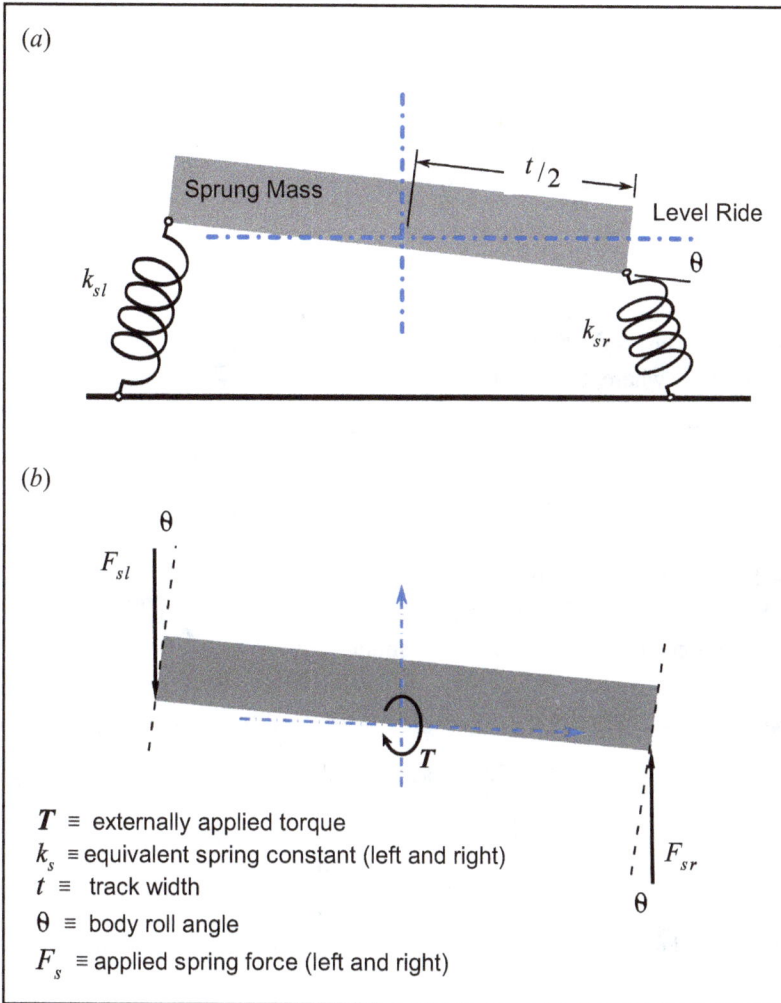

Figure (5.21): SPRUNG MASS (a) GEOMETRY and (b) FREE-BODY-DIAGRAM IN ROLL

To account for roll stiffness in our four-wheel vehicle model we will place a torsion spring at each roll center. The amount of torsional stiffness required to model the suspension springs is determined by examining what happens when an external torque is placed on the sprung mass above the suspension. A simple schematic of a sprung mass experiencing an external torque is shown in Figure (5.21), along with a free-body-diagram. If the sprung mass and suspension are in static equilibrium, then

$$\sum M_0 = -T + F_{sl}\frac{t}{2}\cos\theta + F_{sr}\frac{t}{2}\cos\theta = 0$$

or

$$T = \left(F_{sl} + F_{sr}\right)\frac{t}{2}\cos\theta \qquad (5.66)$$

If we also restrict our attention to linear springs, then the spring forces are related to their respective spring constants and spring deflection by

$$F_{sl} = k_{sl}\,\Delta x \qquad (5.67) \qquad \text{and} \qquad F_{sr} = k_{sr}\,\Delta x \qquad (5.68)$$

Since the sprung mass system under consideration is symmetrically placed above the suspension, the deflection magnitude of the left spring is the same as that of the right spring. The spring deflection in the foregoing spring force equations is written as $\Delta x = x_2 - x_1$ where, from Figure (5.22),

$$x_2 = \sqrt{\left(\frac{t}{2}\sin\theta + x_1\right)^2 + \left[\frac{t}{2}(1 - \cos\theta)\right]^2} \qquad (5.69)$$

Roll angles of practical interest are generally limited to five degrees or less. Under these conditions we can make the customary small angle approximations, i.e., $\sin\theta \approx \theta$ and $\cos\theta \approx 1$. Then Eq (5.67) simplifies to $x_2 = \frac{t}{2}\theta + x_1$, which results in a spring deflection of $\Delta x = \frac{t}{2}\theta$. Inserting the previous result along with Eqs (5.67) and (5.68) into Eq (5.66) gives us a relationship for the torque as a function of roll angle, i.e.,

$$T = \left(k_{sl} + k_{sr}\right)\frac{t^2}{4}\,\theta \qquad (5.70)$$

An equivalent suspension roll stiffness, k_θ, is defined as $k_\theta \equiv \dfrac{T}{\theta}$. Then, based on Eq (5.70), the roll stiffness for our suspension system is

$$k_\theta = \left(k_{sl} + k_{sr}\right)\frac{t^2}{4} \qquad (5.71)$$

If an anti-roll bar was a member of the original suspension design, then it's torsional stiffness, k_b, must be added to the right-hand-side of Eq (5.71) in order to determine the total roll stiffness:

$$k_\theta = \left(k_{sl} + k_{sr}\right)\frac{t^2}{4} + k_b \qquad (5.72)$$

To complete the four-wheel model, Eq (5.72) is individually evaluated for the front and the rear suspensions.

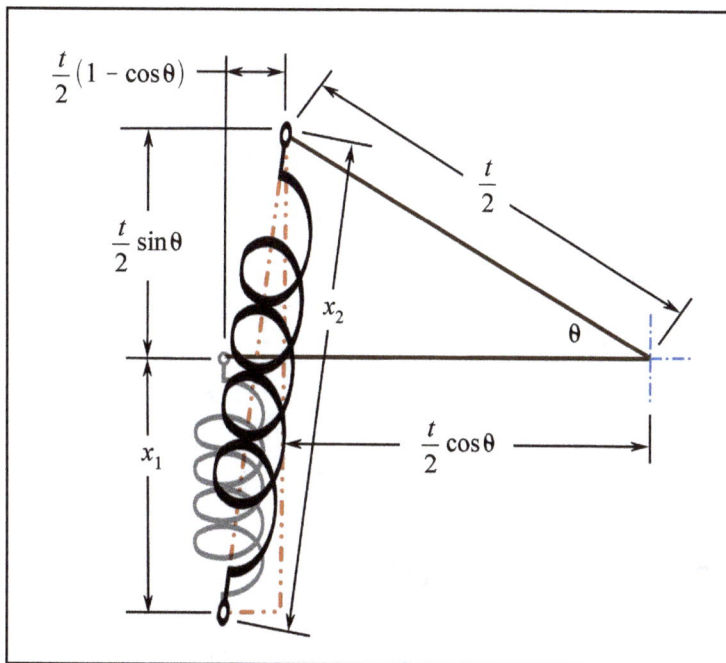

Figure (5.22): *LEVEL RIDE AND DEFLECTED SPRING GEOMETRY*

The basic procedure just described is used to evaluate the roll stiffness for more complicated suspension designs as well. Although in our simple example we ended up with a constant roll stiffness, this is not always the case. The spring geometry of some suspension systems can produce torque relationships that are non-linear functions of roll angle.

(5.4) FOUR-WHEEL STEADY-STATE CORNERING MODEL

The two-wheel cornering model is an excellent tool in understanding the importance and influence of tire cornering stiffness and sprung weight distribution on corning performance. But, this model has many limitations. Two limitations we want to overcome are the inability to account for the effects of body roll, and non-linear tire characteristics. As you might imagine, to accomplish our goal we will need a more sophisticated four-wheel cornering model.

(5.4.1) EQUILIBRIUM EQUATIONS

A schematic of an elementary four-wheel cornering model is shown in Figure (5.23). This model is more advanced than the two-wheel model, but still incorporates simplifications: (1) the unsprung weight is considered to be of minor importance and is therefore neglected in the mathematical analysis, and (2) through out the cornering process the front and rear suspension roll centers are always located in the xz-plane and remain at a fixed location. The equations which describe steady-state cornering behavior can be derived from static equilibrium by replacing the affect of centrifugal acceleration

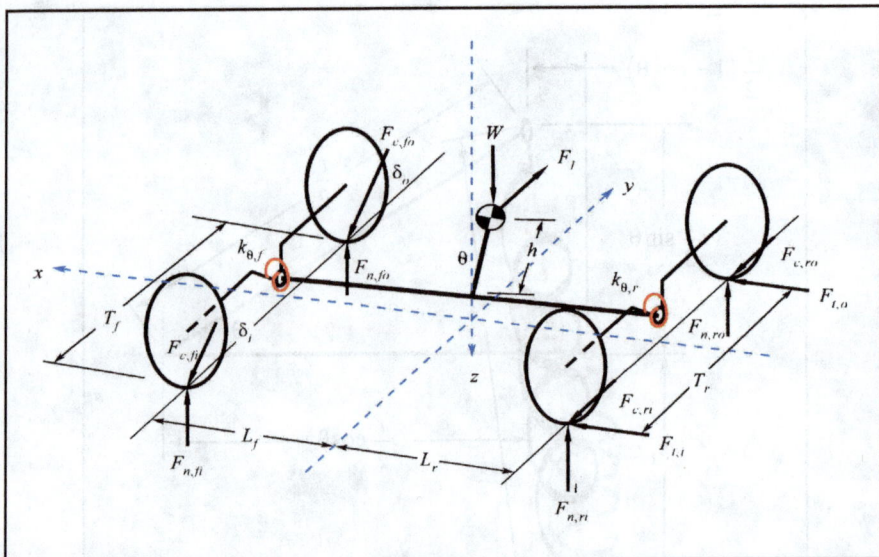

Figure (5.23): FOUR-WHEEL CORNERING MODEL FOR LEFT-HAND TURN

with an applied inertia force[7]. We should also point out that the inertia force does not usually act in the *yz*-plane, but rather at the sideslip or "drift" angle, β, as illustrated in Figure (5.24). Under these conditions[8], the basic equilibrium equations are:

$$\sum F_x = -F_{c,fo}\sin\delta_o - F_{c,fi}\sin\delta_i + F_{t,o} + F_{t,i} + F_I\sin\beta = 0 \qquad (5.73)$$

$$\sum F_y = -F_{c,fo}\cos\delta_o - F_{c,fi}\cos\delta_i - F_{c,ro} - F_{c,ri} + F_I\cos\beta = 0 \qquad (5.74)$$

$$\sum F_z = -F_{n,fo} - F_{n,fi} - F_{n,ro} - F_{n,ri} + W = 0 \qquad (5.75)$$

and

$$\sum M_x = \frac{1}{2}\left(F_{n,fi} - F_{n,fo}\right)T_f + \frac{1}{2}\left(F_{n,ri} - F_{n,ro}\right)T_r$$

$$+ Wh\theta + F_I H\cos\beta = 0 \qquad (5.76)$$

---

[7] *Known as d'Alembert's principle.*

[8] *The sideslip angle, β, is defined in Figure (5.2). A positive value indicates the vehicle is pointing towards the turn center, whereas an a vehicle with negative sideslip points away from the turn center. The steer angles, δ_i and δ_o, are assumed positive for a left-hand turn.*

Chapter 5: CORNERING PERFORMANCE 247

$$\sum M_y = \left(F_{n,fi} + F_{n,fo}\right)L_f - \left(F_{n,ri} + F_{n,ro}\right)L_r$$

$$- F_l H \sin\beta + W h \gamma \theta = 0 \tag{5.77}$$

The tractive forces, $F_{t,i}$ and $F_{t,o}$, shown in Figure (5.23), bear closer examination. In describing an actual vehicle we should have included tire rolling resistance and aerodynamic drag forces in our vehicle diagram. However, in order to minimize the complexity of our simple four-wheel cornering model they have been neglected. With this in mind, the tractive forces in Eqs (5.73) and (5.78) are those

$$\sum M_z = -\left(F_{c,fi} \cos\delta_i + F_{c,fo} \cos\delta_o\right)L_f + \left(F_{c,ri} + F_{c,ro}\right)L_r$$

$$+ \frac{1}{2}\left(F_{c,fo} \sin\delta_o - F_{c,fi} \sin\delta_i\right)T_f - \frac{1}{2}\left(F_{t,o} - F_{t,i}\right)T_r$$

$$+ F_l h \theta \sin\beta = 0 \tag{5.78}$$

which from a mathematical point of view would be required to maintain equilibrium during cornering, and generally do not have significant interest in the study of cornering performance. Consequently, we will no longer concern ourselves with these forces. By neglecting all tractive forces we are also able to use "pure cornering" tire models, such as the Pacejka '94 model discussed in Chapter 2, which are slightly easier to use.

It should be noted that two vehicle center of gravity parameters, h and H, are used in Eqs (5.76) and (5.77). By definition the distance h is measured from the roll axis to the center of gravity (CG), whereas H is measured from the ground surface reference plane. The distance H is fairly easy to determine either by experimental or analytical means, which we will consider to be a know design parameter. The other CG distance h is related to H through the roll axis geometry, i.e.,

$$h = H - \frac{R_{c,f}L_r + R_{c,r}L_f}{L_r + L_f} \tag{5.79}$$

where $R_{c,f}$ and $R_{c,r}$ define the front and rear suspension roll center locations above the ground surface (xy-plane). The angle, γ, found in Eq (5.77) is the roll axis tilt angle. It is also based on the suspension roll center locations, and is determined from

$$\gamma = \arctan\left(\frac{R_{c,r} - R_{c,f}}{L_r + L_f}\right)$$

or, by using the small angle approximation,

$$\gamma \cong \frac{R_{c,r} - R_{c,f}}{L_r + L_f} \tag{5.80}$$

Eqs (5.73) through (5.80) describe the necessary conditions required to maintain external equilibrium. Internal equilibrium must be maintained as well. The inertia force acting at the sprung mass center of gravity causes the body to roll about it's roll axis. This motion is resisted by the roll stiffness of the front and rear suspensions.

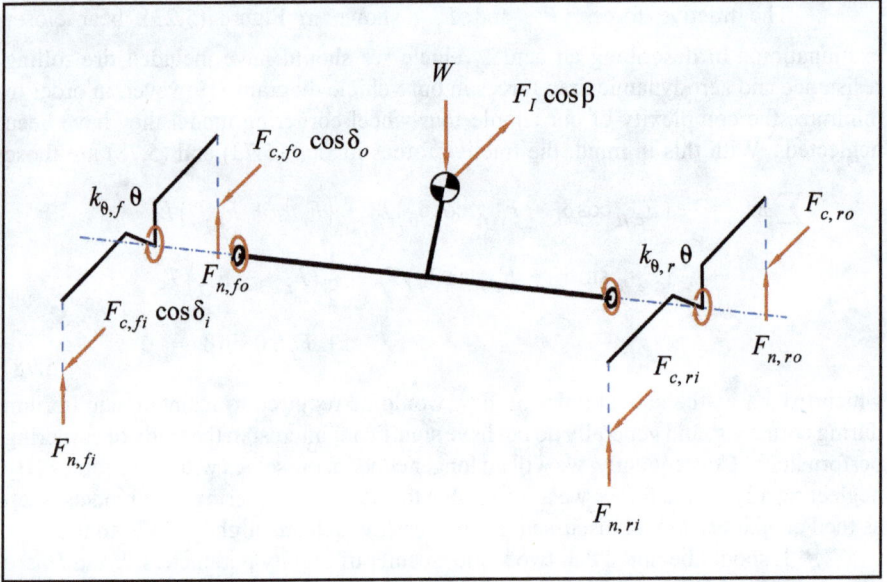

Figure (5.24): FREE-BODY-DIAGRAM OF SIMPLE FOUR-WHEEL VEHICLE IN ROLL

A free-body-diagram depicting the front and rear axles, and the sprung portion of the vehicle is given in Figure (5.24). Closer examination of this figure yields three additional steady-state cornering conditions:

(i) front axle static equilibrium

$$\sum M_{RC_f} = k_{\theta,f}\theta + \left(F_{c,fi}\cos\delta_i + F_{c,fo}\cos\delta_o\right)R_{c,f}$$

$$- \frac{1}{2}\left(F_{n,fo} - F_{n,fi}\right)T_f = 0 \tag{5.81}$$

(ii) rear axle static equilibrium

$$\sum M_{RC_r} = k_{\theta,r}\theta + \left(F_{c,ri} + F_{c,ro}\right)R_{c,r} - \frac{1}{2}\left(F_{n,ro} - F_{n,ri}\right)T_r = 0 \tag{5.82}$$

and (iii) sprung mass roll equilibrium

$$\sum M_{Roll\,Axis} = \left(F_I\cos\beta\right)\left(h\,\cos\gamma\right)\cos\theta$$

$$+ \left(W\cos\gamma\right)\left(h\cos\gamma\right)\sin\theta - \left(k_{\theta,f} + k_{\theta,r}\right)\theta = 0$$

or, using small angle approximations,

$$F_I h \cos\beta + W h \theta - \left(k_{\theta,f} + k_{\theta,r}\right)\theta = 0 \qquad (5.83)$$

In addition to the static equilibrium equations, there are geometric constraints. Figure (5.25) illustrates the basic cornering geometry, as viewed looking down on the ground surface reference plane, of the left-front (inside) tire for the four-wheel model.

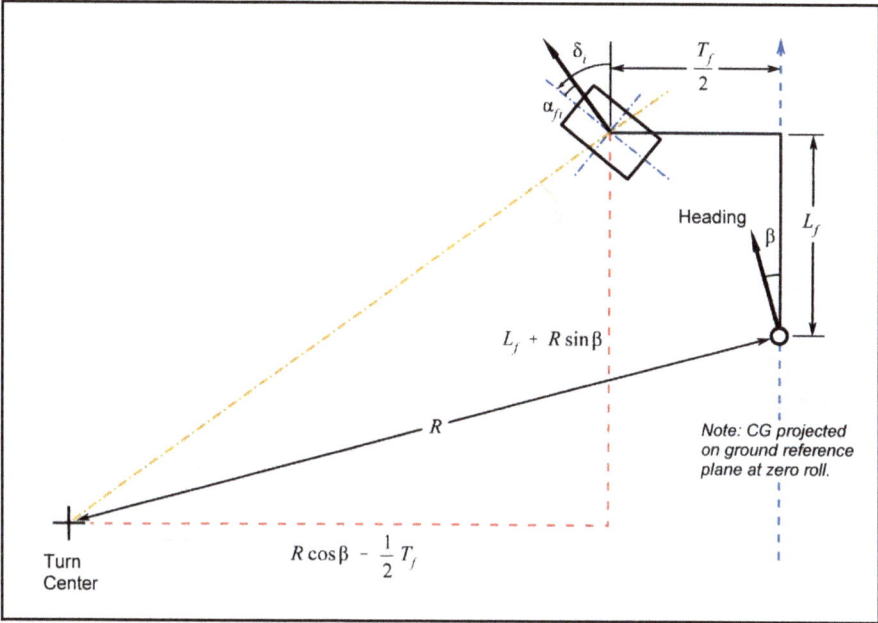

Figure (5.25): LEFT-FRONT (INSIDE) TIRE CORNERING GEOMETRY

When we studied the two-wheel cornering model we found that the tire slip angles and the vehicle sideslip angle had to accommodate a unique turning center. This same principle also holds true for the four-wheel cornering model. The relationship derived from Figure (2.25) for the left-front (inside) tire slip angle as a function of vehicle sideslip angle and steer angle is:

$$\alpha_{fi} = \delta_i - \arctan\left(\frac{L_f + R \sin\beta}{R \cos\beta - \dfrac{1}{2} T_f}\right) \qquad (5.84)$$

and, following a similar geometric analysis, the right-front (outside) relationship is

$$\alpha_{fo} = \delta_o - \arctan\left(\frac{L_f + R\sin\beta}{R\cos\beta + \frac{1}{2}T_f}\right) \qquad (5.85)$$

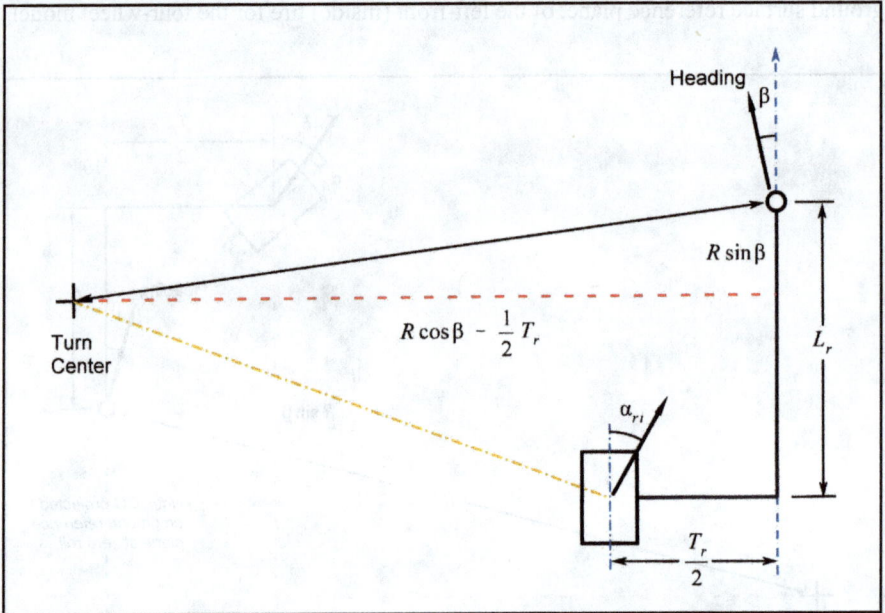

Figure (5.26): LEFT-REAR (INSIDE) TIRE CORNERING GEOMETRY

Figure (5.26) depicts the cornering geometry for the left-rear (inside) tire, which yields the tire slip angle:

$$\alpha_{ri} = \arctan\left(\frac{L_r - R\sin\beta}{R\cos\beta - \frac{1}{2}T_r}\right) \qquad (5.86)$$

In a similar fashion the right-rear (outside) tire slip angle can be expressed as

$$\alpha_{ro} = \arctan\left(\frac{L_r - R\sin\beta}{R\cos\beta + \frac{1}{2}T_r}\right) \qquad (5.87)$$

    Finally, to complete the set of cornering equations we must account for the influence of the tires on cornering behavior. In general, the lateral cornering force can be expressed as a function of the tire slip angle and the normal force exerted on the tire. Unfortunately the elastic non-linear behavior of real tires prevents us from being able to write a simple equation for the lateral force. The precise functional relationship for the lateral force depends on which tire model (refer to Chapter 2) we select. At this point the best we can do is write generic empirical correlations as:

$$F_{c,fi} = \text{fn}\left(\alpha_{fi}, F_{n,fi}\right) \tag{5.88}$$

$$F_{c,fo} = \text{fn}\left(\alpha_{fo}, F_{n,fo}\right) \tag{5.89}$$

$$F_{c,ri} = \text{fn}\left(\alpha_{ri}, F_{n,ri}\right) \tag{5.90}$$

and

$$F_{c,ro} = \text{fn}\left(\alpha_{ro}, F_{n,ro}\right) \tag{5.91}$$

(5.4.2) SOLUTION OF EQUILIBRIUM EQUATIONS

We certainly have a "mess" of equilibrium equations to deal with. Before we attempt a solution lets review the meaning of the symbols contained in these equations and decide whether we should consider them as "known" or "unknown" quantities. The will help us in developing and organizing a solution strategy. The parameters governing the four-wheel model are summarized in Table (5.1).

The known quantities include a set of chassis design parameters, which as suspension designers we would specify, and three independent variables. The decision on which variables should be considered independent comes from the driver-vehicle interaction. If we assume for a moment that we are test driving our vehicle on a skid pad, we have the freedom to change the steering wheel position altering the steer angles[9], and the engine throttle which controls the engine speed. Then from the vehicles point of view, the steer angles and forward velocity can be varied independently from the vehicle system. The forward velocity is not directly contained in the equilibrium equations, but it is related to the cornering inertia force and turning radius by the customary equation

$$F_I = \frac{W}{g}\frac{V^2}{R} \tag{5.92}$$

We finally end up with 16 unknowns in 16 equations. It is not very practical to attempt a closed form solution of the equilibrium equations. Especially when the set of equations contain trigonometric functions and non-linear tire models. So, instead, we will develop a numerical solution. The direction our numerical solution takes now is one of an iterative nature. We start by guessing expected values for the turn radius and yaw angle. The body roll angle and the inertia force can then be calculated from Eqs (5.83) and (5.92), respectively. Substituting these values into Eqs (5.75), (5.76) and (5.77) produces a set of three equations which contain four unknown normal forces.

---

[9] *It should be noted that the inside steer angle cannot be varied independently of the outer steer angle. A unique geometrical relationship exists between the two as governed by the steering linkage mechanism.*

Table (5.1): *FOUR-WHEEL CORNERING MODEL PARAMETERS*

KNOWN QUANTITIES	
Chassis Design Parameters	
W	sprung weight
H, L_f, L_r	CG location
T_f, T_r	front and rear track width
R_{cf}, R_{cr}	front and rear roll center location
$k_{\theta f}, k_{\theta r}$	front and rear roll stiffness
Independent Variables	
δ_i, δ_o	inside and outside steer angle
V	forward velocity
UNKNOWN QUANTITIES	
Cornering Performance Parameter	
R	turning radius
Dependent Variables	
$F_{c,fi}, F_{c,fo}, F_{c,ri}, F_{c,ro}$	lateral forces
$F_{n,fi}, F_{n,fo}, F_{n,ri}, F_{n,ro}$	normal forces
$\alpha_{fi}, \alpha_{fo}, \alpha_{ri}, \alpha_{ro}$	tire slip angles
θ	body roll angle
β	yaw angle
F_I	cornering inertia force

To solve for all of the normal forces we need to construct one more equation from the remaining equilibrium equations. First, rewrite Eq (5.81) as

$$F_{c,fi}\cos\delta_i + F_{c,fo}\cos\delta_o = \frac{1}{2}\left(F_{n,fo} - F_{n,fi}\right)\frac{T_f}{R_{cf}} - \frac{k_{\theta f}}{R_{cf}}\theta$$

and Eq (5.82) as

$$F_{c,ri} + F_{c,ro} = \frac{1}{2}\left(F_{n,ro} - F_{n,ri}\right)\frac{T_r}{R_{cr}} - \frac{k_{\theta r}}{R_{cr}}\theta$$

Using the two previous equations to eliminate the cornering forces from Eq (5.74) yields the forth normal force equation:

$$\frac{1}{2}\left(F_{n,fo} - F_{n,fi}\right)\frac{T_f}{R_{cf}} + \frac{1}{2}\left(F_{n,ro} - F_{n,ri}\right)\frac{T_r}{R_{cr}}$$

$$= \left(\frac{k_{\theta f}}{R_{cf}} + \frac{k_{\theta r}}{R_{cr}}\right)\theta + F_I\cos\beta \tag{5.93}$$

Eqs (5.75), (5.76), (5.77) and (5.93) can be expressed as a single matrix equation of the form

$$\boldsymbol{A}\boldsymbol{F} = \boldsymbol{R} \tag{5.94}$$

where

$$\boldsymbol{A} = \begin{bmatrix} 1 & 1 & 1 & 1 \\ -\dfrac{1}{2}T_f & \dfrac{1}{2}T_f & -\dfrac{1}{2}T_r & \dfrac{1}{2}T_r \\ L_f & L_f & -L_r & -L_r \\ -\dfrac{1}{2}\dfrac{T_f}{R_{cf}} & \dfrac{1}{2}\dfrac{T_f}{R_{cf}} & -\dfrac{1}{2}\dfrac{T_r}{R_{cr}} & \dfrac{1}{2}\dfrac{T_r}{R_{cr}} \end{bmatrix} \tag{5.95}$$

$$\boldsymbol{F} = \left\{ \begin{array}{c} F_{n,fi} \\ F_{n,fo} \\ F_{n,ri} \\ F_{n,ro} \end{array} \right\} \tag{5.96}$$

and

$$R = \left\{ \begin{array}{c} W \\[2mm] W h \theta + F_I H \cos \beta \\[2mm] F_I H \sin \beta - W h \gamma \theta \\[2mm] \left(\dfrac{k_{\theta f}}{R_{cf}} + \dfrac{k_{\theta r}}{R_{cr}} \right) \theta + F_I \cos \beta \end{array} \right\} \tag{5.97}$$

The solution for the normal force vector matrix is found from

$$F = A^{-1} R \tag{5.98}$$

The next step is to evaluate the tire slip angles from Eqs (5.84) through (5.87), and the tire cornering forces from the tire model expressed by Eqs (5.88) through (5.91). At this point it is desirable to obtain a better estimate of the sideslip angle for our assumed value of turning radius. It should be recognized that the normal forces and tire slip angles have been evaluated based on the guessed value of sideslip angle. These in turn were used in determining the tire cornering forces. Consequently we may think of the cornering forces as functions of sideslip angle. Eq (5.74) can therefore be written as

$$F_{c,fo}(\beta) \cos \delta_o + F_{c,fi}(\beta) \cos \delta_i + F_{c,ro}(\beta) + F_{c,ri}(\beta) - F_I \cos \beta = 0 \tag{5.74}$$

Successive approximations are made on the sideslip angle and the previous solution process repeated until Eq (5.74) is satisfied.

We are now left with the task of finding the correct value of turning radius. Once again we look to the equilibrium equations for a relationship to aid us in refining the turning radius, but we do so by looking at the inertia force. Combining Eqs (5.81) and (5.83) gives us the required inertia force to attain static equilibrium at the front axle:

$$F_{If} =$$

$$\left(\frac{k_{\theta f} + k_{\theta r} - W h}{k_{\theta f} h \cos \beta} \right) \left[\frac{1}{2} \left(F_{n,fo} - F_{n,fi} \right) T_f - \left(F_{c,fi} \cos \delta_i + F_{c,fo} \cos \delta_o \right) R_{cf} \right] \tag{5.99}$$

The required rear axle inertia force is found from Eqs (5.82) and (5.83):

$$F_{Ir} = \left(\frac{k_{\theta f} + k_{\theta r} - Wh}{k_{\theta f} h \cos\beta} \right) \left[\frac{1}{2} \left(F_{n,ro} - F_{n,ri} \right) T_r - \left(F_{c,ri} + F_{c,ro} \right) R_{cr} \right]$$

(5.100)

Unless we made an extraordinary guess, the assumed turning radius, which was used to estimate the inertia force, and the calculated inertia forces by Eqs (5.99) and (5.100) will all be different. The inaccuracy of the guessed turning radius causes an imbalance in the static equilibrium at the front and rear axles which results in

$$F_{If} < F_I < F_{Ir} \quad \text{or} \quad F_{If} > F_I > F_{Ir} .$$

Obviously we want to find the value of inertia force where $F_{If} = F_I = F_{Ir}$. It is convenient to define an inertia force error function as

$$F_{I,err} \equiv \left| F_{If} - F_{Ir} \right|$$

(5.101)

The assumed value of turning radius is either increased or decreased as required to decrease the calculated inertia force error towards zero. A flow diagram of the steps taken to achieve a solution to the four-wheel cornering model are shown in Figure (5.27).

Now that we have a working four-wheel cornering model, we can look at the cornering behavior of a few different vehicle designs. The specifications of the vehicles are found in Table (5.2) in which Vehicle 1 is the baseline design[10]. The cornering simulation tabular results for the baseline design are given in Table (5.3) for two different steering angle scenarios. The challenge is to find a convenient method to interpret data without having to sort through all of the numeric data. In this regard it is convenient to utilize a few of the definitions previously introduced during the study of two-wheel cornering behavior in Section (5.2.3). In particular, the sideslip gain and the curvature gain:

$$G_{slip} = \beta / \delta_i \quad \text{and} \quad G_{curv} = \frac{1/R}{\delta_i} .$$

These two gain functions can be plotted against the lateral acceleration, $a_{lat} = V^2 / R$ where, with reference to the simulation results, V is the forward velocity as inputted to the simulation and R is the turn radius as calculated by the simulation. The sideslip gain

---

[10] *The front and rear track width specifications for the baseline design in Table (5.2) are narrower than what would normally be accepted. These dimensions were chosen for illustrative purposes only; they produce a vehicle which is capable of exhibiting a variety of different cornering behaviors. A better vehicle design would have a front track width closer to 58" and a rear track width around 58.5".*

and curvature gain results for three different steering angle inputs[11] are in shown in Figures (5.28) and (5.29),respectively. How do we characterize the cornering behavior of our baseline design? Looking back at the results of the two-wheel cornering model

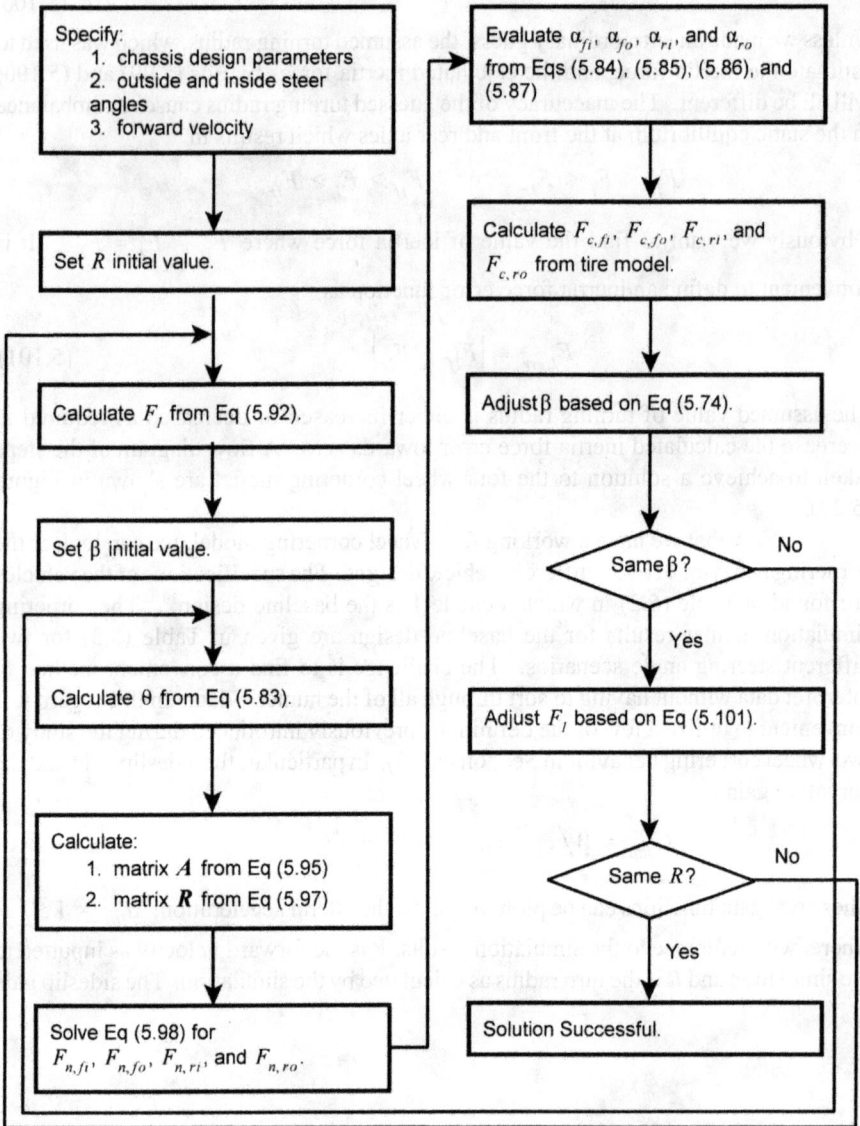

Figure (5.27): SOLUTION DIAGRAM FOR FOUR-WHEEL CORNERING MODEL

---

[11] *In this example the inside steering angles were arbitrarily selected; the outside steering angles were calculated based on Ackermann's steering.*

Table (5.2): CHASSIS DESIGN PARAMETER LISTING FOR FOUR-WHEEL CORNERING MODEL EXAMPLES

Chassis Design Parameters		Vehicle 1	Vehicle 2	Vehicle 3	Vehicle 4
Sprung Weight, W [lb]		1738			
CG Location:	L_f [in]	57.42			
	L_r [in]	50.58			
	H [in]	20.0			*SAME*
Track Width:	T_f [in]	48.0			
	T_r [in]	44.0	*SAME*	*SAME*	
Roll Center:	$R_{c,f}$ [in]	3.0			4.0
	$R_{c,r}$ [in]	3.5			3.5
Front Spring Rate: $k_{s,fi}$, $k_{s,fo}$ [lb/in]		56.7			
Rear Spring Rate: $k_{s,ri}$, $k_{s,ro}$ [lb/in]		73.8			*SAME*
Anti-roll Stiffness: $k_{b,f}$ [ft-lb/deg]		- - -	197	197	
$k_{b,r}$ [ft-lb/deg]		- - -	0	131	

as indicated by Figure (5.8), and comparing those with Figure (5.28), we conclude that our vehicle does not understeer. But it is difficult to draw a conclusion if the vehicle oversteers or has neutral steer characteristics. A more definitive method is to examine the slope of the curvature gain versus lateral acceleration plot. That is,

(i) $\dfrac{dG_{curv}}{da_{lat}} > 0$ indicates oversteer, (ii) $\dfrac{dG_{curv}}{da_{lat}} = 0$ indicates neutral steer, and

(iii) $\dfrac{dG_{curv}}{da_{lat}} < 0$ indicates understeer. Figure (5.29) indicates that our baseline vehicle

has very interesting characteristics. For the three different steering inputs the slope of each curvature gain graph is positive for all forward velocities (lateral accelerations).

Table (5.3): SAMPLE CORNERING RESULTS FOR "VEHICLE 1"

Inside Steer Angle [deg]: 5.0
Outside Steer Angle [deg]: 4.84

Vel [fps]	Yaw [deg]	Roll [deg]	Radius [ft]	Normal Force [lb]				Slip Angle [deg]				Cornering Force [lb]			
				fi	fo	ri	ro	fi	fo	ri	ro	fi	fo	ri	ro
10	-2.08	0.46	104.75	395	420	447	477	-0.22	-0.24	-0.23	-0.22	10	14	14	13
12	-2.05	0.66	104.74	389	425	440	484	-0.25	-0.27	-0.26	-0.25	15	20	20	20
14	-2.02	0.90	104.72	383	432	432	491	-0.28	-0.30	-0.30	-0.29	21	27	26	28
16	-1.98	1.18	104.71	376	439	423	500	-0.32	-0.34	-0.34	-0.33	27	35	34	37
18	-1.93	1.49	104.68	367	448	412	511	-0.37	-0.38	-0.38	-0.37	34	45	41	48
20	-1.88	1.84	104.65	358	457	401	522	-0.42	-0.43	-0.44	-0.42	42	56	50	60
22	-1.82	2.22	104.62	348	468	388	535	-0.48	-0.49	-0.49	-0.48	49	68	58	75
24	-1.76	2.65	104.57	336	479	374	549	-0.54	-0.55	-0.56	-0.54	57	82	67	91
26	-1.69	3.11	104.53	324	492	358	564	-0.61	-0.61	-0.63	-0.61	66	99	76	110
28	-1.61	3.61	104.46	310	506	342	580	-0.68	-0.69	-0.71	-0.69	74	117	85	131
30	-1.53	4.15	104.39	296	520	324	598	-0.77	-0.77	-0.80	-0.77	81	138	93	154
32	-1.44	4.72	104.29	281	536	305	617	-0.86	-0.86	-0.90	-0.86	88	161	101	181
34	-1.33	5.34	104.18	264	552	284	637	-0.97	-0.96	-1.01	-0.97	95	187	107	212
36	-1.22	5.99	104.03	246	570	263	659	-1.08	-1.07	-1.12	-1.09	100	216	112	246
38	-1.09	6.69	103.85	227	589	240	682	-1.21	-1.19	-1.26	-1.22	104	249	115	284
40	-0.94	7.43	103.63	207	609	215	707	-1.35	-1.33	-1.41	-1.37	106	286	115	328
42	-0.77	8.21	103.37	186	630	189	733	-1.51	-1.48	-1.59	-1.54	105	328	113	377
44	-0.58	9.04	103.04	163	652	162	760	-1.70	-1.66	-1.79	-1.73	102	375	106	434
46	-0.36	9.92	102.64	139	676	133	790	-1.91	-1.87	-2.03	-1.96	95	428	95	497

Inside Steer Angle [deg]: 10.0
Outside Steer Angle [deg]: 9.45

Vel [fps]	Yaw [deg]	Roll [deg]	Radius [ft]	Normal Force [lb]				Slip Angle [deg]				Cornering Force [lb]			
				fi	fo	ri	ro	fi	fo	ri	ro	fi	fo	ri	ro
10	-4.29	0.91	52.77	383	432	431	491	-0.21	-0.37	-0.30	-0.28	8	40	27	27
12	-4.23	1.31	52.76	373	443	418	504	-0.27	-0.42	-0.36	-0.34	18	52	38	40
14	-4.16	1.78	52.75	360	457	402	520	-0.34	-0.49	-0.44	-0.41	29	67	50	57
16	-4.08	2.33	52.73	346	472	383	537	-0.42	-0.56	-0.52	-0.49	40	84	63	77
18	-3.99	2.95	52.70	329	489	362	558	-0.51	-0.65	-0.62	-0.58	52	105	75	101
20	-3.88	3.64	52.67	311	508	339	580	-0.62	-0.74	-0.73	-0.68	64	130	88	130
22	-3.76	4.41	52.63	291	529	313	605	-0.74	-0.86	-0.87	-0.81	76	160	99	165
24	-3.62	5.26	52.56	268	553	285	633	-0.88	-0.99	-1.02	-0.95	86	194	109	205
26	-3.45	6.18	52.49	244	578	254	663	-1.04	-1.14	-1.20	-1.12	95	235	115	254
28	-3.25	7.19	52.38	217	606	220	696	-1.23	-1.32	-1.41	-1.31	101	283	118	312
30	-3.02	8.28	52.23	188	635	184	731	-1.45	-1.53	-1.66	-1.55	102	339	114	380
32	-2.73	9.45	52.04	156	667	145	770	-1.73	-1.78	-1.98	-1.84	98	406	102	463

This is an oversteering characteristic, but the vehicle has positive sideslip angles indicating that the vehicle is pointing away from the turn center. The later observation is more characteristic of an understeering vehicle. Looking at the raw data in Table (5.3) we do indeed notice a slight decrease in the turn radius as the forward velocity increases, along with a decrease in the sideslip angle. It is also important to remember that the four-wheel cornering model, just as an actual automobile, has lateral force/tire slip characteristics which are non-linear, and also allows for weight transfer due to body roll. The tires and weight transfer are primarily responsible for the unusual cornering behavior we see in Figures (5.28) and (5.29). In its present state of design, "Vehicle 1" does not corner particularly well. It has oversteer and an excessive amount of body roll,

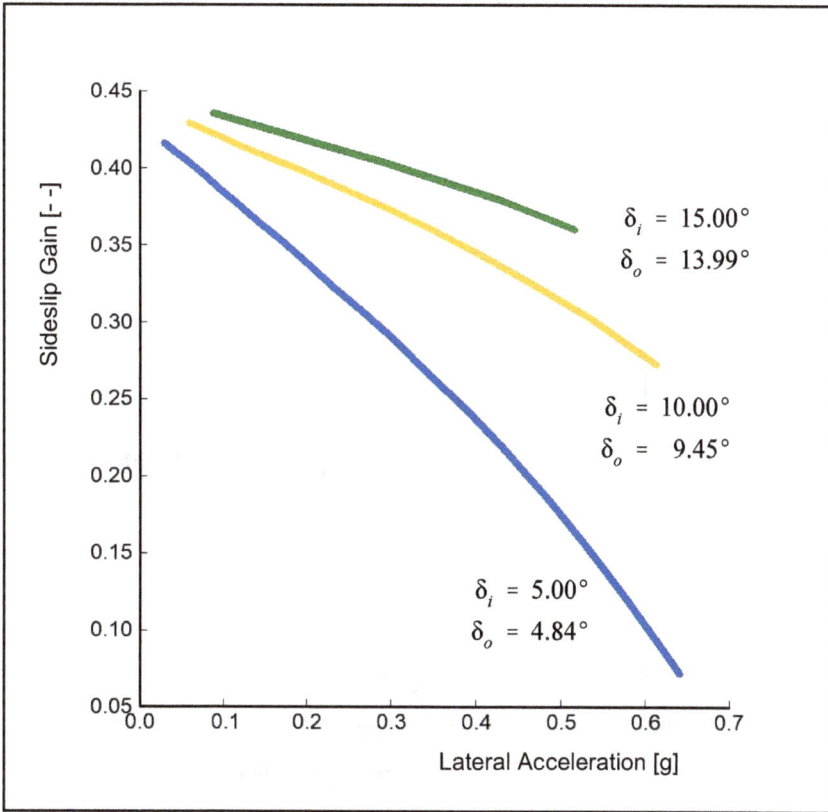

Figure (5.28): SIDESLIP GAIN RESULTS OF "VEHICLE 1"

which most drivers would find difficult to adjust to. Not to mention the problems in designing enough suspension travel to accommodate the body roll.

What design changes can be done to improve cornering performance? Looking at the simulation results in Table (5.3), the excessive body roll certainly stands out as a problem. The solution is to increase the roll stiffness. The front and rear spring rates in the foregoing example were not chosen at random, but were selected to provide an acceptable ride by the methods detailed in Chapter 4. So increasing the spring rates, while improving cornering, would result in a much harsher ride. We can, however, incorporate a torsion spring (referred to as an anti-roll bar, anti-sway bar, or stabilizer bar) into the suspension system which will increase the roll stiffness without affecting the ride. A typical design and installation of an anti-roll bar is shown in Figure (5.30).

Now that we have decided to include an anti-roll bar in our design, how stiff (torsional spring rate) should it be? We can obtain a fairly good estimate of the combined front and rear roll stiffness by introducing another design parameter called the roll gradient. The roll gradient is defined as the change in roll angle with respect to lateral acceleration, i.e., $\dfrac{d\theta}{d\left(a_{lat}/g\right)}$. Recall Eq (5.83),

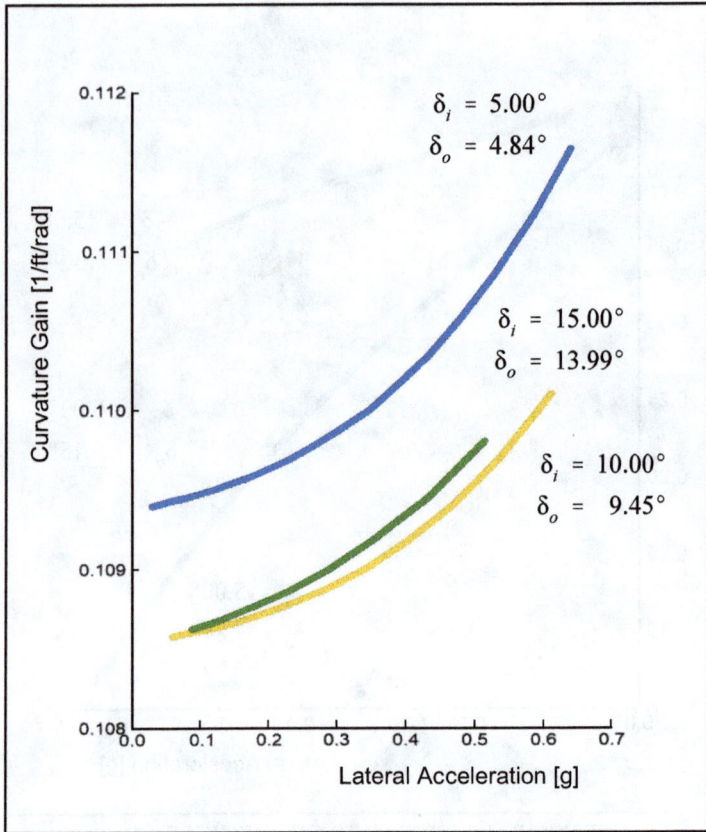

Figure (5.29): CURVATURE GAIN RESULTS OF "VEHICLE 1"

$$F_I \, h \cos\beta \, + \, Wh\theta \, - \, \left(k_{\theta,f} + k_{\theta,r}\right)\theta \, = \, 0 \qquad (5.83)$$

which, using the lateral acceleration, can also be written as

$$W \frac{a_{lat}}{g} \, h \, \cos\beta \, = \, \left[\left(k_{\theta f} + k_{\theta r}\right) - W h\right]\theta$$

Differentiating the previous equation with respect to the lateral acceleration results in

$$W \, h \cos\beta \, - \, W \frac{a_{lat}}{g} \, h \, \sin\beta \, \frac{d\beta}{d\left(\dfrac{a_{lat}}{g}\right)} \, = \, \left[\left(k_{\theta f} + k_{\theta r}\right) - W h\right] \frac{d\theta}{d\left(\dfrac{a_{lat}}{g}\right)}$$

For purposes of estimating roll stiffness we may assume, from practical experience, that $\cos\beta \approx 1$ and $\sin\beta \approx 0$; therefore the foregoing equation becomes

Chapter 5: CORNERING PERFORMANCE

$$k_{\theta f} + k_{\theta r} = W h \left[1 + \frac{1}{d\theta \Big/ d\left(\dfrac{a_{lat}}{g}\right)} \right] \qquad (5.102)$$

In essence, the right-hand-side of Eq (5.102) consists of only two unknowns: the rolling moment, Wh, and the roll gradient. The rolling moment is dependent on the vehicle design specifications. For instance, in our baseline example $W = 1738\,\text{lb}$ and from Eq (5.79):

$$h = H - \frac{R_{cf} L_r + R_{cr} L_f}{L_w}$$

$$= 20.0 - \frac{(3.0)(50.58) + (3.5)(57.42)}{108.0} = 16.73 \text{ in } \{1.395 \text{ ft}\}$$

therefore, $Wh = 2424\,\text{ft-lb}$. The combined front and rear roll stiffness can then be calculated by specifying a desired roll gradient. Table (5.4) lists representative values of roll gradient based on cornering or roll stiffness. This table is useful in establishing a starting point, and also gives us a subjective feel for how a driver might perceive our vehicle. To continue on with our current suspension design problem, we will specify a "firm" roll gradient. Then, based on Table (5.4), the required combined roll stiffness is

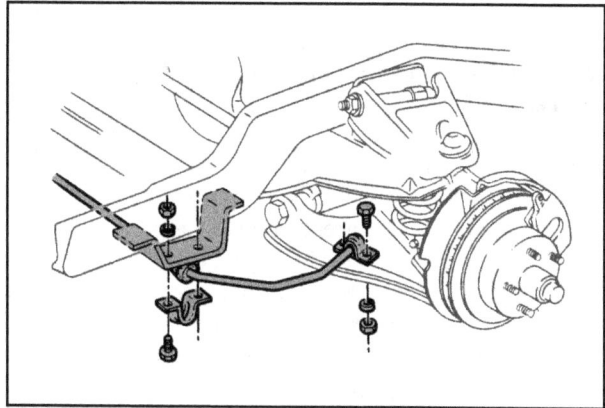

Figure (5.30): EXAMPLE ANTI-ROLL BAR INSTALLATION ON AN INDEPENDENT FRONT SUSPENSION

$$k_{\theta f} + k_{\theta r} = 2424 \left[1 + \frac{1}{5}\frac{180}{\pi} \right] = 30{,}200\,\frac{\text{ft-lb}}{\text{rad}} \left\{ 527\frac{\text{ft-lb}}{\text{deg}} \right\}$$

It should be noted that the suspension springs offer some resistance to roll. To calculated the required anti-roll bar torsional stiffness we must deduct the portion of roll stiffness due to the springs from the combined total above. Based on Eq (5.72) we can express the combined roll stiffness as

$$k_{\theta f} + k_{\theta r} =$$

$$k_{sf}\frac{T_f^2}{2} + k_{sr}\frac{T_r^2}{2} + \left(k_{bf} + k_{br}\right)$$
$$(5.103A)$$

which can be rewritten to yield

$$k_{bf} + k_{br} =$$

$$\left(k_{\theta f} + k_{\theta r}\right) - \left[k_{sf}\frac{T_f^2}{2} + k_{sr}\frac{T_r^2}{2}\right]$$
$$(5.103B)$$

Substituting the calculated combined roll stiffness along with vehicle design parameters from Table (5.4) into Eq (5.103B) produces

$$k_{bf} + k_{br} = 328\ \frac{\text{ft-lb}}{\text{deg}}.$$

Table (5.4): REPRESENTATIVE ROLL GRADIENTS

Roll Stiffness	Roll Gradient [deg/g]
Very Soft	8.5
Soft	7.5
Semi-Soft	7.0
Semi-Firm	6.0
Firm	5.0
Very Firm	4.2
Extremely Firm	3.0
Hard	1.5

Adapted from WF Milliken and DL Milliken, RACE CAR VEHICLE DYNAMICS, SAE, Warrendale,PA, 1995, pg 584.

Unfortunately, the equations we have do not tell us how the anti-roll stiffness should be split between the front and rear anti-roll bars. Logically, if we divide the total roll stiffness equally between the front and rear suspension, then we should maintain the cornering behavior, i.e., understeer, oversteer, etc., that we currently have. In our current baseline design we have oversteer, which we would like to eliminate in favor of a slight amount of understeer. With this in mind, it is desirable to have more roll stiffness in the front suspension relative to the rear suspension. As a starting point in the anti-roll bar design process, we will arbitrarily select a front-to-rear split of approximately 60% / 40% roll stiffness distribution. Then,

$$k_{bf} = 197\ \frac{\text{ft-lb}}{\text{deg}} \qquad \text{and} \qquad k_{br} = 131\ \frac{\text{ft-lb}}{\text{deg}}.$$

For discussion purposes only, we will install an anti-roll bar on just the front suspension to ascertain it's affect on cornering performance. This suspension change/addition to our baseline vehicle is cataloged in Table (5.2) as "Vehicle 2." The cornering results of the modified design are tabulated in Table (5.5) and graphically presented as the curvature gain in Figure (5.31). "Vehicle 2" shows a significant improvement in body roll angle, but there is a drastic change in cornering behavior. We went from a vehicle design which has oversteering characteristics to a new design that now has a substantial amount of understeer.

Looking at Table (5.5), the maximum forward speed listed for "Vehicle 2" is 34 fps. Why is this the maximum? The answer to this question lies in the tabulated values of normal force acting on the front-inside tire. If we allowed the simulation to continue, we would find that the front-inside normal force becomes negative for speeds

Table (5.5): CORNERING RESULTS FOR "VEHICLE 2", "VEHICLE 3" AND "VEHICLE 4"

– Vehicle 2 –

Vel [fps]	Drift [deg]	Roll [deg]	Radius [ft]	Normal Force [lb]				Slip Angle [deg]				Cornering Force [lb]			
				fi	fo	ri	ro	fi	fo	ri	ro	fi	fo	ri	ro
16	4.09	1.03	52.74	326	492	420	501	-0.42	-0.56	-0.52	-0.48	37	87	67	72
18	3.99	1.31	52.73	304	514	409	511	-0.51	-0.64	-0.61	-0.57	48	109	83	93
20	3.89	1.61	52.73	280	539	396	522	-0.62	-0.74	-0.72	-0.67	57	137	100	117
22	3.77	1.95	52.75	253	567	383	535	-0.75	-0.86	-0.85	-0.79	66	169	118	145
24	3.63	2.32	52.77	224	597	368	549	-0.89	-0.99	-0.98	-0.92	72	207	135	178
26	3.47	2.72	52.83	192	630	352	564	-1.06	-1.15	-1.14	-1.07	75	253	152	214
28	3.29	3.15	52.94	158	664	335	581	-1.25	-1.33	-1.32	-1.23	73	307	169	256
30	3.08	3.61	53.11	122	701	317	598	-1.49	-1.55	-1.52	-1.42	65	369	183	303
32	2.83	4.08	53.37	83	740	298	617	-1.77	-1.81	-1.75	-1.64	49	442	196	355
34	2.55	4.58	53.77	44	780	279	636	-2.10	-2.11	-2.02	-1.88	25	526	205	412

– Vehicle 3 –

Vel [fps]	Drift [deg]	Roll [deg]	Radius [ft]	Normal Force [lb]				Slip Angle [deg]				Cornering Force [lb]			
				fi	fo	ri	ro	fi	fo	ri	ro	fi	fo	ri	ro
16	4.08	0.75	52.74	346	471	401	520	-0.42	-0.56	-0.52	-0.48	40	84	65	74
18	3.99	0.95	52.72	330	488	385	535	-0.51	-0.65	-0.61	-0.57	52	105	79	97
20	3.89	1.18	52.70	312	507	367	552	-0.62	-0.74	-0.73	-0.68	64	130	94	124
22	3.76	1.43	52.67	291	529	347	571	-0.74	-0.86	-0.85	-0.80	76	159	108	155
24	3.63	1.70	52.65	269	552	325	592	-0.88	-0.99	-1.00	-0.93	87	193	122	192
26	3.47	1.99	52.61	245	577	301	615	-1.04	-1.14	-1.17	-1.09	95	233	134	234
28	3.28	2.31	52.58	219	604	276	640	-1.22	-1.31	-1.36	-1.27	101	281	143	284
30	3.07	2.66	52.54	190	633	248	667	-1.44	-1.51	-1.59	-1.48	103	335	149	342
32	2.80	3.03	52.52	160	664	219	696	-1.71	-1.76	-1.86	-1.74	100	400	151	409
34	2.49	3.42	52.49	127	696	188	727	-2.02	-2.05	-2.19	-2.04	90	474	146	486
36	2.10	3.84	52.49	92	730	155	760	-2.43	-2.43	-2.59	-2.42	72	560	134	575
38	1.59	4.27	52.58	56	766	122	794	-2.96	-2.92	-3.11	-2.90	45	659	114	675
40	0.89	4.70	52.92	18	800	90	830	-3.71	-3.62	-3.80	-3.54	10	766	89	780

– Vehicle 4 –

Vel [fps]	Drift [deg]	Roll [deg]	Radius [ft]	Normal Force [lb]				Slip Angle [deg]				Cornering Force [lb]			
				fi	fo	ri	ro	fi	fo	ri	ro	fi	fo	ri	ro
16	4.08	0.73	52.74	345	472	402	518	-0.42	-0.56	-0.52	-0.48	40	84	65	74
18	3.99	0.92	52.72	329	490	387	533	-0.51	-0.65	-0.61	-0.57	52	105	79	96
20	3.89	1.14	52.70	310	509	369	550	-0.62	-0.74	-0.73	-0.68	64	130	94	123
22	3.77	1.38	52.68	290	530	350	568	-0.74	-0.86	-0.85	-0.80	76	160	109	154
24	3.63	1.65	52.65	267	554	328	589	-0.88	-0.99	-1.00	-0.93	86	194	123	191
26	3.47	1.93	52.63	243	579	305	611	-1.04	-1.14	-1.17	-1.09	94	234	135	233
28	3.28	2.24	52.60	216	606	280	635	-1.23	-1.31	-1.36	-1.27	100	282	145	282
30	3.07	2.58	52.58	187	636	253	662	-1.44	-1.51	-1.59	-1.48	102	337	152	339
32	2.81	2.93	52.56	157	667	225	690	-1.71	-1.76	-1.85	-1.73	98	401	154	404
34	2.50	3.31	52.56	124	700	195	720	-2.03	-2.05	-2.17	-2.03	87	476	151	480
36	2.11	3.71	52.61	89	734	163	751	-2.43	-2.43	-2.57	-2.40	69	563	140	567
38	1.61	4.13	52.73	52	770	131	785	-2.95	-2.91	-3.07	-2.87	41	661	123	663
40	0.94	4.54	53.16	15	804	100	819	-3.69	-3.60	-3.73	-3.48	6	766	100	765

Inside Steer Angle: 10.0 deg
Outside Steer Angle: 9.45 deg

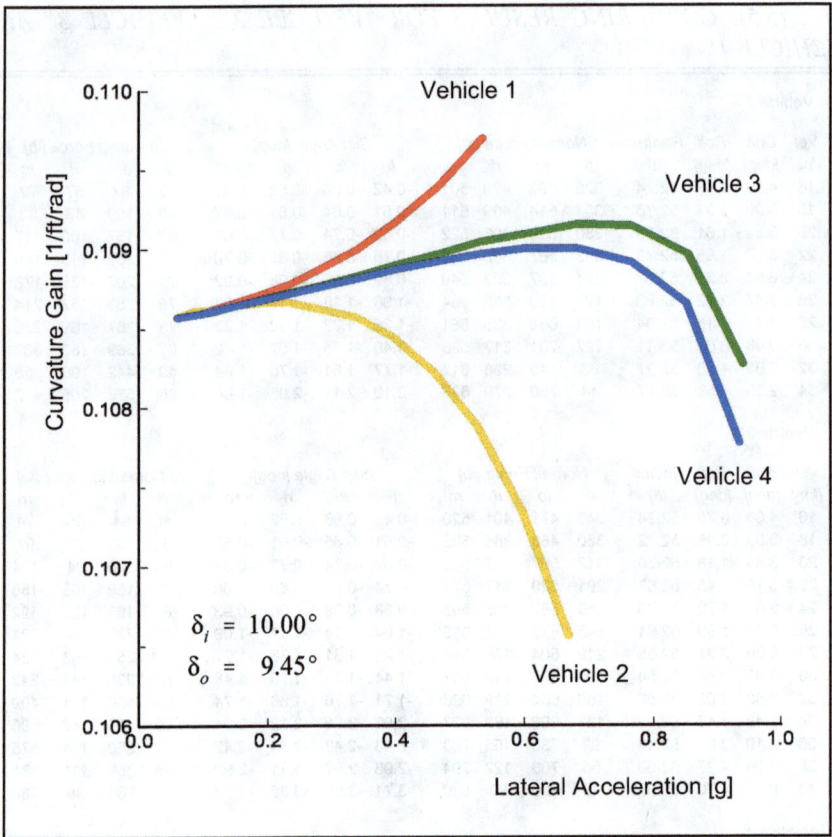

Figure (5.31): CURVATURE GAIN IMPROVEMENT RESULTS

greater than about 34 fps. This indicates a condition where the front-inside tire lifts off the road surface. In this case we no longer have a four-wheeled vehicle, but rather a three-wheeled vehicle with an attached spare tire. To eliminate the wheel lifting problem we need to obtain a better balance, i.e., smaller difference, in the front and rear suspension roll stiffness. The obvious solution to this problem is to install the rear suspension anti-roll bar. The results of this modification are found under the "Vehicle 3" label.

The cornering characteristics demonstrated by "Vehicle 3" are definitely an improvement over those of "Vehicle 1" or "Vehicle 2", but now we have a suspension design that exhibits variable steer characteristics. The vehicle starts out with oversteer, goes to neutral steer at about 0.7 g and then starts to understeer until the lateral acceleration reaches a value of about 0.94. A further increase in vehicle speed once again causes the front-inside tire to lift off of the road surface.

The selection of anti-roll stiffness rates used in the suspension design of "Vehicle 3" is probably not optimum. A more complete treatise of the vehicle cornering characteristics with anti-roll bars would include examining the behavior of a range of steer angles. The general goal of a successful suspension design is to maximize the vehicle speed while cornering, and to obtain a slight amount of understeer to maintain stability. At this point adjusting the anti-roll bar stiffness rates to improve the cornering performance is somewhat of a trial and error process. For instance, if it is desirable to increase the amount of understeer without increasing the total roll stiffness, Table (5.6) indicates that the roll stiffness distribution should be shifted from the rear suspension to the front. The positive or negative effect from this shift can be ascertained by re-running the four-wheel cornering simulation. Based on the results, further adjustment is made in the roll stiffness distribution is made until desired results are obtained. We could spend more time developing anti-roll stiffness rate specifications to improve cornering performance of our current example, but at this juncture in our study of corning behavior we can better utilize our time by examining other aspects of suspension design.

The slope of the roll axis and the location of the suspension roll centers also have an impact of cornering performance. Raising the roll center decreases the lever arm of the sprung mass about the roll axis, thereby reducing the roll couple. This creates an effect similar to that caused by increasing the suspension anti-roll bar stiffness. The exact effect of altering the roll center is difficult to deduce from the steady-state cornering equations. But, in general terms, the roll center height determines, in part, the distribution of normal force primarily from the inside to the outside wheel. Raising the roll center height at one axle plane causes a more even distribution of normal force at the opposition axle plane. This means, for instance, by

Table (5.6): RELATIVE EFFECT OF UNDERSTEER/OVERSTEER ADJUSTMENTS

Increase Understeer (Decrease Oversteer)	Increase Oversteer (Decrease Understeer)
Shift CG location toward rear axle plane.	Shift CG location toward front axle plane.
Increase front roll stiffness and/or decrease rear roll stiffness.	Decrease front roll stiffness and/or increase rear roll stiffness.
Raise front roll center height and/or lower rear roll center height.	Lower front roll center height and/or raise rear roll center height.
Increase aerodynamic down force at rear axle and/or decrease aerodynamic down force at front axle.	Decrease aerodynamic down force at rear axle and/or increase aerodynamic down force at front axle.
Change tire selection: increase front slip angles and/or decrease rear slip angles.	Change tire selection: decrease front slip angles and/or increase rear slip angles.

raising the roll center at the front suspension we will design more understeer into the vehicle. The effect of changing roll center location is most noticeable at the higher end of lateral acceleration. These effects are summarized in Table (5.6).

Lets return to our example design problem. Looking at Figure (5.31), the curvature gain plot for "Vehicle 3" shows oversteer characteristics for lateral accelerations less than about 0.7 g. Our goal now is to adjust the roll center location(s) to produce a more neutral steering vehicle. This is accomplished by offsetting the current oversteer behavior with an appropriate increase in understeer. Generally it is best, and least confusing, to make only one suspension design modification at a time. Rather than changing both roll center locations we will just increase the front roll center by one inch. The cornering results of this latest suspension change, labeled as "Vehicle 4", are listed in Table (5.5) and are also graphed in Figure (5.31). The results do indeed show a more neutral steering car.

One should bear in mind, just in case it is not obvious, that while making roll center adjustment in the cornering simulation is easy, making changes to an existing vehicle is not. For an existing vehicle, altering the roll center height usually involves a major re-design of suspension linkages and suspension pick-up points on the chassis. It is also necessary to take precautions to limit the roll center height. Too high of location leads to jacking.

The last modification of our current "Vehicle 4" design is to examine the affect of changing tires on cornering performance. In this instance we will look at replacing

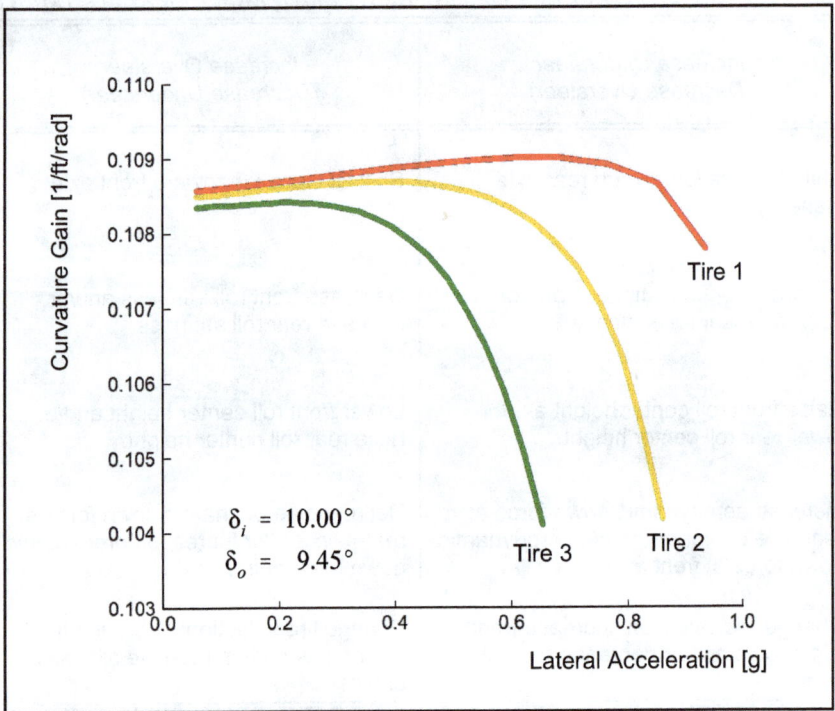

Figure (5.32): CURVATURE GAIN OF "VEHICLE 4" WITH DIFFERENT TIRES

the current tires, labeled "Tire 1" in Figure (5.32), with a set of tires ("Tire 2") that have half the cornering stiffness, and with a set ("Tire 3") of one-quarter cornering stiffness. At first glance we notice quite a bit of change. We started out with a slightly oversteering vehicle and ended up with a vehicle which has considerable understeer with poorer lateral acceleration performance. In fact, for the "Tire 3" selection, the vehicle starts out with a slight amount of oversteer but very quickly changes to understeer which continues until the lateral acceleration reaches a value of about 0.65 g. At this point the vehicle is near the control limit, i.e., maximum slip angle before the tire starts to skid. Increasing the lateral acceleration further causes a very non-linear decrease in curvature gain for an understeering vehicle. (A dramatic increase in curvature gain would be observed for an oversteering car.) This cornering behavior is caused by the non-linear relationship between tire lateral cornering force and slip angle when one or more of the vehicle's tires approach the limit of adhesion. A small increase in cornering force, required to offset a small increase in lateral acceleration, generates a large increase in slip angle. In turn, the larger slip angle forces an understeering vehicle into an attitude that greatly increases the turning radius. In our example, the turning radius increases fast enough to offset increasing forward velocity such that the lateral acceleration experienced by the vehicle becomes nearly constant. Hence the rapid drop-off in the curvature gain shown in Figure (5.32). Our conclusion is that the selection of tires becomes an integral part of the suspension design. The remaining facet of cornering performance is to examine the affect of steer angles and steering geometry.

(5.5) VEHICLE STEERING

The previous section is concerned mainly with the cornering forces generated by the tires and an understanding of the dynamics involved in executing a cornering maneuver. Equally important in the cornering process is the steering system, through which the intended direction of the vehicle is controlled by the driver. In this section we will undertake a study of geometry for front-wheel steered vehicles.

The foundation of any steering linkage and mechanism design is the geometric analysis. Simply stated, the first goal of the geometric analysis is to determine an appropriate relationship between the outside steer angle and the inside steer angle. The steer angle relationship can then be used as "input-output" criterion for the geometric design of a four-bar steering linkage.

(5.5.1) JEANTAUD STEERING LINKAGE

The earliest attempt to devise a method for steering linkage design is generally credited to Rudoff Ackermann; however, he was not the inventor[2]. The concept was actually developed for horse-drawn carriages by Georg Langensperger (Munich, 1816). Ackermann was Langensperger's patent agent in London. After obtaining a patent for Langensperger, Ackermann published a pamphlet regarding steering improvements for four-wheel carriages.

The principle of Ackermann steering states that at any particular instant while traveling through a corner, all four wheels move in an arc about a common point or turn center. This basic geometry was introduced in Figure (5.1). Under these conditions, the relationship between the inside steer angle and the outside steer angle is

$$\delta_i = \arctan\left(\frac{L_w \tan\delta_o}{L_w - T_f \tan\delta_o}\right) \tag{5.104}$$

The previous equation is referred to as Ackermann steering, or the Ackermann effect.

Around 1881 Charles Jeantaud began to re-examine the concept of Ackermann steering for electric car applications. The result of his investigations is an attempt at a graphical method of steering linkage design which will produce an Ackermann effect. At that time this method was known in France as the Jeantaud diagram, illustrated in Figure (5.33). Basically, the steering arm can have any length, but must be located

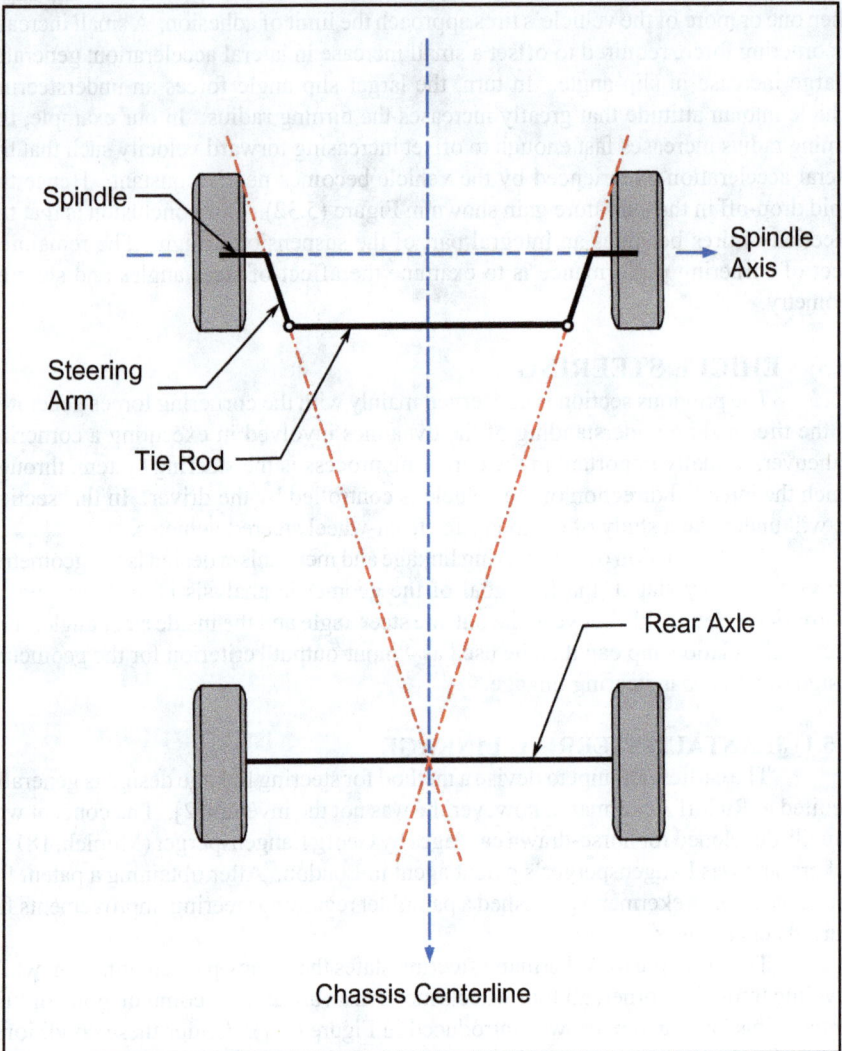

Figure (5.33): JEANTAUD STEERING LINKAGE DIAGRAM

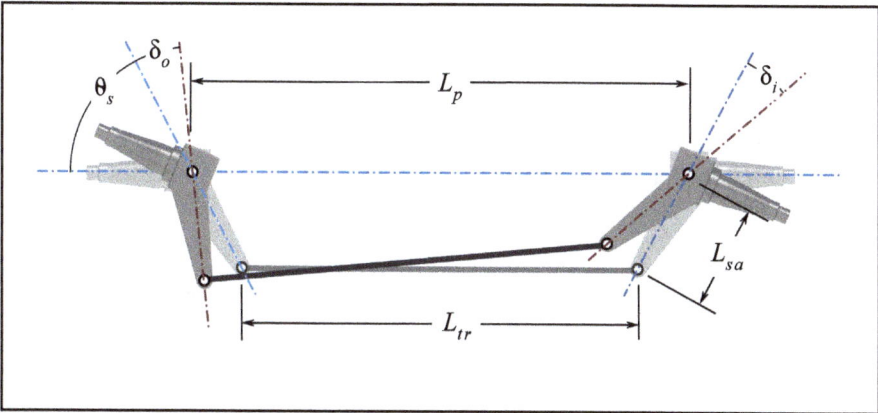

Figure (5.34): TWO-DIMENSIONAL STEERING LINKAGE GEOMETRY

along an imaginary line connecting the front wheel pivot point to the point created by the intersection of the chassis centerline with the rear axle axis. It should be pointed out that Jeantaud's design is two-dimensional. That is, any affect of suspension movement in bounce, or rebound, on steer angle(s) is neglected.

The behavior of any two-dimensional steering linkage, including Jeantaud's design, can be studied by treating the linkage as a symmetrical four-bar mechanism as shown in Figure (5.34). The relationship between the steer angles and the lengths of the links is given by a variation of Freudenstein's equation[3]:

$$R_1 - R_2 \left[\cos\left(\theta_s - \delta_i\right) + \cos\left(\theta_s + \delta_o\right) \right] + \cos\left(2\theta_s - \delta_i + \delta_o\right) = 0 \qquad (5.105)$$

where

$$R_1 = \frac{L_p^2 + 2L_{sa}^2 - L_{tr}^2}{2L_{sa}^2} \qquad (5.106)$$

$$R_2 = \frac{L_p}{L_{sa}} \qquad (5.107)$$

and

$$\theta_s = \arccos\left(\frac{L_p - L_{tr}}{2L_{sa}}\right) \qquad (5.108)$$

It should be noted that in the specific study of Jeantaud's steering linkage, the tie rod length is not arbitrary. It length depends on the wheelbase, distance between steered wheel pivot points, and steering arm length, i.e.,

Chapter 5: CORNERING PERFORMANCE

$$L_{tr} = L_p - 2 L_{sa} \sin \left[\arctan \left(\frac{L_p}{2 L_w} \right) \right]$$ (5.109)

Eq (5.105) can be developed into an expression that describes the inside steer angle as a function of linkage lengths and outside steer angle. First, rewrite Eq (5.105) as

$$R_1 - R_2 \cos(\theta_s + \delta_o) + \left[\cos(2\theta_s + \delta_o) - R_2 \cos\theta_s \right] \cos\delta_i$$
$$+ \left[\sin(2\theta_s + \delta_o) - R_2 \sin\theta_s \right] \sin\delta_i = 0$$ (5.110)

For convenience, define

$$A_1 = R_1 - R_2 \cos(\theta_s + \delta_o)$$ (5.111)

$$A_2 = \cos(2\theta_s + \delta_o) - R_2 \cos\theta_s$$ (5.112)

and

$$A_3 = \sin(2\theta_s + \delta_o) - R_2 \sin\theta_s$$ (5.113)

Using these definitions, Eq (5.110) becomes

$$A_1 + A_2 \cos\delta_i = - A_3 \sin\delta_i$$ (5.114)

Squaring both sides of Eq (5.114) results in

$$A_1^2 + 2 A_1 A_2 \cos\delta_i + A_2^2 \cos^2\delta_i = A_3^2 \sin^2\delta_i$$
$$= A_3^2 \left(1 - \cos^2\delta_i \right)$$

or, upon rearranging terms

$$\left(A_2^2 + A_3^2 \right) \cos^2\delta_i + 2 A_1 A_2 \cos\delta_i + \left(A_1^2 - A_3^2 \right) = 0$$ (5.115)

Solving Eq (5.115) for the inside steer angle leads to:

$$\delta_i = \arccos \left(\frac{- A_1 A_2 - A_3 \sqrt{A_2^2 - A_1^2 + A_3^2}}{A_2^2 + A_3^2} \right)$$ (5.116)

It is interesting to study the Jeantaud steering linkage because it does not provide correct Ackermann geometry for all steered wheel positions. As the example in Figure (5.35) illustrates, the linkage only matches the Ackermann curve at small angles of steer. The deviation of the actual inside-outside steer angle curve from the theoretical Ackermann curve is called steering linkage error. Further optimization of Jeantaud's linkage is necessary in order to reduce the amount of linkage error; however, this is not practical. If you recall, Ackermann's principle is based on tires that exhibit

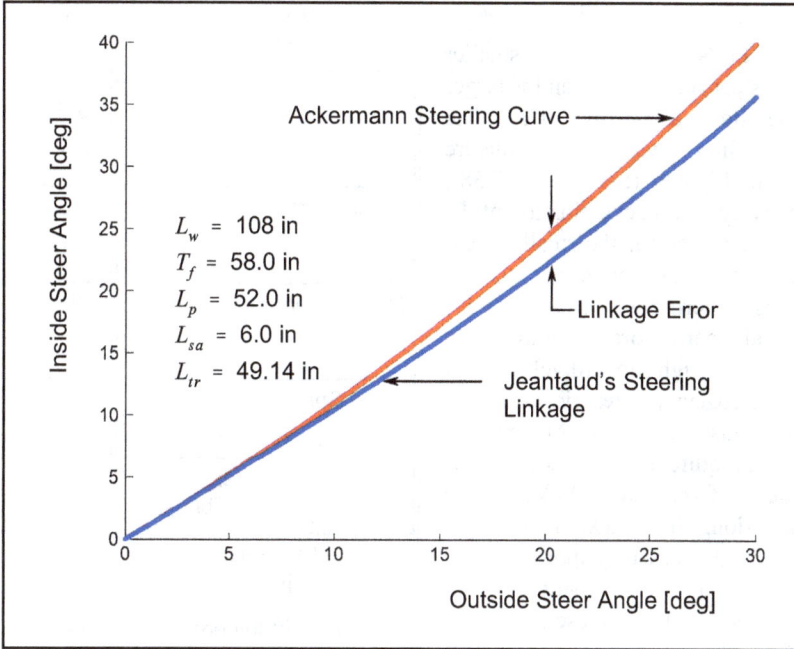

Figure (5.35): EXAMPLE COMPARISON OF ACKERMANN STEERING GEOMETRY TO JEANTAUD'S STEERING LINKAGE

no slip. This means that even the Ackermann steering curve is in error, especially at large angles of steer. Before we attempt to optimize the steering linkage, our first priority should be to develop a more accurate steering design curve.

(5.5.2) STEERING WITH TIRE SLIP

A unique feature of Ackermann steering is the location of the turn center, which is always located in the rear axle plane. This is not the case when the affects of tire slip are included in vehicle cornering models. Although the Ackermann steering angle relationship is a good starting point, small adjustments are made to improve handling characteristics. For example, consider the vehicle which has chassis design parameters as listed in Table (5.7). At an outside steer angle of $10°$ the inside steer angle calculated from Eq (5.104) is $11.031°$. As shown in Figure (5.36), we can change the handling characteristics of our vehicle by making slight changes to the calculated inside steer angle. For our vehicle when we increase the inside steer angle we incorporate more understeer, and, conversely, when the steer angle is decreased more oversteer is obtained. In this case, a slight decrease in the Ackermann steer angle tends to provide us with more neutral steer characteristics. This is generally the case, and is somewhat erroneously referred to as anti-Ackermann steering adjustment. However, the deviation from the theoretical Ackermann curve is not linear.

Let's consider a smaller outside steer angle of 2° and a larger outside steer angle of 30°. The corresponding curvature gain results are shown in Figures (5.37) and (5.38), respectively. Upon examination of the results we notice for the smaller steer angle that the Ackermann specification provides cornering characteristics close to neutral steer before transitioning to understeer at high forward velocity. A this smaller angle any decrease in Ackermann steering causes oversteer. At the larger steer angle quite a bit of understeer is noticed for the Ackermann specification. The Ackermann angle needs to be decreased by approximately 2° to obtain more neutral steer characteristics. From these results we can draw the general conclusion that very little deviation from the

Table (5.7): CHASSIS DESIGN PARAMETER LISTING FOR STEERING COMPARISON EXAMPLES

Chassis Design Parameters	Vehicle 5
Sprung Weight, W [lb]	1738
CG Location: L_f [in]	57.42
L_r [in]	50.58
H [in]	20.0
Track Width: T_f [in]	58.5
T_r [in]	58.0
Roll Center: $R_{c,f}$ [in]	4.0
$R_{c,r}$ [in]	2.5
Front Spring Rate: $k_{s,fi}$, $k_{s,fo}$ [lb/in]	56.7
Rear Spring Rate: $k_{s,ri}$, $k_{s,ro}$ [lb/in]	73.8
Anti-roll Stiffness: $k_{b,f}$ [ft-lb/deg]	205
$k_{b,r}$ [ft-lb/deg]	- - -

Ackermann specification is required at small angles, but must be progressively decreased as the outside steer angle increases.

Returning to our previous example, summarized in Figure (5.35) and Table (5.7), discrete values of outside steer angle are selected. Adjustment of the inside steer angle is then made in accordance with the foregoing anti-Ackermann approach in obtaining desired handling performance. The exact amount of deviation is left to the judgement of the design engineer. The Ackermann steering curve along with the discrete adjusted values are shown in Figure (5.39).

When examining the four-bar linkage equation, Eq (5.105), which we may think of as the steering linkage performance equation, we notice in Figure (5.34) that there are three geometrical parameters involved in the design: L_p, the distance between the steering knuckle pivot axis; L_{sa}, the steering arm length; and L_{tr}, the tie-rod length. As a practical matter, the dimension L_p is fixed based on the design of the front suspension. The steering arm length is also fixed as the steer arm component must fit within the confines of the wheel rim. This leaves only the tie-rod length as the parameter to be adjusted to match the steering linkage performance to the desired inside-outside steer angle points; an example of which is shown in Figure (5.39).

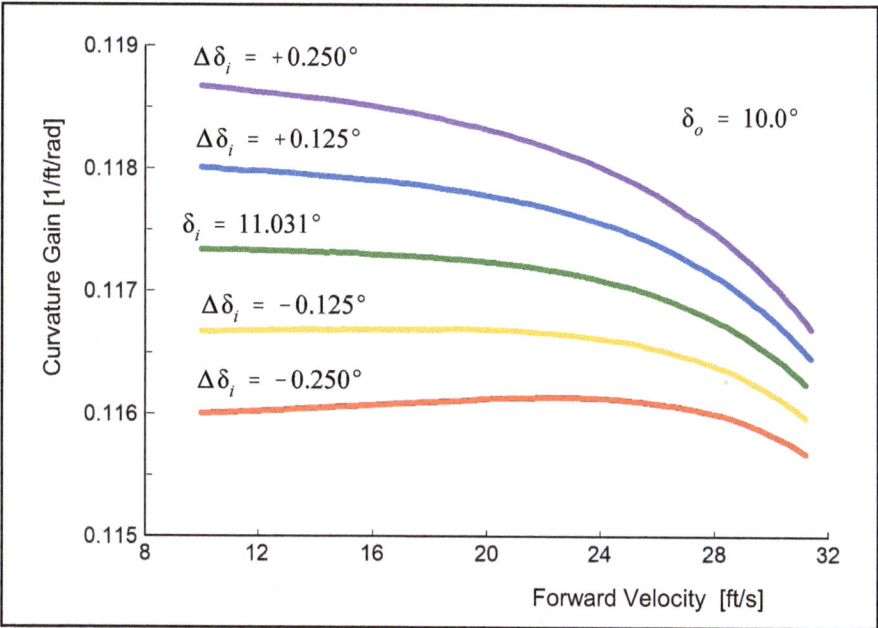

Figure (5.36): INSIDE STEER ANGLE ADJUSTMENT ON CURVATURE GAIN FOR OUTSIDE STEER ANGLE OF 10°

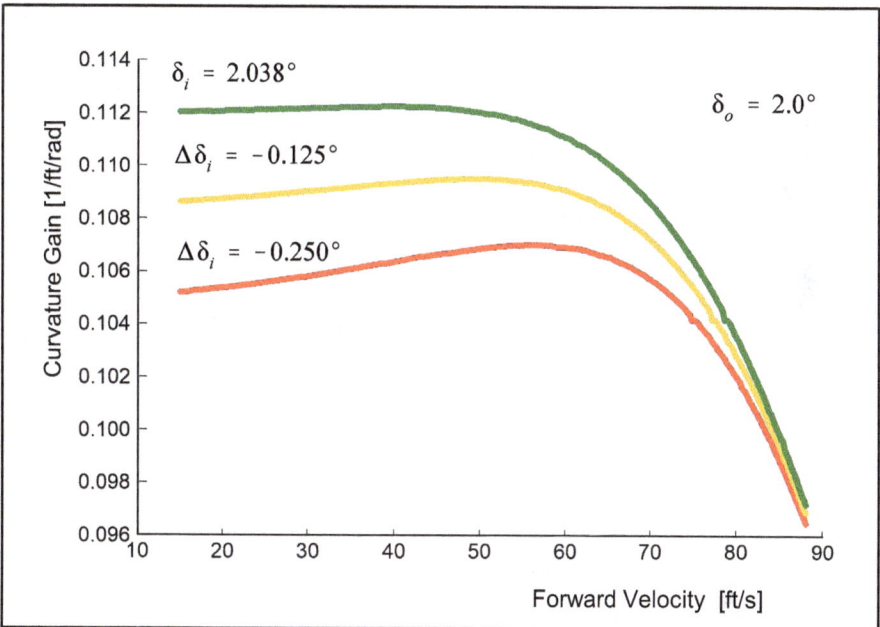

Figure (5.37): INSIDE STEER ANGLE ADJUSTMENT ON CURVATURE GAIN FOR OUTSIDE STEER ANGLE OF 2°

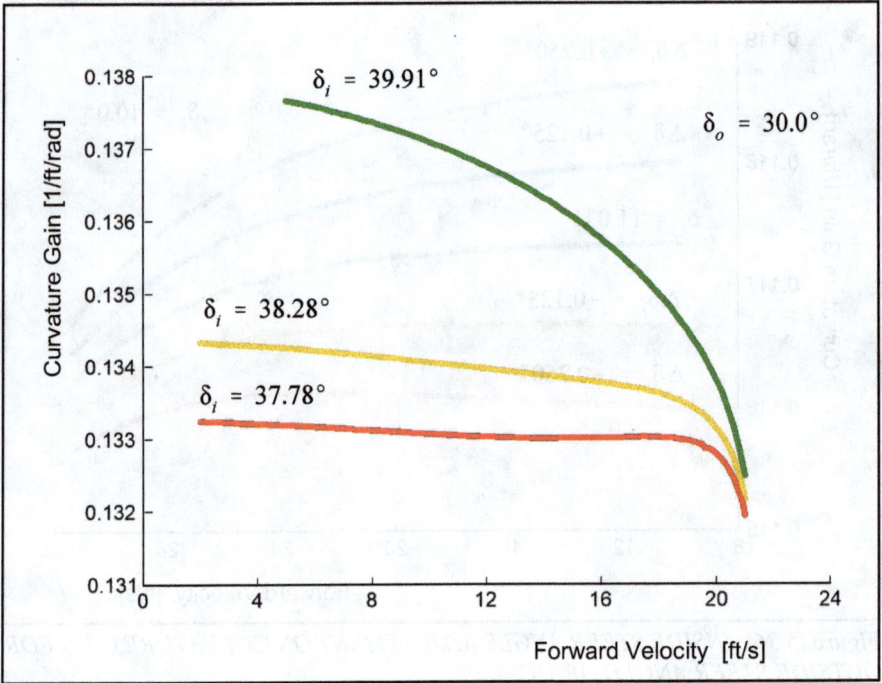

Figure (5.38): INSIDE STEER ANGLE ADJUSTMENT ON CURVATURE GAIN FOR OUTSIDE STEER ANGLE OF 30°

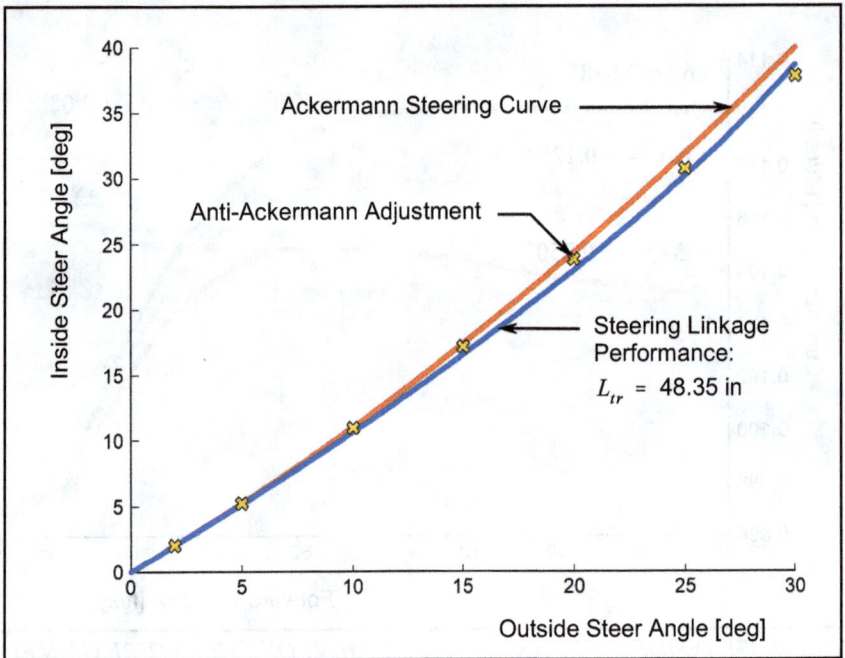

Figure (5.39): EXAMPLE STEERING LINKAGE PERFORMANCE

Chapter 5: CORNERING PERFORMANCE

(5.5.3) RACK AND PINION STEERING

The last example we will examine is the inclusion of rack and pinion steering into the front suspension design. In this design the tie-rod component of the four-bar steering linkage is replaced by a rack and pinion steering assembly, demonstrated by the example shown in Figure (5.40). We begin by asking the question: "How do the rack

Figure (5.40): TYPICAL RACK AND PINION STEERING INSTALLATION

and pinion design dimensions affect steering geometry?" A two-dimensional geometric representation of a rack and pinion steering system is shown in Figure (5.41). Two of the geometrical parameters involved in the design: L_p, the distance between the steering

Figure (5.41): TWO-DIMENSIONAL RACK AND PINION STEERING GEOMETRY

knuckle pivot axis and L_{sa}, the steering arm length, are the same as the four-bar steering system. The tie-rod length, L_{tr}, is slightly different as it only pertains to only one side of the rack, although left and right sides are the same. Two new design dimensions are needed to complete the steering system geometry: L_r, the rack length from pivot-to-pivot, and L_{rp}, the rack installation offset from the front axle plane. The independent feature of the rack and pinion steering system is the rack displacement, ΔR, as it is controlled by the driver through the steering wheel and pinion gear.

Using the xy-coordinate system shown in Figure (5.41), the coordinates of the pivot points (designated as 1, 2, 3 and 4) are:

$$\text{Pivot 1: } x_1 = L_{sa} \sin(\theta - \delta_i) \qquad\qquad y_1 = L_{sa} \cos(\theta - \delta_i)$$

$$\text{Pivot 2: } x_2 = L_{rp} \qquad\qquad y_2 = \frac{1}{2}(L_p - L_r) + \Delta R$$

$$\text{Pivot 3: } x_3 = L_{rp} \qquad\qquad y_3 = \frac{1}{2}(L_p + L_r) + \Delta R$$

$$\text{Pivot 4: } x_4 = L_{sa} \sin(\theta + \delta_o) \qquad\qquad y_4 = L_p - L_{sa} \cos(\theta + \delta_o)$$

The geometric constraint of the tie-rod component is given by

$$L_{tr}^2 = (x_2 - x_1)^2 + (y_2 - y_1)^2 \qquad\qquad (5.117a)$$

as well as

$$L_{tr}^2 = (x_4 - x_3)^2 + (y_4 - y_3)^2 \qquad\qquad (5.117b)$$

Substituting the pivot coordinate locations into Eqs (5.117a) and (5.117b), along with quite a bit of mathematical manipulation, leads to the relationships:

(i) inside steer angle

$$\theta - \delta_i = \arcsin\left(\frac{-B_i + \sqrt{B_i^2 - 4A_i C_i}}{2A_i}\right) \qquad\qquad (5.118)$$

where

$$A_i = L_{rp}^2 + \left[\frac{1}{2}(L_p - L_r) + \Delta R\right]^2$$

$$B_i = -2E_i L_{rp}$$

$$C_i = E_i^2 - \left[\frac{1}{2}(L_p - L_r) + \Delta R\right]^2$$

and

$$E_i = \frac{L_{sa}^2 + L_{rp}^2 - L_{tr}^2 + \left[\frac{1}{2}(L_p - L_r) + \Delta R\right]^2}{2 L_{sa}}$$

(*ii*) outside steer angle

$$\theta + \delta_o = \arccos\left(\frac{-B_o - \sqrt{B_o^2 - 4 A_o C_o}}{2 A_o}\right) \tag{5.119}$$

where

$$A_o = L_{rp}^2 + \left[\frac{1}{2}(L_p - L_r) - \Delta R\right]^2$$

$$B_o = 2 E_o\left[\frac{1}{2}(L_p - L_r) - \Delta R\right]$$

$$C_o = E_o^2 - L_{rp}^2$$

and

$$E_o = \frac{L_{tr}^2 - L_{sa}^2 - L_{rp}^2 - \left[\frac{1}{2}(L_p - L_r) - \Delta R\right]^2}{2 L_{sa}}$$

The angle θ is the steering arm angle at $0°$ steer, or zero rack displacement, and is found by substituting $\Delta R = 0$ into either Eq (5.118) or (5.119).

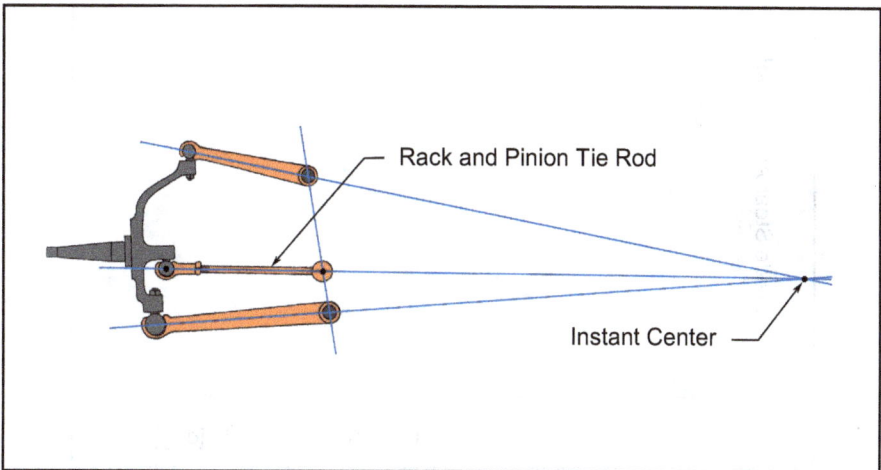

Figure (5.42): RACK AND PINION TIE ROD GEOMETRY FOR DOUBLE A-ARM FRONT SUSPENSION

Chapter 5: CORNERING PERFORMANCE

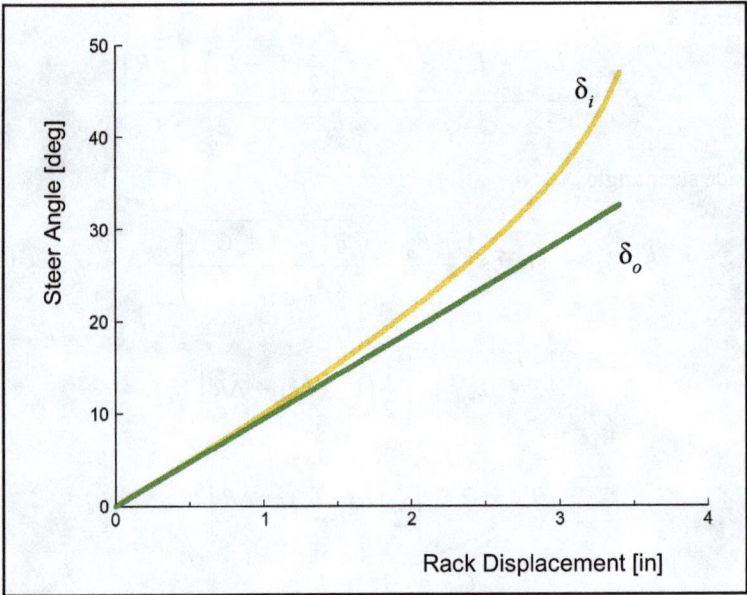

Figure (5.43): EXAMPLE INSIDE/OUTSIDE STEER ANGLE PERFORMANCE VERSUS RACK DISPLACEMENT

Similar to our previous four-bar steering system analysis, we think of Eqs (5.18) and (5.19) as rack and pinion steering performance equations. As before, the geometrical parameters L_p and L_{sa} are constrained due to the design of the suspension

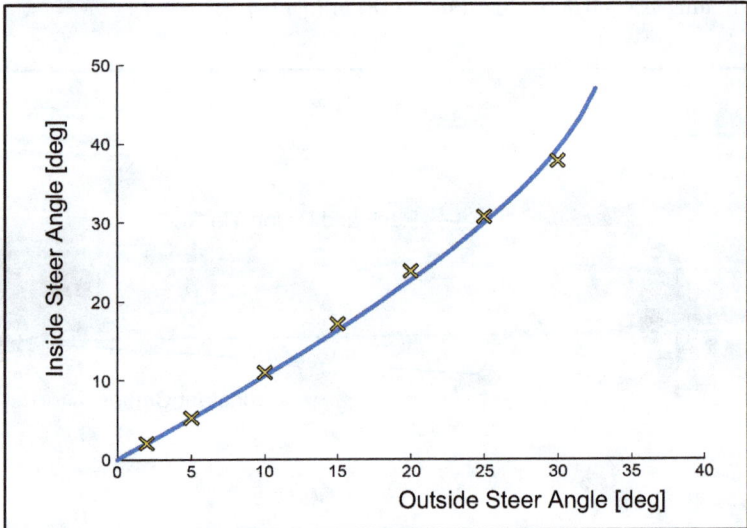

Figure (5.44): EXAMPLE RACK AND PINION PERFORMANCE GRAPH

system. The rack offset, L_{rp}, is also constrained by practical considerations as it must be located within the confines of the chassis and provide space for equipment such as the engine, or perhaps the drivetrain. There is also a practical consideration of the overall rack length, L_r , as well. To clarify this statement, consider the double a-arm front suspension system shown in Figure (5.42). In order to minimize bump steer, that is toe-in/toe-out change that occurs during bounce and rebound, the tie-rod has one end fixed at a location in the plane that connects the upper a-arm pivots and lower a-arm pivots with the instantaneous centerline. The left-hand-side tie-rod pivot location is symmetric with the right-hand-side and thus creates a fixed length that defines the overall rack length. As before, this leaves the tie-rod length, L_{tr}, as the only geometric parameter to adjust to obtain desirable inside/outside steer angle performance.

As an example, lets rack and pinion geometry parameters of: L_p = 52 in, L_{sa} = 6 in, L_r = 28 in, L_{rp} = 4 in, and L_{tr} = 10 in. In this case, the inside and outside steer angles created by moving the rack are shown in Figure (5.43). The rack and pinion steering geometry dimensions in this example were not chosen at random, but were specified for use in the previous vehicle example summarized in Table (5.7). A cross-correlation of inside steer angle against outside steer angle can be created from the rack and pinion performance graph of Figure (5.43). This correlation together with the desired anti-Ackermann steering points is shown in Figure (5.44). As with the four-bar steering linkage, there is linkage error that occurs, and with further refinement, may be reduced. However, it should be realized that much of this section is limited to two-dimensional analysis. The three-dimensional influence of suspension movement (bounce and rebound), body roll, and changes in roll center location have not been accounted for. I will leave it up to the reader to advance the current analysis. *Happy motoring!*

Appendix A:
VEHICLE DYNAMICS TERMINOLOGY

In the discussion of vehicle dynamics and suspension systems it is beneficial to have a set of standard terms which we can use to describe the features of suspension systems as well as features of vehicle performance. The following material discusses the more commonly encountered terms, accompanied by their definitions as recommended by the Society of Automotive Engineers.

(A.1) WHEEL ALIGNMENT PARAMETERS

The function of wheel alignment parameters is to accurately describe the geometric orientation of the wheels relative to the vehicle chassis. These parameters are (a) inclination angle, (b) camber, (c) caster angle or caster offset, and (d) wheel toe. The phrase "wheel alignment" as encountered in automotive maintenance normally refers to the initial values of the alignment variables when the vehicle is in a stationary position. However, the characteristics of a particular suspension design may be such that the alignment parameters change with vehicle movement over a bump or during cornering. Consequently these parameters must be thought of as geometric variables rather than fixed quantities.

(A.1.1) INCLINATION AXIS AND SCRUB RADIUS

The inclination axis generally refers to a design aspect of the front suspension. It is also known as the "steering inclination axis" or "kingpin inclination axis." The inclination axis is an imaginary line about which the wheel pivots when steered. The axis illustrated in Figure (A.1) is determined by drawing a line through two points defined by the rotation centers of the upper and lower ball joints. The inclination axis of rear double A-arm independent suspensions is determined from pivot point centers at the connection of the wheel hub assembly to the A-arm linkage. On strut type suspensions, front or rear, the strut centerline defines the inclination axis.

The amount of inclination is an angle which is used in mathematical models of suspension systems. The angle is expressed as the number of degrees away from the wheel-centered vertical plane.

The intersection of the inclination axis with the road surface defines a distinct point. This is the point which the tire contact patch, that is the area of rubber in contact with the road surface, moves about when the tire is turned. If the tire centerline is thought of as a geometric plane dividing the tire in half, then the intersection of this plane with the road surface would create a line. The perpendicular distance (shortest) from the line to the contact patch turning center is called the scrub radius. The scrub radius as depicted in Figure (A.1) is defined as having a positive magnitude since the contact patch turning point is inboard of the wheel. Outboard turning centers are negative.

The scrub radius influences the vehicle stability, turning radius, and steering effort. Vehicles with rear wheel drive are normally designed with a slight positive scrub radius built into the front suspension. A positive scrub radius assists the vehicle in maintaining a straight stable path, but at the expense of increased steering effort required in turning the wheels. Front wheel drive cars, on the other hand, are designed with negative scrub radius to provide stability during braking. Another benefit of positive scrub radius is a slight improvement in turning radius. On a four wheeled vehicle with positive scrub radius the inside tire tends to move toward the rear tire as the wheel is turned during a cornering maneuver; the outer tire moves the opposite. The resulting wheel stagger produces a wheelbase which is different from one side to the other. This makes the vehicle turn more sharply in the corner.

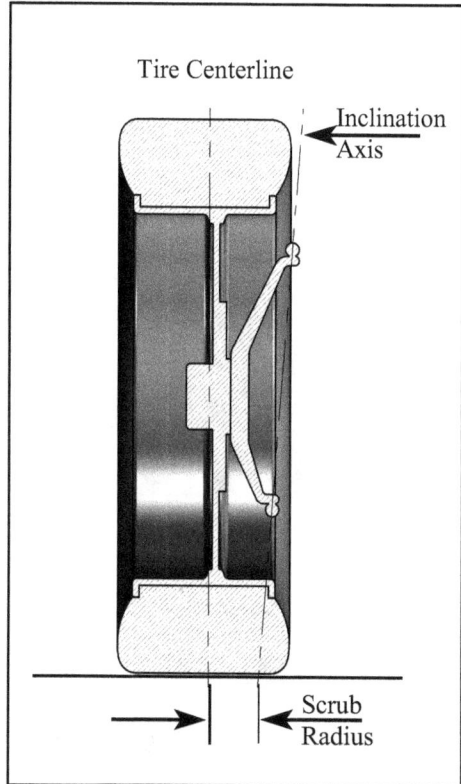

Figure (A.1): INCLINATION AXIS and SCRUB RADIUS ILLUSTRATION

(A.1.2) CAMBER

Camber is defined, or measured, as the angle between the wheel-centered vertical axis and the wheel centerline when viewed from the front of the vehicle. As shown in Figure (A.2), camber is either positive or negative depending upon which direction the wheel centerline is tilted. A positive value is assigned to camber angles when the top of the tire is leaning away from the chassis, while negative camber angles indicate that the tire is leaning in.

Originally, camber was a design parameter of early suspension systems. The camber angle was fixed such that a majority of the wheel load was transferred to the inner wheel bearing of front suspensions. Due to advances in metallurgy and improvements in bearing design, this is no longer the case. Wheel camber of modern front and rear suspensions is important from the standpoint of optimizing tire performance. For maximum cornering speed, acceleration, and braking, the suspension

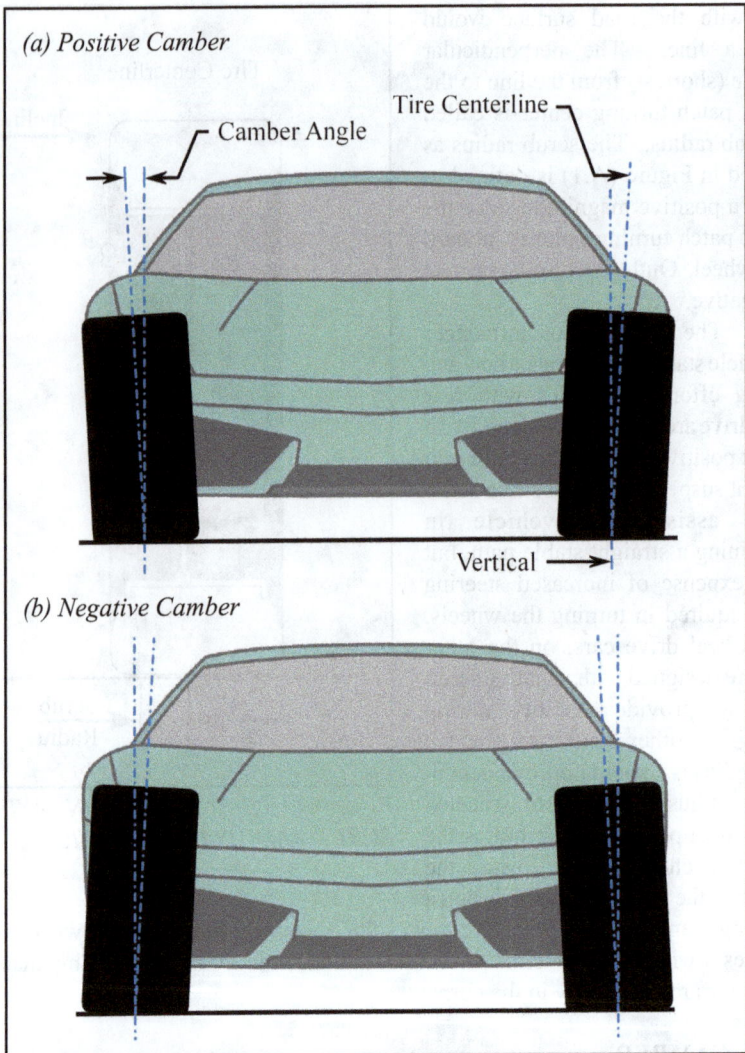

Figure (A.2): CAMBER ANGLE DEFINITION

system should be designed to maintain camber angles close to zero while undergoing various maneuvers.

(A.1.3) CASTER ANGLE AND CASTER OFFSET

Caster angle is the amount of fixed rotational displacement of the steering inclination axis about the spindle/knuckle centerline. Positive caster is defined as a rearward tilt of the inclination axis. A pictorial description is shown in Figure (A.3). Caster offset is the linear displacement, as measured in a side elevation view of the vehicle, from the tire contact patch center to the point defined by the intersection of the inclination axis with the ground plane.

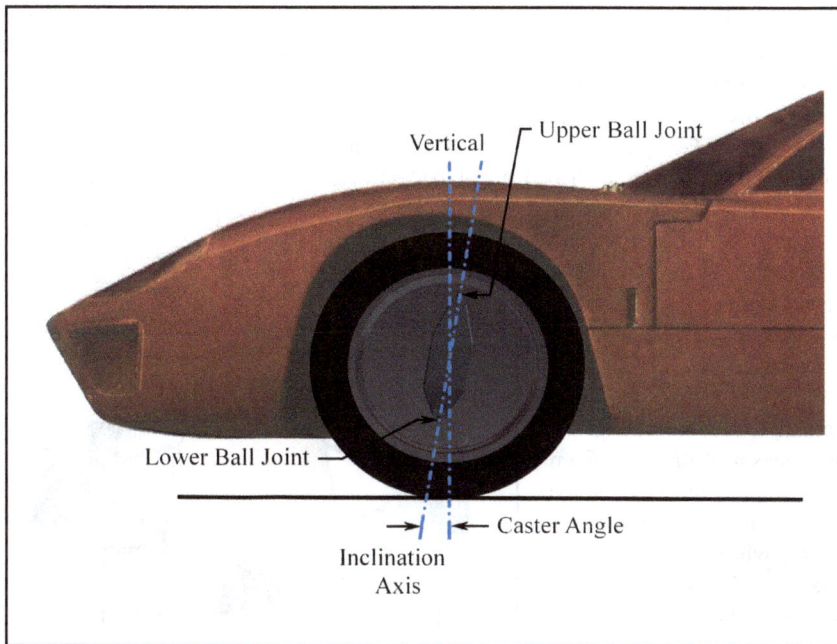

Figure (A.3): CASTER ANGLE DEFINITION {Positive Shown}

Positive caster is generally designed into the front suspension. Its purpose is to place the center of the contact patch behind the point defined by the intersection of the inclination axis with the road surface. This condition generates a self-aligning torque which tends to point the tire toward the center of the road when a deviation in the road surface occurs.

A simple analogy describing the affect of positive caster on vehicle performance is the caster wheels found on a dolly. When the dolly is pushed down the hallway for example, the caster wheel will always follow behind the pivot axis of the caster wheel in a stable condition, i.e., the contact point always swings to the opposite side of the wheel direction.

(A.1.4) TOE AND TOE ANGLE

"Toe" describes the alignment of tires relative to the longitudinal axis of the chassis. Wheel toe alignment applies primarily to all
types of front suspension systems and to independent rear suspensions. Positive toe, called toe-in, indicates that both the left and right tires of the front suspension, for instance, are pointing inward as the vehicle is steered in a straight forward direction. If the tires point away from the chassis centerline, then the tires are said to be "toed-out."

Toe-in is an important alignment parameter which influences the handling characteristics of a vehicle in two ways: (1) it provides steering stability, i.e., the ability of the vehicle to maintain a straight path without being unduly affected by wind gusts or road surface irregularities; and, (2) it affects the transient response of the front tires to a change in steer angle during cornering. It is important to understand that the characteristics of pneumatic tires are such that toe-in induces a rolling resistance. As a result, too much toe-in inhibits the forward motion of the vehicle, and also causes excessive tire wear.

Figure (A.4) depicts the procedure by which the wheel toe is determined. First, a circumferential line is scribed at, or near, the center of tire. Next, the distances between the two scribed lines are measured at the front and backside of the tire tread. The measurements are made above the ground at a height equivalent to the spindle axis. The wheel toe is then calculated by subtracting the front measured distance from the rear distance. In some instances it is convenient to describe wheel toe in terms of a toe angle. The toe angle is defined as the angle between the tire centerline, as viewed from above the vehicle looking down at the road surface, and the longitudinal centerline of the chassis. The toe angle can be calculated from the toe measurement using the following equation:

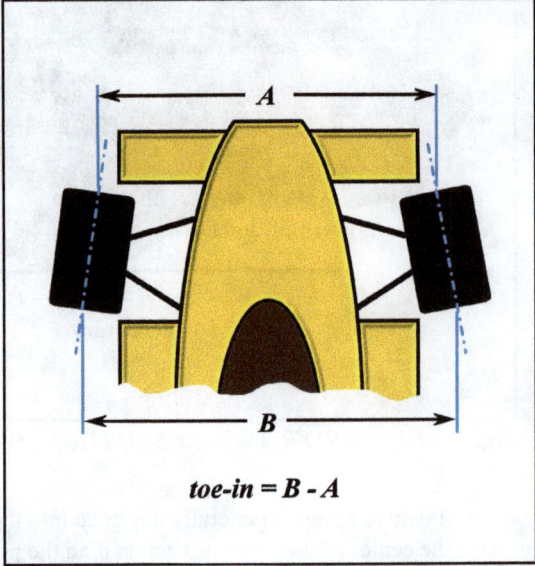

$$toe\text{-}in = B - A$$

Figure (A.4): METHOD OF WHEEL TOE MEASUREMENT

$$toe\ angle = \arcsin\left(\frac{1/2\ toe\text{-}in}{tire\ diameter}\right)$$

(A.2) SPRUNG AND UNSPRUNG WEIGHT

Simply stated the sprung weight is that portion of the total vehicle weight which is isolated from the road surface by a spring or set of springs. The unsprung weight is determined by subtracting the sprung weight from the total weight.

An important suspension performance parameter is the sprung to unsprung weight ratio. A low sprung/unsprung weight ratio inhibits the vehicles ability to hold the road. At modest ground speeds the tires are able to follow the road contour relatively well; but, as the ground speed increases, the effect of the low ratio causes the unsprung weight of the vehicle to push the sprung weight up. This is directly opposite of the desired effect of having the sprung weight push the unsprung weight (mainly the tires) down firmly against the road. Consequently, the tires tend to leave the road surface over dips. Under severe conditions, e.g., washboard road, this can result in an unstable ride leading to loss of control. In addition to diminished vehicle performance, the bouncing or jarring motion of the vehicle caused by the relatively large unsprung weight may cause discomfort to the driver. That is, the driver may have the subjective opinion that the vehicle has a "rough" ride.

(A.3) VEHICLE MOVEMENT PARAMETERS

In developing a vehicle kinematic analysis it is convenient to use a vehicle

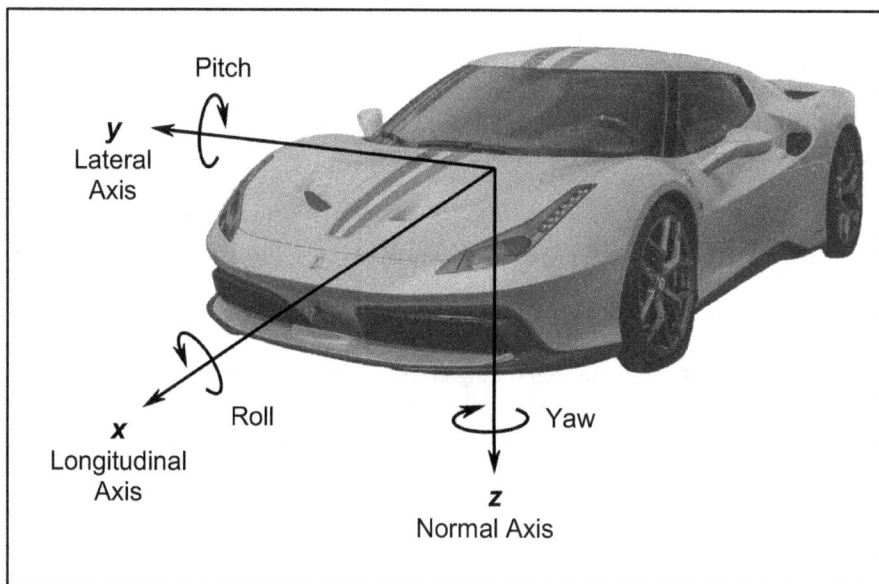

Figure (A.5): VEHICLE COORDINATE SYSTEM

coordinate system, such as the one pictured in Figure (A.5). This vehicle coordinate system is a right-handed, relative coordinate system which is usually, but not necessarily, placed at the center of gravity of the sprung weight, or geometrically centered within the body. The *x-axis*, or *longitudinal axis*, points in forward direction and is essentially parallel with the road surface. The *y-axis*, or *lateral axis*, also lies in the longitudinal plane but points towards the drivers right; and finally, the *z-axis*, or *normal axis*, points downward. Three additional parameters are required to describe the rotation of the vehicle: yaw, pitch, and roll.

(A.3.1) YAW

The direction a vehicle points, or heads, while cornering is usually different from the path the vehicle actually travels. The angular displacement, relative to a fixed coordinate system, of this deviation is called yaw. In vehicle cornering studies there are instances where it is more convenient to describe vehicle rotation about the normal axis relative to the vehicle heading. In this case the angular displacement is referred to as sideslip, or drift.

(A.3.2) BOUNCE AND PITCH

The term "bounce" is used to describe the oscillatory linear motion of the vehicle body and chassis, i.e., the sprung weight, along the normal axis. "Pitch" describes the fore and aft oscillatory rotational motion about the lateral axis. Bounce and pitch motions are introduced into the vehicle body (sprung weight) through surface irregularities in the road. Bounce and pitch terminology do not apply to the unsprung weight since under normal driving conditions the wheels are in contact with the road. Instead the linear movement of the suspension system relative to the vehicle is described by jounce (up) and rebound (down). For example, when a car travels over a bump the suspension is in jounce, the springs are compressed and the shock absorbers collapse as the wheels and axles move towards the chassis. After traveling over the bump the suspension returns to its original position by rebound, the wheels and axles move away from the chassis.

(A.3.3) ROLL

Roll describes the vehicle body/chassis movement during a cornering maneuver. Roll movement is actually a combination of lateral translation along the y-axis and a rotation about the longitudinal axis. The latter movement is called body roll and is expressed as an angle.

The inertia force induced during cornering tends to move the sprung weight center of gravity about a roll axis. The roll axis is geometrically determined by passing a line through the front roll center and the rear roll center of the vehicle. The suspension roll center is a characteristic of a particular design. It is essentially an instant center defined from kinematic relationships obtained by treating the suspension system as a mechanical linkage or mechanism.

(A.4) UNDERSTEER, OVERSTEER, and NEUTRAL STEER

Three terms are used to describe the cornering behavior of a vehicle while maneuvering through a turn: understeer, oversteer, and neutral steer. At first it seems only logical that a car would naturally follow the intended path as steered by the driver. But, the vehicle does not usually respond in this manner. The dynamics of cornering are rather involved, but, basically, the elastic behavior of pneumatic tires allows a twisting deformation of the tire caused indirectly by the centrifugal forces encountered during cornering. Here we make a distinction between the direction the wheel rim is pointing and the direction the tire is pointing where it makes contact with the road surface. Since there is a twisting deformation, the two directions need not be the same. As a result, the tire deformation occurring in the rear wheels will also steer the vehicle. Thus, it is the combination of the front wheel steering and the rear wheel steering which actually guides the vehicle. If the driver does not compensate for the rear steering, the car will deviate from the desired path. A large amount of deviation can have serious consequences in that it generally leads to loss of control, i.e., the car will either leave the confines of the road boundary or spin out. Obviously, a successful maneuver requires that the actual path match the radius of the turn; therefore, the driver must compensate by changing the steered position of the front wheels. If the driver must compensate by steering further into the turn, the car is said to understeer. On the other hand, if the driver steers out of the turn, then the car oversteers. Figure (A.6) graphically illustrates these two situations. Neutral steer refers to a vehicle which can execute a constant radius turn without steering correction by the driver.

Drivers with average ability can intuitively perform necessary corrections to maintain control of an understeering vehicle. This is the primary reason that many suspension designers consider a vehicle which understeers to be stable. Consequently the majority of production passenger cars are intentionally designed to understeer. Oversteer can be unstable. If an oversteering car is left uncorrected while cornering the car will start to turn a smaller radius. This increases the centrifugal force which in turn increases the oversteering effect. The process of turning a smaller radius continues until the car loses traction and finally spins out of control. An oversteering car does have the advantage over other cars of cornering faster on a low speed course; but, it generally requires a skilled driver to handle an oversteering car. Although neutral steer is theoretically possible, in practice it is very difficult to attain. This is due mainly to factors which cannot be controlled by the driver, such as the tire temperature and tire wear patterns. In some instances, especially noticeable on

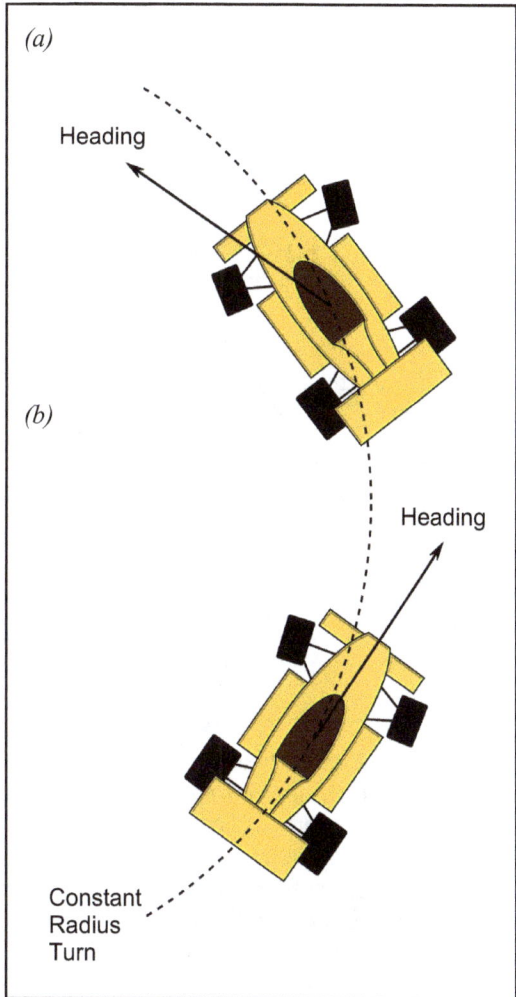

Figure (A.6): (a) OVERSTEERING and (b) UNDERSTEERING VEHICLE

lightweight vehicles, an understeering vehicle can change after a period of time to an oversteering vehicle as a shift in weight distribution occurs from the fuel level changing in the fuel tank.

A quantitative description of understeer or oversteer is most easily defined by the curvature gain, which is the reciprocal of turn radius per steer angle. Increasing curvature gain is indicative of an oversteering vehicle and, conversely, an understeering vehicle exhibits decreasing curvature gain characteristics.

Appendix B:
ESTIMATION OF VEHICLE WEIGHT CHARACTERISTICS

During the design process of a new vehicle, it is necessary to have estimates of weight characteristics such as sprung and unsprung weights, center of gravity location and moment of inertia about the lateral and normal axes, as well as the moment of inertia of rotating components connected to the wheel and hub assemblies. All of these characteristics are required order to predict vehicle behavior during acceleration (braking), ride evaluation, and cornering maneuvers. In the event the vehicle does not yet exist we must rely on estimates of the vehicle's weight characteristics.

The information presented in this appendix provides us with a methodology to estimate the value of sprung and unsprung weights, the location of the center of gravity, as well as an estimate of required moments of inertia.

(B.1) CENTER OF GRAVITY ESTIMATION

For the majority of fundamental vehicle dynamic studies, a two-dimensional representation (i.e., side view) of the vehicle is sufficient. In this case, the scheme used to determine the center of gravity (or mass center) starts by defining two planes: the ground (horizontal) surface and either the front or rear axle (vertical) plane. For purposes of our discussion we will chose the front axle plane. The next step is to define a standard Cartesian coordinate system based on the horizontal and vertical planes.

The general strategy used in estimating the center of gravity (or mass center) is to divide the vehicle components or group of components up into discrete units of weight. The individual weight units are then placed at an estimated location on the xy-plane. A general plot is shown in Figure (B.1). The vehicle center of gravity is also place on the xy-plane, but at an arbitrary location.

The vehicle weight is calculated by summing the individual weight items:

$$W_{CG} = W_1 + W_2 + \cdots + W_i + \cdots + W_n = \sum_{i=1}^{n} W_i \qquad (B.1)$$

The CG coordinates are then calculated from:

$$x_{CG} = \frac{1}{W_{CG}} \sum_{i=1}^{n} x_i W_i \qquad (B.2)$$

and

$$y_{CG} = \frac{1}{W_{CG}} \sum_{i=1}^{n} y_i W_i \qquad (B.3)$$

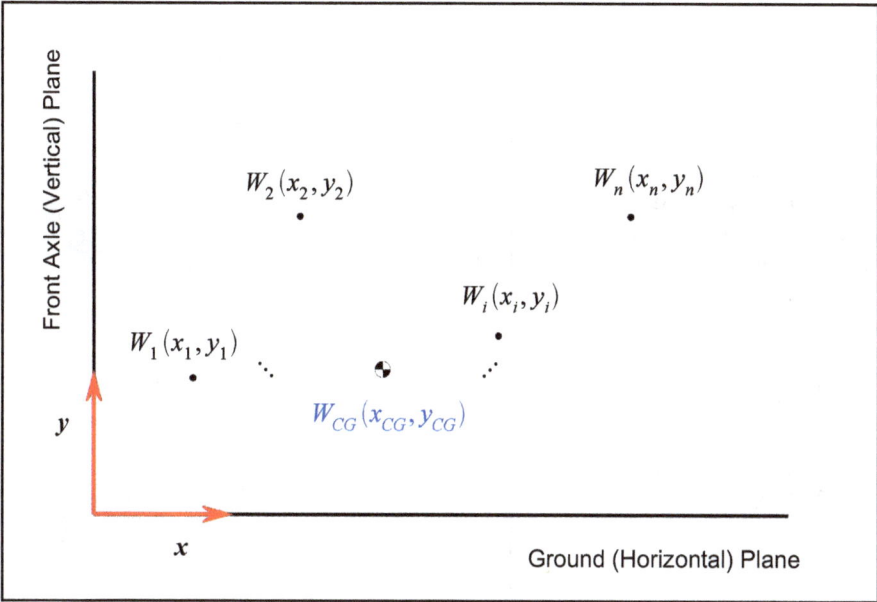

Figure (B.1): VEHICLE DISCRETE COMPONENT WEIGHT DISTRIBUTION

It should be noted that Eqs (B.1) through (B.3) can evaluated from two different points of view. If the vehicle center of gravity is to be calculated, then all of the vehicle weight components along with their location need to be included. However, if the location of the sprung mass center of gravity is being sought, the unsprung weight components should be omitted.

(B.2) MOMENT OF INERTIA AND DYNAMIC INDEX

Two values of moment of inertia need to be estimated. One about the lateral axis passing through the sprung mass center of gravity, which is required for bounce and pitch motion studies, and the other about the normal axis passing through the vehicle center of gravity, which is used in modeling vehicle cornering behavior. In either case, the moment of inertia basic equation is

$$I = \int_{0}^{m} r^2 \, dm \qquad (B.4)$$

As noted above, it is convenient to divide the vehicle up into discrete mass (weight) components as shown in Figure (B.1). In order to estimate moments of inertia, we will also need to include a spin axis in order to evaluate the radial distance r. A graphical representation is shown in Figure (B.2). Using this concept the mass moment of inertia as defined by Eq (B.4) can be approximated by

$$I = \sum_{i=i}^{n} r_i^2 m_i \qquad \text{(B.5)}$$

Under the condition of a two-dimensional representation of our vehicle, the individual radial displacements about the lateral axis passing through the sprung mass center of gravity is found from

$$r_i =$$

$$\sqrt{(x_i - x_{CG})^2 + (y_i - y_{CG})^2}$$

$$\text{(B.6)}$$

The radial displacement values about the normal axis passing through the vehicle center of gravity is

$$r_i = \sqrt{(x_i - x_{CG})^2} \qquad \text{(B.7)}$$

However, it should be noted that the value of x_{CG} is not the same

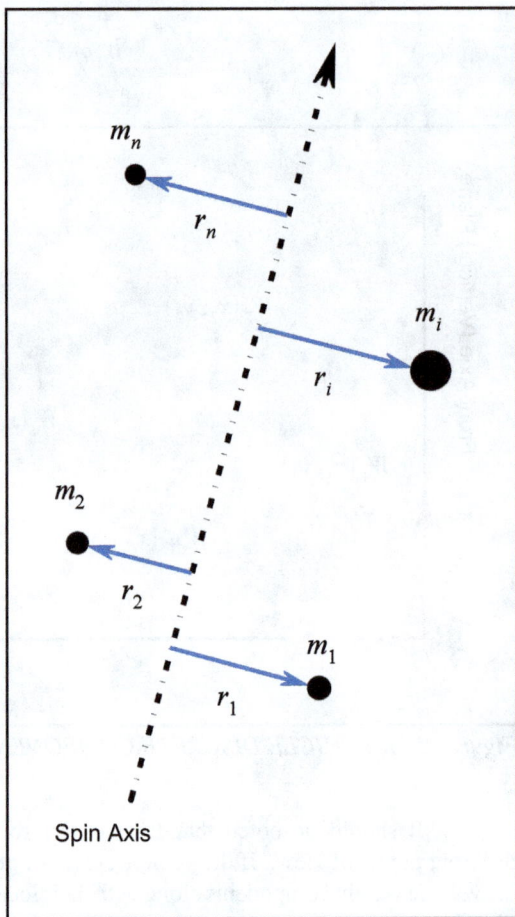

Figure (B.2): DISCRETE MASS DISTRIBUTION ABOUT SPIN AXIS

in Eqs (B.6) and (B.7) as the two represent different centers of gravity.

The lateral axis moment of inertia, I_{lat}, is estimated from Eq (B.5) using the radial values obtained from Eq (B.6). In a similar manner, the normal axis moment of inertia, I_{norm}, is estimated from Eqs (B.5) and (B.7).

When studies of vehicle bounce and pitch, as well as cornering, it is convenient in the description of dynamic governing equations to use the concept of equivalent radial position, ρ. The equivalent radial position is the position that if the entire mass were to be placed, the resulting moment of inertia would be identical to that of the system. For example, consider the lateral moment of inertia. From Eqs (B.5) and (B.6)

$$I_{lat} = \sum_{i=i}^{n} \left[(x_i - x_{CG})^2 + (y_i - y_{CG})^2 \right] m_i = \rho^2 m \qquad (B.8)$$

The equivalent radial distance related to the lateral moment of inertia is then

$$\rho = \sqrt{\frac{\left[(x_i - x_{CG})^2 + (y_i - y_{CG})^2 \right] m_i}{m}} \qquad (B.9)$$

(B.2.1) ROTATING WHEEL MOMENT OF INERTIA

Determining the moment of inertia for a rotating wheel assembly is quite an undertaking; however, a simple strategy may be used to arrive at an estimate suitable for vehicle dynamic studies. In this case it is convenient to lump all of the rotating components into a single disc of identical weight rotating about the hub spindle (spin axis). Based on this consideration, the general equation describing the polar moment of inertia is

$$I = \frac{1}{2} m r^2 = \frac{1}{2g} \left(W_{tire} + W_{rim} + W_{hub} + W_{disc} + \cdots \right) r_{tire}^2 \qquad (B.10)$$

(B.3) EXAMPLE APPLICATION

Using a worksheet or spreadsheet is a convenient method to catalog the components of a vehicle and can also be used to estimate weights, weight distribution, and moments of inertia. Lets consider a very, very simple off-road vehicle as illustrated

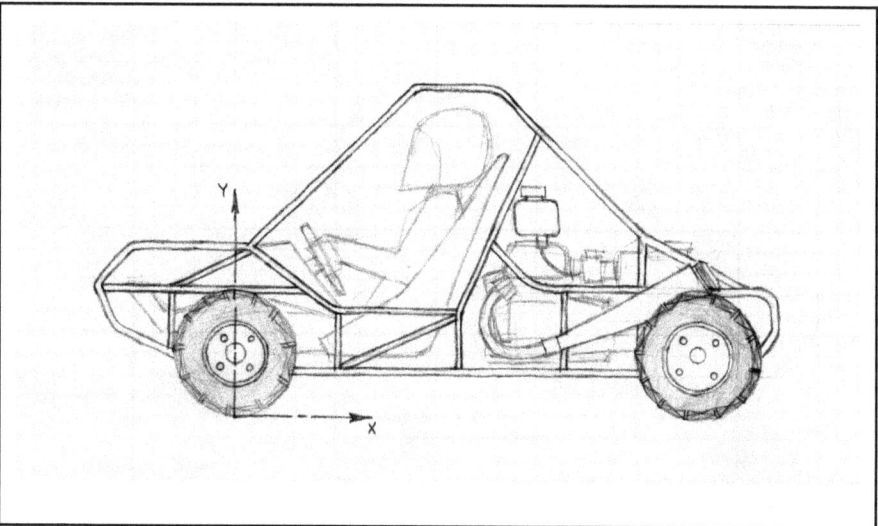

Figure (B.3): SIMPLE OFF-ROAD VEHICLE SKETCH

in the sketch of Figure (B.3). The first step in creating a worksheet is to specify basic geometric vehicle parameters, based on the intended use of the vehicle, such as wheelbase and tire diameters. Then, looking at the sketch and contemplating the actual construct of the prototype, divide the vehicle up into subsystems or subassemblies, as shown in Table (B.1). Within each subassembly, all required components are listed along with an estimated weight and an estimated location (based on the coordinate system defined in the sketch). Finally, using the foregoing equations, estimated values of weights, CG locations, and equivalent moment radii can be calculated. We should point out that as the vehicle design progresses, the worksheet will need to be updated with changes and modifications.

	72
[in]	22
[in]	22

	Weight [lb]			Location		Moment of Inertia			
	sprung	unsprung		x	y	r-lat [in]	I-lat	r-norm [in]	I-r
		nonrotating	rotating						
bly									
			15.0						
			5.0						
			2.0						
kle			7.0						
		1.5							
		0.5							
			6.0						
		6.5							
		0.5							
total	0.0	9.0	35.0						

N OF VEHICLE WEIGHT CHARACTERISTICS

subtotal	0.0	9.0	35.0					
total	*0.0*	*25.5*	*76.0*	*0.0*	*10.5*	*37.83*	*0*	*37.76*
l rod ends	0.3	0.3						
	1.0	1.0						
nd shock	3.5	3.5						
e	0.3	0.3						
linkage	1.0	1.0						
subtotal	6.0	6.0	0.0					
subtotal	6.0	6.0	0.0					
total	*12.0*	*12.0*	*0.0*	*0.0*	*12.0*	*37.65*	*452*	*37.65*

Assembly

		18.0
		8.0
earings		4.0
& knuckle		6.5

MATION OF VEHICLE WEIGHT CHARACTERISTICS

				3.0					
				1.5					
				5.5					
		6.0							
		2.5							
total	0.0	15.0	40.0						
total	0.0	15.0	40.0						
total	*0.0*	*30.0*	*80.0*	*72.0*	*10.5*	*34.95*	*0*	*34.62*
sion								
ds	0.3	0.3						
	1.5	1.5						
k	3.5	3.5						
	0.3	0.3						
total	5.5	5.5	0.0					
total	5.5	5.5	0.0					
total	*11.0*	*11.0*	*0.0*	*72.0*	*12.0*	*34.76*	*382*	*34.49*
n								

N OF VEHICLE WEIGHT CHARACTERISTICS

ts	85.0								
	7.5								
total	*92.5*				*48.0*	*8.0*	*13.13*	*1215*	*12.1*
	2.5								
arrier	20.0								
ousing	10.0								
brgs	9.0								
yokes	4.5	4.5							
	2.0	1.0							
	0.5								
total	*48.5*	*5.5*	*0.0*	*68.0*	*12.0*	*30.78*	*1493*	*30.5*
	180.0								
t	15.0								
	9.0								
	3.0								
column	10.0								
gauges	10.0								

MATION OF VEHICLE WEIGHT CHARACTERISTICS

total	*227.0*	*0.0*	*0.0*	*30.0*	*22.0*	*9.68*	*2197*	*10.86*
	6.5							
	15.0							
	8.0							
	24.0							
	6.0							
	75.0							
	3.0							
total	137.5	0.0	0.0					
front	*75.6*			*0.0*	*12.0*	*37.65*	*2847*	*37.65*
rear	*61.9*			*72.0*	*14.0*	*34.59*	*2140*	*34.42*

84.0
156.0

]   528.5

37.46

ON OF VEHICLE WEIGHT CHARACTERISTICS

| [in] | 15.83 |

: [lb]	768.5
ion:	
[in]	37.58
[in]	14.22

Moment Radius
| [in] | 20.30 |
| m [in] | 25.39 |

MATION OF VEHICLE WEIGHT CHARACTERISTICS

Appendix C:
DYNAMICS REVIEW

This appendix serves as a review of the principles of dynamic analysis which are used in the prediction of vehicle behavior and performance.

(C.1) KINEMATIC RELATIONSHIPS BETWEEN FIXED AND MOVING REFERENCE FRAMES

Figure (C.1) depicts two difference reference frames. One is a fixed (Newtonian) Cartesian coordinate system, designated as **XYZ**, and the other is a moving (non-Newtonian) reference frame that is free to translate and rotate, designated as **xyz**. The absolute position of point "**P**", i.e., referenced to the fixed coordinate system, can be related to the moving coordinate system and the relative position by the relationship:

$$\vec{r}_{P/O}(t) = \vec{r}_{A/O}(t) + \vec{r}_{P/A}(t) \tag{C.1}$$

The absolute velocity of point "**P**" can be found by taking the time derivative of the position vector defined by Eq (C.1):

$$\vec{V}_{P/O}(t) = \frac{d}{dt}\left(\vec{r}_{P/O}(t)\right)$$

$$= \frac{d}{dt}\left(\vec{r}_{A/O}(t)\right) + \frac{d}{dt}\left(\vec{r}_{P/A}(t)\right) \tag{C.2}$$

The first term on the right-hand-side of Eq (C.2) is the translational velocity of the **xyz**-coordinate system, i.e.,

$$\vec{V}_{A/O}(t) = \frac{d}{dt}\left(\vec{r}_{A/O}(t)\right) \tag{C.3}$$

The second term requires more explanation. We start by noting that the position vector $\vec{r}_{P/A}(t)$ can be written in terms of components as

$$\vec{r}_{P/A}(t) = r_x(t)\,\hat{i}_x(t) + r_y(t)\,\hat{i}_y(t) + r_z(t)\,\hat{i}_z(t) \tag{C.4}$$

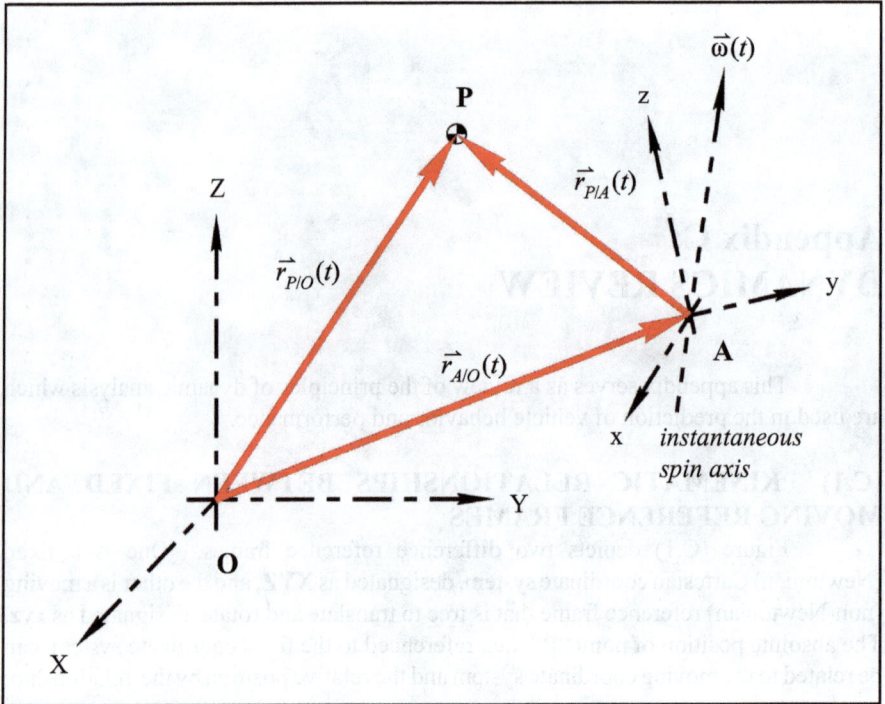

Figure (C.1): REFERENCE FRAME ORIENTATION

Differentiating Eq (C.4) with respect to time results in

$$\frac{d}{dt}\left(\vec{r}_{P/A}\right) = \frac{d}{dt}\left(r_x\,\hat{\boldsymbol{i}}_x + r_y\,\hat{\boldsymbol{i}}_y + r_z\,\hat{\boldsymbol{i}}_z\right)$$

$$= \frac{dr_x}{dt}\,\hat{\boldsymbol{i}}_x + \frac{dr_y}{dt}\,\hat{\boldsymbol{i}}_y + \frac{dr_z}{dt}\,\hat{\boldsymbol{i}}_z + r_x\frac{d\hat{\boldsymbol{i}}_x}{dt} + r_y\frac{d\hat{\boldsymbol{i}}_y}{dt} + r_z\frac{d\hat{\boldsymbol{i}}_z}{dt} \qquad (C.5)$$

The first three terms on the right-hand-side of Eq (C.5) mathematically defines the linear velocity of "**P**" relative to the moving **xyz**-coordinate system, i.e.,

$$\vec{v}(t) = \frac{dr_x}{dt}\,\hat{\boldsymbol{i}}_x + \frac{dr_y}{dt}\,\hat{\boldsymbol{i}}_y + \frac{dr_z}{dt}\,\hat{\boldsymbol{i}}_z \qquad (C.6)$$

The time derivatives of the unit vectors can be found from geometric interpretation. The vector diagram in Figure (C.2) shows that for a small time increment, Δt, the new unit vector in the **x**-direction is

$$\hat{\boldsymbol{i}}_x(t + \Delta t) = \hat{\boldsymbol{i}}_x(t) + \vec{\omega}(t)\,\Delta t \times \hat{\boldsymbol{i}}_x(t) \qquad (C.7)$$

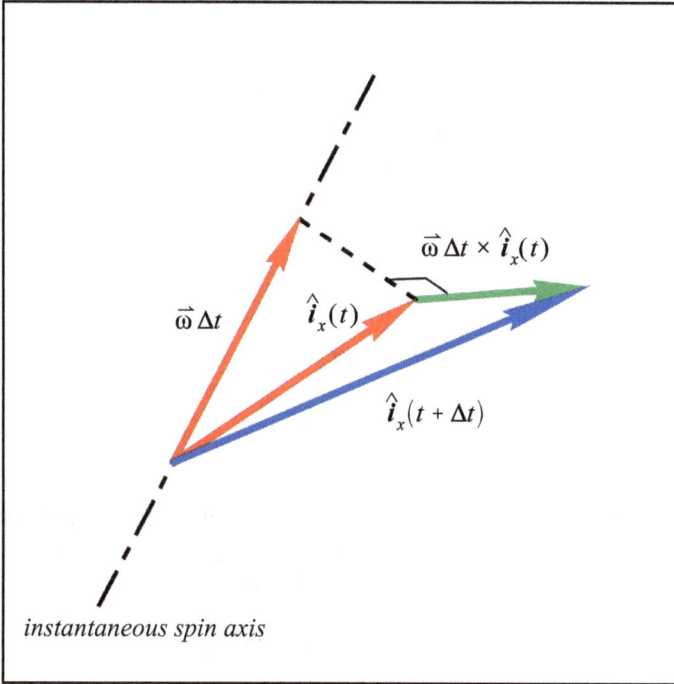

Figure (C.2): UNIT VECTOR GEOMETRY DESCRIPTION

The new unit vector, $\hat{i}_x(t + \Delta t)$, can also be estimated via Taylor series expansion, i.e.,

$$\hat{i}_x(t + \Delta t) \;=\; \hat{i}_x(t) \;+\; \frac{d\hat{i}_x(t)}{dt}\,\Delta t \;+\; \frac{d^2\hat{i}_x(t)}{dt^2}\,\frac{\Delta t^2}{2} \;+\; \cdots \tag{C.8}$$

Assuming Δt is small, terms of order Δt^2 and higher can be neglected in the previous equation.

Substituting Eq (C.8) into Eq (C.7) produces

$$\hat{i}_x(t) \;+\; \frac{d\hat{i}_x(t)}{dt}\,\Delta t \;=\; \hat{i}_x(t) \;+\; \vec{\omega}(t)\,\Delta t \times \hat{i}_x(t) \tag{C.9}$$

or

$$\frac{d\hat{i}_x(t)}{dt} \;=\; \vec{\omega}(t) \times \hat{i}_x(t) \tag{C.10}$$

Similarly

$$\frac{d\hat{i}_y(t)}{dt} \;=\; \vec{\omega}(t) \times \hat{i}_y(t) \tag{C.11}$$

and

$$\frac{d\hat{i}_z(t)}{dt} = \vec{\omega}(t) \times \hat{i}_z(t) \tag{C.12}$$

Using Eqs (C.10), (C.11), and (C.12), the last three terms on the right-hand-side of Eq (C.7) can be written as

$$r_x \frac{d\hat{i}_x}{dt} + r_y \frac{d\hat{i}_y}{dt} + r_z \frac{d\hat{i}_z}{dt} = r_x\left(\vec{\omega} \times \hat{i}_x\right) + r_y\left(\vec{\omega} \times \hat{i}_y\right) + r_z\left(\vec{\omega} \times \hat{i}_z\right)$$

$$= \vec{\omega} \times \left(r_x\hat{i}_x + r_y\hat{i}_y + r_z\hat{i}_z\right)$$

$$= \vec{\omega}(t) \times \vec{r}_{P/A}(t) \tag{C.13}$$

The resulting expression from the previous equation is the mathematical definition of the tangential velocity of point "**P**" relative to the moving **xyz**-coordinate system. From Eqs (C.6) and (C.13) the relative velocity of "**P**" can be written as

$$\frac{d}{dt}\left(\vec{r}_{P/A}(t)\right) = \vec{v}(t) + \vec{\omega}(t) \times \vec{r}_{P/A}(t) \tag{C.14}$$

Finally, using Eqs (C.1.3) and (C.14), the absolute velocity of point "**P**" becomes

$$\vec{V}_{P/O}(t) = \vec{V}_{A/O}(t) + \vec{v}(t) + \vec{\omega}(t) \times \vec{r}_{P/A}(t) \tag{C.15}$$

The absolute acceleration, $\vec{A}_{P/O}(t)$, of point "**P**" is found by taking the time derivative of the velocity vector given by Eq (C.15), i.e.,

$$\vec{A}_{P/O}(t) = \frac{d}{dt}\left(\vec{V}_{P/O}(t)\right)$$

$$= \frac{d}{dt}\left(\vec{V}_{A/O}(t)\right) + \frac{d}{dt}\left(\vec{v}(t)\right) + \frac{d}{dt}\left(\vec{\omega}(t) \times \vec{r}_{P/A}(t)\right) \tag{C.16}$$

The first term on the right-hand-side of Eq (C.16) is the acceleration of the moving coordinate system, i.e.,

$$\vec{A}_{A/O}(t) = \frac{d}{dt}\left(\vec{V}_{A/O}(t)\right)$$

The remaining terms in Eq (C.16) can be expanded in two parts:
- Part (*i*)

$$\frac{d}{dt}\left(\vec{v}(t)\right) = \frac{d}{dt}\left(v_x\,\hat{i}_x + v_y\,\hat{i}_y + v_z\,\hat{i}_z\right)$$

$$= \frac{dv_x}{dt}\,\hat{i}_x + \frac{dv_y}{dt}\,\hat{i}_y + \frac{dv_z}{dt}\,\hat{i}_z + v_x\frac{d\hat{i}_x}{dt} + v_y\frac{d\hat{i}_y}{dt} + v_z\frac{d\hat{i}_z}{dt}$$

(C.18)

The linear, relative acceleration of "**P**" is represented by the terms

$$\vec{a}(t) = \frac{dv_x}{dt}\,\hat{i}_x + \frac{dv_y}{dt}\,\hat{i}_y + \frac{dv_z}{dt}\,\hat{i}_z$$

(C.19)

By virtue of Eqs (C.10), (C.11), and (C.12), the second group of three terms in Eq (C.18) can be expressed as

$$v_x\frac{d\hat{i}_x}{dt} + v_y\frac{d\hat{i}_y}{dt} + v_z\frac{d\hat{i}_z}{dt} = v_x\left(\vec{\omega}\times\hat{i}_x\right) + v_y\left(\vec{\omega}\times\hat{i}_y\right) + v_z\left(\vec{\omega}\times\hat{i}_z\right)$$

$$= \vec{\omega}\times\left(v_x\,\hat{i}_x + v_y\,\hat{i}_y + v_z\,\hat{i}_z\right)$$

$$= \vec{\omega}(t)\times\vec{v}(t)$$

(C.20)

Therefore,

$$\frac{d}{dt}\left(\vec{v}(t)\right) = \vec{a}(t) + \vec{\omega}(t)\times\vec{v}(t)$$

(C.21)

- Part (*ii*)

$$\frac{d}{dt}\left(\vec{\omega}(t)\times\vec{r}_{P/A}(t)\right) = \dot{\vec{\omega}}(t)\times\vec{r}_{P/A}(t) + \vec{\omega}(t)\times\frac{d}{dt}\left(\vec{r}_{P/A}(t)\right)$$

(C.22)

Substituting Eq (C.14) into the previous equation results in

$$\frac{d}{dt}\left(\vec{\omega}(t)\times\vec{r}_{P/A}(t)\right) = \dot{\vec{\omega}}(t)\times\vec{r}_{P/A}(t)$$

$$+ \vec{\omega}(t)\times\left[\vec{v}(t) + \vec{\omega}(t)\times\vec{r}_{P/A}(t)\right]$$

$$= \dot{\vec{\omega}}(t)\times\vec{r}_{P/A}(t)$$

$$+ \vec{\omega}(t)\times\vec{v}(t) + \vec{\omega}(t)\times\left[\vec{\omega}(t)\times\vec{r}_{P/A}(t)\right]$$

(C.23)

Back substitution of Eqs (C.17), (C.21) and (C.23) into Eq (C.16) produces the general expression for the absolute acceleration of point "**P**":

$$\vec{A}_{P/O}(t) = \vec{A}_{A/O}(t) + \vec{a}(t) + \vec{\omega}(t) \times \vec{r}_{P/A}(t)$$
$$+ 2\vec{\omega}(t) \times \vec{v}(t) + \vec{\omega}(t) \times \left[\vec{\omega}(t) \times \vec{r}_{P/A}(t) \right]$$

(C.24)

(C.2) INERTIA FORCE AND DYNAMIC EQUILIBRIUM

Consider a point **P** in a moving **xyz**-coordinate system which has a relative velocity of zero. When viewed from the origin of the **xyz**-coordinate system this point would appear to be motionless. Noting that the relative velocity and acceleration are both zero, i.e., $\vec{v}(t) = 0$ and $\vec{a}(t) = 0$, the absolute acceleration, from Eq (C.24), for point **P** is given by

$$\vec{A}_{P/O} = \vec{A}_{A/O} + \vec{\omega} \times \vec{r}_{P/A} + \vec{\omega} \times \left(\vec{\omega} \times \vec{r}_{P/A} \right)$$

(C.25)

The motion of point **P** is governed by Newton's Second Law: $\Sigma \vec{F} = m\vec{A}_{P/O}$ which, by virtue of Eq (C.25), when applied to a fixed point in a moving coordinate system becomes:

$$\Sigma \vec{F} = m \left[\vec{A}_{A/O} + \vec{\omega} \times \vec{r}_{P/A} + \vec{\omega} \times \left(\vec{\omega} \times \vec{r}_{P/A} \right) \right]$$

(C.26)

When examining point **P** relative to the moving coordinate system, it is convenient to introduce an inertia force defined as:

$$\vec{F}_{inertia} = -m \left[\vec{A}_{A/O} + \vec{\omega} \times \vec{r}_{P/A} + \vec{\omega} \times \left(\vec{\omega} \times \vec{r}_{P/A} \right) \right]$$

(C.27)

Using this definition, Eq (C.26) reduces to

$$\Sigma \vec{F} + \vec{F}_{inertia} = 0$$

(C.28)

Eq (C.28) is referred to as the equation of dynamic equilibrium.

(C.3) REVIEW OF PARTICLE KINEMATICS

At some point in the study of vehicle cornering performance it is usually necessary to use kinematic relationships which describe the motion of the vehicle's center of gravity (**CG**). The most efficient approach in this case is to employ a curvilinear coordinate system. Figure (C.3) shows the curvilinear motion of the **CG** in a fixed **XY**-plane, along with symbolic terminology used in this section.

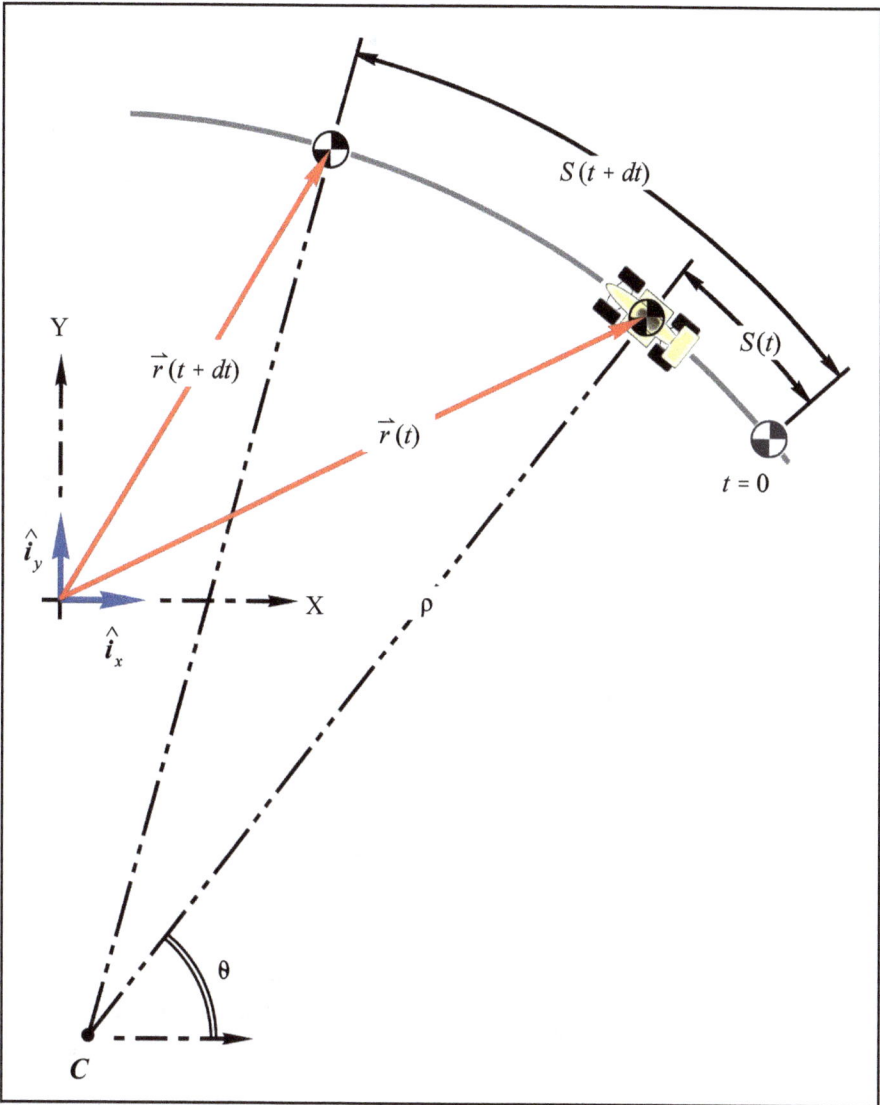

Figure (C.3): VEHICLE CURVILINEAR MOTION AND GEOMETRY

The **CG** velocity is universally defined as

$$\vec{V} = \frac{d\vec{r}}{dt} \qquad (C.29)$$

Using the chain rule of differentiation, the velocity can also be expressed as

$$\vec{V} = \frac{d\vec{r}}{dS}\frac{dS}{dt} \qquad (C.30)$$

The term $\dfrac{dS}{dt}$ is the forward speed of the vehicle and can be written in short-hand

notation simply as V. The remaining term, $\dfrac{d\vec{r}}{dS}$, which also has a physical significance,

can be deduced from geometrical interpretation. The trick to discovering this significance is to first write the position vector, $\vec{r}(t + dt)$, and the displacement, $S(t + dt)$, as a Taylor series expansion:

$$\vec{r}(t + dt) = \vec{r}(t) + \frac{d\vec{r}}{dt} dt + \frac{d^2\vec{r}}{dt^2} \frac{dt^2}{2!} + \cdots \qquad (C.31)$$

$$S(t + dt) = S(t) + \frac{dS}{dt} dt + \frac{d^2S}{dt^2} \frac{dt^2}{2!} + \cdots \qquad (C.32)$$

Neglecting terms of dt^2 and greater, the previous equations reduce to approximately

$$\vec{r}(t + dt) \approx \vec{r}(t) + d\vec{r} \qquad (C.33)$$

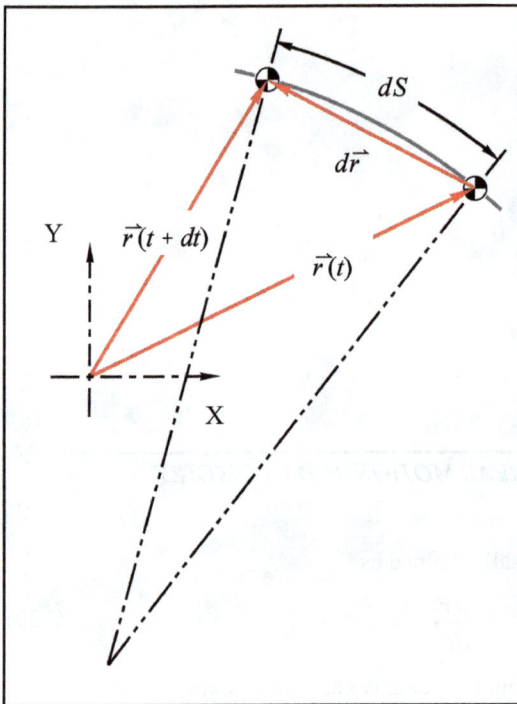

Figure (C.4): DIFFERENTIAL POSITION VERSUS DIFFERENTIAL PATH DISPLACEMENT

$$S(t + dt) \approx S(t) + dS \qquad (C.34)$$

Eqs (C.33) and (C.34) can be expressed graphically as shown in Figure (C.4). The geometry in Figure (C.4) shows that as the included angle becomes very small, or as $dt \to 0$, the magnitude of $d\vec{r}$ approaches dS. Based on this observation we conclude that the quantity

$\dfrac{d\vec{r}}{dS}$ defines a unit vector

tangent to the vehicle path pointing in the direction of travel. Not surprisingly, the resulting vector is referred to as a unit tangential vector, $\hat{i}_T$. From Eq. (C.30) the **CG** velocity ultimately becomes

$$\vec{V} = V \hat{i}_T \qquad (C.35)$$

The **CG** acceleration is defined as

$$\vec{a} = \frac{d\vec{V}}{dt} \qquad (C.36)$$

or from Eq (C.35)

$$\vec{a} = \frac{dV}{dt}\hat{i}_T + V\frac{d\hat{i}_T}{dt} \qquad (C.37)$$

The time derivative of the tangent unit vector is not necessarily zero since it may vary in direction with time. To aid us in developing an expression for the tangential unit vector derivative, we must first relate the tangential unit vector to the fixed coordinate system:

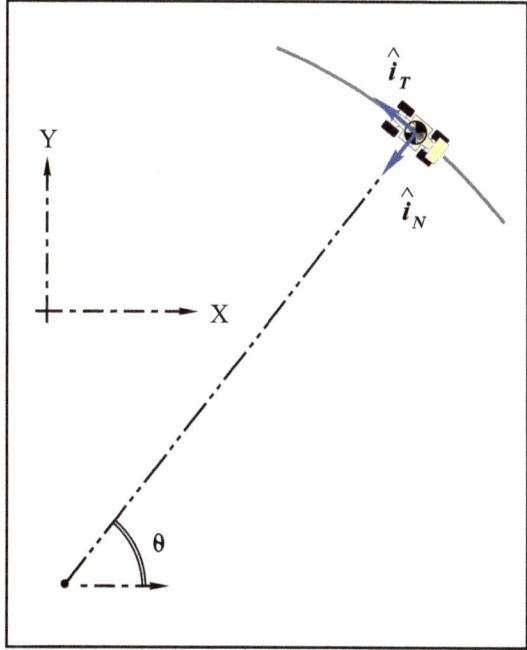

Figure (C.5): TANGENTIAL AND NORMAL UNIT VECTOR DEFINITIONS

$$\hat{i}_T = -\sin\theta \, \hat{i}_x + \cos\theta \, \hat{i}_y \qquad (C.38)$$

Differentiating Eq (C.38) with respect to time results in:

$$\frac{d\hat{i}_T}{dt} = -\cos\theta \frac{d\theta}{dt} \hat{i}_x - \sin\theta \frac{d\hat{i}_x}{dt} - \sin\theta \frac{d\theta}{dt} \hat{i}_y + \cos\theta \frac{d\hat{i}_y}{dt} \qquad (C.39)$$

However, the unit vectors associated with the fixed reference frame are invariant with time; therefore, the time derivatives of these unit vectors is zero. Then Eq. (C.39) reduces to:

$$\frac{d\hat{i}_T}{dt} = \frac{d\theta}{dt}\left(-\cos\theta \, \hat{i}_x - \sin\theta \, \hat{i}_y\right) \qquad (C.40)$$

The terms in parenthesis define a unit normal vector, as shown in Figure (C.5). By noting that the angular speed, ω, is defined as $\frac{d\theta}{dt}$. , Eq (C.40) becomes

$$\frac{d\hat{i}_T}{dt} = \omega \, \hat{i}_N \qquad (C.41)$$

Appendix C: DYNAMICS REVIEW

Finally the **CG** acceleration, Eq (C.37), simplifies to

$$\vec{a} = \frac{dV}{dt} \hat{i}_T + V\omega \, \hat{i}_N \qquad\qquad (C.42)$$

The magnitudes of the terms on the right-hand-side are, respectively, the tangential acceleration and the normal acceleration.

REFERENCES

Chapter 1

[1] J.P. Norbye, THE CAR AND ITS WHEELS - A GUIDE TO MODERN SUSPENSION SYSTEMS, Tab Books Inc., Blue Ridge Summit, PA, 1980, pp. 221-224.

Chapter 2

[1] C. Campbell, THE SPORTS CAR, ITS DESIGN AND PERFORMANCE, 3rd ed., Robert Bentley, Inc., Cambridge, MA, 1974, pg. 160.

[2] L.J.K. Setright, AUTOMOBILE TYRES, Chapman and Hall, London, 1972, pg. 25.

[3] K. Jost, "Tire Materials and Construction," AUTOMOTIVE ENGINEERING, Vol. 100, No. 10, 1992, pp. 23-28.

[4] K.G. Budinski, ENGINEERING MATERIALS: PROPERTIES AND SELECTION, 3rd ed,, Prentice-Hall, New Jersey, 1989, pg. 127.

[5] T.D. Gillespie, FUNDAMENTALS OF VEHICLE DYNAMICS, S.A.E., Warrendale, PA,1992, pg. 344.

[6] J.Y. Wong, THEORY OF GROUND VEHICLES, John Wiley & Sons, New York, 1978, pp. 15-18.

[7] J.J. Taborek, "Mechanics of Vehicles," MACHINE DESIGN, May 30 - Dec 26, 1957.

[8] S.P. Timoshenko and D.H. Young, THEORY OF STRUCTURES, 2nd ed., McGraw-Hill, New York, 1965, pp. 601-602.

[9] E. Bakker, L. Nyborg, and H.B. Pacejka, "Tyre Modeling for Use in Vehicle Dynamic Studies," S.A.E paper no. 870421, 1987.

[10] E. Bakker, H.B. Pacejka, and L. Lider, "A New Tire Model with an Application in Vehicle Dynamic Studies," S.A.E. paper no. 890087, 1989.

[11] H.B. Pacejka and E. Bakker, "The Magic Tyre Formula," PROCEEDINGS OF THE 1ST INTERNATIONAL COLLOQUIUM ON

TYRE MODELS FOR VEHICLE DYNAMIC ANALYSIS, Swets & Zeitlinger B.V., Amsterdam/Lisse, 1993.

[12] H.B. Pacejka, TIRE AND VEHICLE DYNAMICS, Butterworth-Heinemann, Woburn, MA, 2002, pp. 172-191.

[13] C.T. Sun and Y.P. Lu, VIBRATION DAMPING OF STRUCTURAL ELEMENTS, Prentice-Hall, New Jersey, 1995, pp. 12-17.

Chapter 3
[1] C. Campbell, THE SPORTS CAR, ITS DESIGN AND PERFORMANCE, 3rd ed., Robert Bentley, Inc., Cambridge, MA, 1974, pg. 228.

[2] H. Heisler, Chapter 14: Vehicle Body Aerodynamics, ADVANCED VEHICLE TECHNOLOGY, 2nd ed., 2002, pp 584-634.

Chapter 4
[1] J.P. Norbye, THE CAR AND ITS WHEELS - A GUIDE TO MODERN SUSPENSION SYSTEMS, Tab Books, Blue Ridge Summit, PA, 1980, p 74.

[2] H.E. von Gierke and D.E. Goldman, "Effects of Shock and Vibration on Man," SHOCK AND VIBRATION HANDBOOK, 3 ed, C.M. Harris editor, McGraw-Hill, New York, 1988, pp 44-1 - 44-58.

[3] B.D. Van Deusen, "Human Response to Vehicle Vibration," paper no. 680090, SAE Trans, Vol. 77, 1969, pp 328-345.

[4] W.E. Woodson, HUMAN FACTORS DESIGN HANDBOOK, McGraw-Hill, New York, 1981, pp 860-861.

[5] D. Dieckmann, "Einfluss Vertihaler Mechanischer Schwingungen auf den Menschen," Internat. Z. Angew Physiol., Vol. 16, 1947, pp 519-564.

[6] J.Y. Wong, THEORY OF GROUND VEHICLES, John Wiley & Sons, New York, 1978, p 265.

[7] W.T. Thompson, THEORY OF VIBRATIONS WITH APPLICATIONS, 4th ed., Prentice Hall, New Jersey, 1993.

[8] C. Campbell, THE SPORTS CAR: ITS DESIGN AND PERFORMANCE, 3rd ed., Robert Bentley Inc., Cambridge, MA, 1974, pp. 198-199.

[9] T.D. Gillespie, FUNDAMENTALS OF VEHICLE DYNAMICS, Soc. of Auto. Engrs., Inc., Warrendale, PA, 1992.

[10] M. Olley, "National Influences on American Passenger Car Design," Proc. Inst. Auto. Engrs., Vol. XXXII, 1937-38.

[11] R.G.S. White, VEHICLE DYNAMICS, course notes in Mechanical Engineering, University of Illinois, Champaign-Urbana, IL, circa 1972.

[12] T.T. Soong and M. Grigoriu, "Calculus of Stochastic Process," RANDOM VIBRATION OF MECHANICAL AND STRUCTURAL SYSTEMS, Prentice Hall, New Jersey, 1992, pp 95 - 159.

[13] B.D. Van Deusen, "Analytical Techniques for Designing Riding Quality into Automotive Vehicles," paper no. 670021, SAE Trans, Vol. 76, 1968.

[14] C.J. Dodds and J.D. Robson, "The Description of Road Surface Roughness," Journal Sound Vib, Vol. 32, No. 2, 1973, pp 175 - 183.

[15] R.P. LaBarre, R.T. Forbes and S. Andrew, "The Measurement and Analysis of Road Surface Roughness," MOTOR INDUSTRY RESEARCH ASSOCIATION, Report No. 1970/5, 1969.

Chapter 5
[1] W.F. Milliken and D.L. Milliken, RACE CAR VEHICLE DYNAMICS, Soc of Auto Engrs, Warrendale, PA, 1995, pp 301-313.

[2] J.P. Norbye, THE CAR AND ITS WHEELS - A GUIDE TO MODERN SUSPENSION SYSTEMS, Tab Books Inc., Blue Ridge Summit, PA, 1980, pp. 252-253.

[3] G.H. Martin, KINEMATICS AND DYNAMICS OF MACHINES, McGraw-Hill, New York, 1969, pp. 330-331.

* 9 7 8 1 8 4 8 9 0 4 3 0 9 *